D1085245

STP 1186

Thermomechanical Fatigue Behavior of Materials

Huseyin Sehitoglu, editor

ASTM Publication Code Number (PCN)
04-011860-30

ASTM
1916 Race Street
Philadelphia, PA 19103

Library of Congress Cataloging-in-Publication Data

Thermomechanical fatigue behavior of materials / Huseyin Sehitoglu.
(STP ; 1186)
"ASTM publication code number (PCN) 04-011860-30."
Includes bibliographical references and index.
ISBN 0-8031-1871-6
1. Alloys—Thermomechanical properties. 2. Composite materials—Thermomechanical properties. 3. Fracture mechanics.
I. Sehitoglu, Huseyin, 1957– . II. Series: ASTM special technical publication ; 1186.
TA483.T47 1993
620.1′126—dc20 93-21663
CIP

Copyright © 1993 AMERICAN SOCIETY FOR TESTING AND MATERIALS, Philadelphia, PA. All rights reserved. This material may not be reproduced or copied, in whole or in part, in any printed, mechanical, electronic, film, or other distribution and storage media, without the written consent of the publisher.

Photocopy Rights

Authorization to photocopy items for internal or personal use, or the internal or personal use of specific clients, is granted by the AMERICAN SOCIETY FOR TESTING AND MATERIALS for users registered with the Copyright Clearance Center (CCC) Transactional Reporting Service, provided that the base fee of $2.50 per copy, plus $0.50 per page is paid directly to CCC, 27 Congress St., Salem, MA 01970; (508) 744-3350. For those organizations that have been granted a photocopy license by CCC, a separate system of payment has been arranged. The fee code for users of the Transactional Reporting Service is 0-8031-1871-6/93 $2.50 + .50.

Peer Review Policy

Each paper published in this volume was evaluated by three peer reviewers. The authors addressed all of the reviewers' comments to the satisfaction of both the technical editor(s) and the ASTM Committee on Publications.

The quality of the papers in this publication reflects not only the obvious efforts of the authors and the technical editor(s), but also the work of these peer reviewers. The ASTM Committee on Publications acknowledges with appreciation their dedication and contribution to time and effort on behalf of ASTM.

Printed in Ann Arbor, MI
September 1993

Foreword

This publication, *Thermomechanical Fatigue Behavior of Materials,* contains the papers presented at the symposium of the same name held in San Diego, CA on 14-16 Oct. 1991. The symposium was sponsored by ASTM Committee E-9 on Fatigue. Huseyin Sehitoglu, University of Illinois, Urbana, IL, served as chairman of the symposium and is editor of the publication.

Contents

Overview

Background

Thermo-mechanical fatigue (TMF) problems are encountered in many applications, such as high-temperature engines, structural components used in high-speed transport, contact problems involving friction, and interfaces in computer technology. Thermo-mechanical fatigue provides a challenge to an analyst as well as to an experimentalist. The analyst is faced with describing the constitutive representation of the material under TMF, which is compounded by complex internal stresses, aging effects, microstructural coarsening, and so forth. The evolution of microstructure and micromechanisms of degradation differ from that encountered in monotonic deformation or in isothermal fatigue. Experimentalists conducting TMF tests need to ensure simultaneous control of temperature and strain waveforms, and minimization of temperature gradients to enable uniform stress and strain fields. Failure to meet these requirements may result in fortuitous results.

This symposium was organized to provide a means of disseminating new research findings in thermo-mechanical fatigue behavior of materials. The need for the symposium grew naturally from the activities of the E9.01.01 Task Group on Thermomechanical Fatigue. There have been numerous developments in understanding thermo-mechanical damage mechanisms over the last decade. The last ASTM symposium on TMF was held in 1975, and since then, the role of oxidation damage is now better recognized, the asymmetry of creep damage is well accepted, and microstructural evolution is established as a contributor to stress-strain response and to damage behavior. Moreover, the experimental techniques to study TMF evolved significantly over the last decade. Computer control of strain and temperature waveforms, high-temperature strain, and temperature measurement techniques were refined considerably. Researchers are gaining a better understanding of damage at the micro-level with sophisticated microscopy tools probing to ever lower size scales. At the same time, with refined numerical models and improved computer power, it is possible to conduct more realistic simulations of material behavior. The last decade has seen increased emphasis on composite materials designed to withstand high operating temperatures and severe TMF environments. Both the experiments and their interpretation are difficult on these highly anisotropic materials with complex internal stress and strain fields.

The purpose of thermomechanical fatigue studies is twofold. First, to gain a deeper understanding of defect initiation and growth as influenced by the underlying microstructure or discrete phases, and second, to obtain useful engineering relationships and mathematical models for macroscopic behavior, allowing the design and evaluation of engineering systems. The first goal is sought by materials scientists and mechanicians conducting basic research, while the second goal is pursued by engineers and designers who are integrating this basic information and experimental data to develop structural models. It is desirable that basic research in this field be guided by the needs and requirements set by designers in their search for better performance.

The papers presented in this special technical publication (STP) have the aim of addressing both the basic research and the design issues in thermomechanical fatigue. The authors have been active researchers in high-temperature fatigue and have all made notable contributions in their specific areas of interest. In addition to U.S. researchers, the contributions from overseas researchers are noteworthy and encouraging.

Summary of the Papers

It is now widely accepted that a materials' TMF behavior be studied under the in-phase case (where maximum temperature and maximum strain coincides) and the out-of-phase case (where maximum temperature and minimum strain coincide). These two loading types represent strain-temperature histories that often produce different damage mechanisms. The reader will find these terms used repeatedly in this publication. A mini-summary of the 14 papers included in this STP follows.

Dr. Remy and colleagues have elucidated the dramatic contribution of oxidation on fatigue crack growth in thermomechanical fatigue by comparing preoxidized and virgin samples. Mr. Zauter and colleagues demonstrated dynamic strain aging and dynamic recovery effects in austenitic stainless steels under thermomechanical fatigue. Similar behavior was seen in Hastelloy X studied by Castelli et al. who proposed a constitutive equation to describe the aging phenomena. Kadioglu and Sehitoglu studied the MarM247 alloy and calculated internal stresses caused by oxide spikes and refined an early model proposed by the senior author. Miller et al. proposed microcrack propagation laws suitable for TMF loadings incorporating creep, fatigue and oxidation effects. Thermomechanical fatigue of In-738 was considered by Bernstein et al. who proposed a life model incorporating time, temperature, and strain effects. Single crystal and directionally solidified nickel alloy was considered by Guedou and Honnorat who also examined coated alloys. Kalluri and Halford studied the Haynes 188 under various TMF cycle shapes demonstrating creep and oxidation damages. Halford et al. discussed the thermomechanical fatigue damage mechanisms in several unidirectional metal-matrix composites. Analysis of local stresses and strains for same class of materials has been achieved in the work of Coker et al. Experiments demonstrating deviations from linear summation of creep and fatigue damages in TMF have been conducted by McGaw. Characterization of crack growth through temperature and stress gradients has been considered by Sakon et al. The shear stress-strain behavior of solder materials in TMF has been studied as a function of cycle time in Hacke et al.

Future Needs

Advanced monolitic materials and their composites will provide challenges to experimentalists and analysts working on thermomechanical fatigue. Beyond the need for TMF resistance in applications listed earlier, studies of thermomechanical fatigue and fracture in the electronics industry and in manufacturing operations involving thermomechanical processing are other areas likely to attract attention in the future.

I would like to express my gratitude to all authors, reviewers, and ASTM staff for their contribution to the publication of this STP. A follow up symposium is planned in two years, which will highlight new developments in this field.

Huseyin Sehitoglu

Symposium chairperson and editor;
University of Illinois, Urbana, Ill.

L. Rémy,[1] *H. Bernard,*[2] *J. L. Malpertu,*[3] *and F. Rezai-Aria*[4]

Fatigue Life Prediction Under Thermal-Mechanical Loading in a Nickel-Base Superalloy

REFERENCE: Rémy, L., Bernard, H., Malpertu, J. L., and Rezai-Aria, F., **"Fatigue Life Prediction Under Thermal-Mechanical Loading in a Nickel-Base Superalloy,"** *Thermomechanical Fatigue Behavior of Materials, ASTM STP 1186,* H. Sehitoglu, Ed., American Society for Testing and Materials, Philadelphia, 1993, pp. 3–16.

ABSTRACT: Thermal-mechanical fatigue of IN-100, a cast nickel base superalloy, was previously shown to involve mainly early crack growth using either bare or aluminized specimens. This crack growth was found to be controlled by interdendritic oxidation. A model for engineering life to crack initiation is thus proposed to describe this microcrack growth phase using local stresses in a microstructural volume element at the crack tip. The identification of damage equations involves fatigue crack growth data on compact tension (CT) specimens, interdendritic oxidation kinetics measurements and fatigue crack growth on CT specimens that have been embrittled by previous oxidation at high temperature. The application of this model to life prediction is shown for low cycle fatigue and thermal-mechanical fatigue specimens of bare and coated specimens as well as for thermal shock experiments.

KEYWORDS: life prediction, low-cycle fatigue, thermal-mechanical fatigue, high temperature fatigue, nickel base superalloy, oxidation

Thermal fatigue with or without superimposed creep is the primary life limiting factor for blades and vanes in gas turbines for jet or aircraft engines. Damage modeling under thermal-mechanical cyclic loading is still at an early stage as compared to the developments made for high temperature isothermal fatigue [*1–3*]. A major reason has been the difficulty of simulating thermal stress cycling in the laboratory. During recent years considerable effort has been devoted to develop thermal-mechanical fatigue (TMF) tests to simulate the behavior of a volume element in a structure. Since all test parameters are known (measured or imposed), such tests can be used to check the validity of damage models to be used for actual components.

TMF tests were thus run on conventionally cast superalloy IN-100 used for blades and vanes in jet engines. The conventional low cycle fatigue (LCF) behavior of this alloy was previously studied in the bare condition in our laboratory [*4*]. From computations of real blades under service conditions, the behavior of bare IN-100 was studied under various TMF cycles, which had maximum and minimum temperatures of 1050 and 600°C (1323 and 873 K) as shown in Fig. 1. Cycles I and II had periods of 9.5 and 3 min, respectively, with a strain ration $R_\varepsilon = -1$ and the mechanical strain was set to zero at minimum temperature. Peak strains occur at 900°C (1173 K) in compression on heating and at 700°C (973 K) on cooling.

[1] Centre des Materiaux P. M. Fourt, Ecole des Mines de Paris, URA CNRS, 866, BP87, 91003 Evry Cedex, France.
[2] Peugot S.A., Velizy, France.
[3] Joseph Paris S.A., Nantes, France.
[4] Ecole Polytechnique Federale de Lausanne, Ecublens, Switzerland.

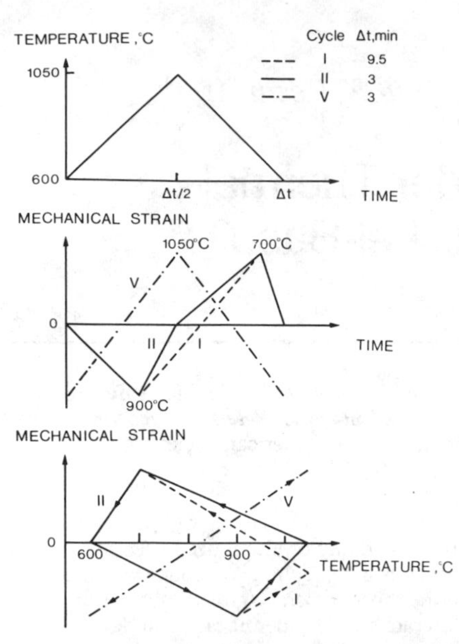

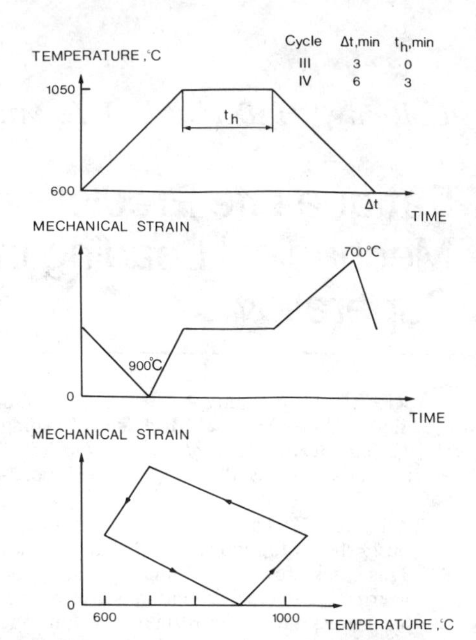

FIG. 1—*Shape of thermal-mechanical fatigue cycles:* (a) *Cycles I, II, and V (with no mean strain) and* (b) *Cycles III and IV (with a zero minimum strain). Each figure shows the plots of temperature versus time (*Δt *is the cycle period), mechanical strain versus time, and mechanical strain versus temperature.*

Cycle III was similar to Cycle II with $R_\varepsilon = 0$ and Cycle IV was Cycle III with a 3-min hold time at maximum temperature and at half the maximum strain. Cycle V was a conventional in-phase cycle where mechanical strain was a maximum (respectively, minimum) at maximum temperature (respectively, minimum) using a period of 3 min All tests were run using hollow cylindrical specimens.

Results were reported in a previous paper [*5*] and some trends are shown in Fig. 2. The TMF life of hollow specimens with a 1-mm wall thickness was conventionally defined as corresponding to a 0.3-mm depth of the major crack. Plastic replicas taken at various fractions of life have shown that the major part of TMF life was spent in the growth of microcracks. The crack growth rate was very sensitive to TMF cycle shape and frequency.

This behavior under TMF cycling is in good agreement with earlier LCF results at 1000°C in air [*4*] since a large frequency-dependence of LCF life had been observed especially in the frequency range $5 \times 10^{-2} - 2$ Hz. The LCF life in air was found to be mainly spent in the propagation of microcracks even at high frequency (1 to 2 Hz). The fatigue life in vacuum was, on the contrary, almost frequency-independent. The marked difference between fatigue lives in air and in vacuum vanished at high frequency (1 to 2 Hz). These results showed clearly a large influence of oxidation on the high temperature fatigue damage of this alloy.

These results were recently completed by TMF tests on aluminized IN-100 specimens, since actual components are coated [*6,7*]. The TMF Cycle II was mainly used, but some more complex cycles were used, including a Cycle II with l-h period instead of 3 min. Aluminized specimens were found to have a longer life than bare specimens for a given cycle shape. Sections of TMF specimens tested to various fractions of life have shown that the TMF life of coated specimens, as that of bare specimens, was mainly controlled by oxidation and involved an important microcrack growth phase.

The microcrack growth phase provided therefore a lower bound of the engineering life to crack initiation.

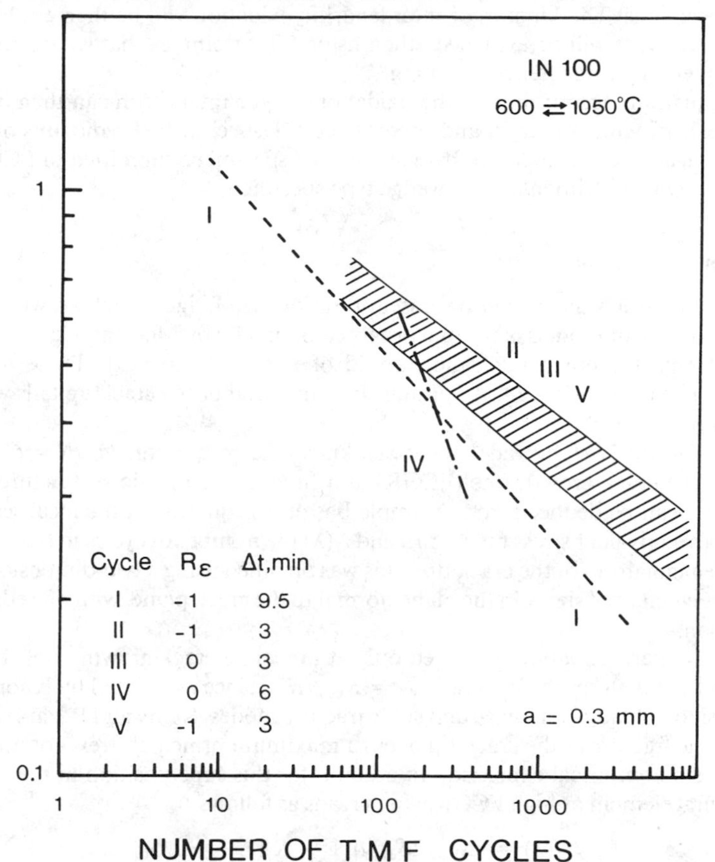

FIG. 2—*Variation of the number of TMF cycles to 0.3-mm crack depth with the mechanical strain range. For sake of simplicity only the trends of results are shown.*

Plastic replicas and metallographic sections of various specimens tested in LCF and in TMF in alloy IN-100 and in other cast superalloys [*4,5,8,9*] have shown that the depth of the major surface crack increases linearly with the number of cycles up to several tenths of a millimetre (at least 0.3 mm) and this constant crack-growth rate regime amounts to 30 to 60% of fatigue life. Then the crack-growth rate increases with crack length as expected from fracture mechanics, and the specimen is actually a structure. The former regime, where microcracks grow at a constant rate, is typical of a volume element behavior and can be taken as the lifetime to initiate an engineering crack several tenths of a millimetre (at least 0.3 mm) in depth.

The paper will therefore describe a fatigue damage model applicable to both LCF and TMF which assumes that fatigue life is spent in the growth of microcracks. The oxidation fatigue interaction will be considered in the following manner: exposure to high temperature during the TMF cycle oxidizes the material at the crack tip and then high stress ranges at medium temperatures give rise to fatigue damage in the material that has been embrittled by oxidation.

The present model is thus different from previous simpler oxidation-fatigue models [*10–12*] such as the one proposed by one of the authors [*11*], which uses a simple summation of pure fatigue and oxidation contributions to crack growth rate and assumes oxide film cracking at each tensile stroke.

A fatigue damage equation will be first fitted to fatigue crack growth data on compact tension (CT) specimens. The kinetics of interdendritic oxidation will be then established. Oxidation embrittlement will be evidenced then using CT specimens that were previously oxidized at high temperature after precracking.

Damage equations accounting for the oxidation-fatigue interaction can then be identified from these experiments on virgin and preoxidized CT specimens. Predictions of the model will be tested against experiment on bare and coated specimens submitted to LCF and TMF and thermal shock experiments using wedge type specimens.

Fatigue Damage Equation

A number of models have been proposed to rationalize fatigue crack growth. One of the most powerful class of models is the one proposed originally by McClintock, who considered a process of repeated crack nucleation ahead of the crack tip [*13*]. These models were reviewed, for instance, in Ref *14*. A volume element ahead of the crack tip fails when a local fracture criterion is reached.

Chalant and Rémy [*14*] showed that the well-known Paris equation $da/dN = C\,\Delta K^m$, which expresses fatigue crack growth rate (FCGR) as a function of the global fracture mechanics parameter ΔK, can be deduced from a simple Basquin's equation at the local scale between the Von Mises equivalent stress range $\Delta\sigma_{eq}$ and $N(\lambda)$ the number of cycles to fracture a microstructural element ahead of the crack tip. This was obtained using a two-dimensional analysis with a square element of size λ in the plane normal to the crack plane, with one edge along the crack direction.

However, the Paris equation is obeyed only at moderate crack growth rates. High FCGR are strongly dependent on the load ratio $R = K_{mim}/K_{max}$ since as pointed by Knott [*15*] there is a superposition of fatigue damage and static fracture modes. Rémy and Rézai-Aria assumed that monotonic fracture at the crack tip obeys a maximum principal stress criterion [*16*] and they proposed an empirical expression to account for this superposition in the fracture of a microstructural element at high FCGR which reads as follows

$$1/N(\lambda) = (\Delta\sigma_{eq}/2S_0)^M/[(1 - R)(\sigma_c - \sigma_{yy}/S_0]^\alpha \tag{1}$$

with

$$R = 1 - \Delta\sigma_{yy}/\sigma_{yy} \qquad \text{for } \Delta\sigma_{yy} \leq \sigma_{yy}$$
$$R = 0 \qquad \text{for } \Delta\sigma_{yy} > \sigma_{yy}$$

and where $\Delta\sigma_{eq}$ is the Von Mises equivalent stress range averaged over the microstructural element at the crack tip, σ_{yy} is the maximum tensile value of the normal stress of the crack tip at a distance λ (only the tensile part of the normal stress range is supposed to contribute to monotonic fracture), S_0, M, α are constants at a given temperature, and σ_c is the critical value of σ_{yy} when $N(\lambda)$ tends to infinity (that is, at monotonic fracture).

Equation 1 was fitted to FCGR measured for different load ratios in the range 10^{-9} to 10^{-5} m/cycle at high frequency (20 or 50 Hz) to minimize environmental effects. $\Delta\sigma_{eq}$, $\Delta\sigma_{yy}$, and σ_{yy} were deduced [*14,16*] from the stress singularity ahead of the crack tip computed by Tracey for plane-strain small scale yielding under monotonic loading [*17,18*]. This finite element analysis was adapted to cyclic loading according to Rice's hypothesis [*19*] using stress and strain ranges instead of stresses and strains and the cyclic stress-strain relationship measured on stabilized loops

$$\Delta\sigma_{eq}/\Delta\sigma_o = (\Delta\varepsilon_{eq}/\Delta\varepsilon_o)^N \tag{2}$$

where $\Delta\sigma_{eq}$, $\Delta\varepsilon_{eq}$ are the Von Mises equivalent stress and total strain ranges, $\Delta\sigma_o$ and N are constant and $\Delta\varepsilon_o = \Delta\sigma_o/3G$ (G is shear modulus with Poisson's coefficient $\nu = 0.3$). λ was defined as the mean secondary dendrite size ($\lambda = 100\ \mu m$, measured edge to edge). $N(\lambda)$ was thus deduced from the crack growth rate $da/dN = \lambda/N(\lambda)$ on CT specimens and from potential drop measurements of the number of cycles to 0.3-mm crack depth as $N(\lambda) = N(0.3\text{ mm})\ \lambda/0.3$.

Figure 3*a* shows $\Delta\sigma_{eq} - N(\lambda)$ curves for IN-100 superalloy deduced from experimental $da/dN - \Delta K$ curves on CT specimens with two load ratios of 0.1 and 0.7 at 1000°C as well as

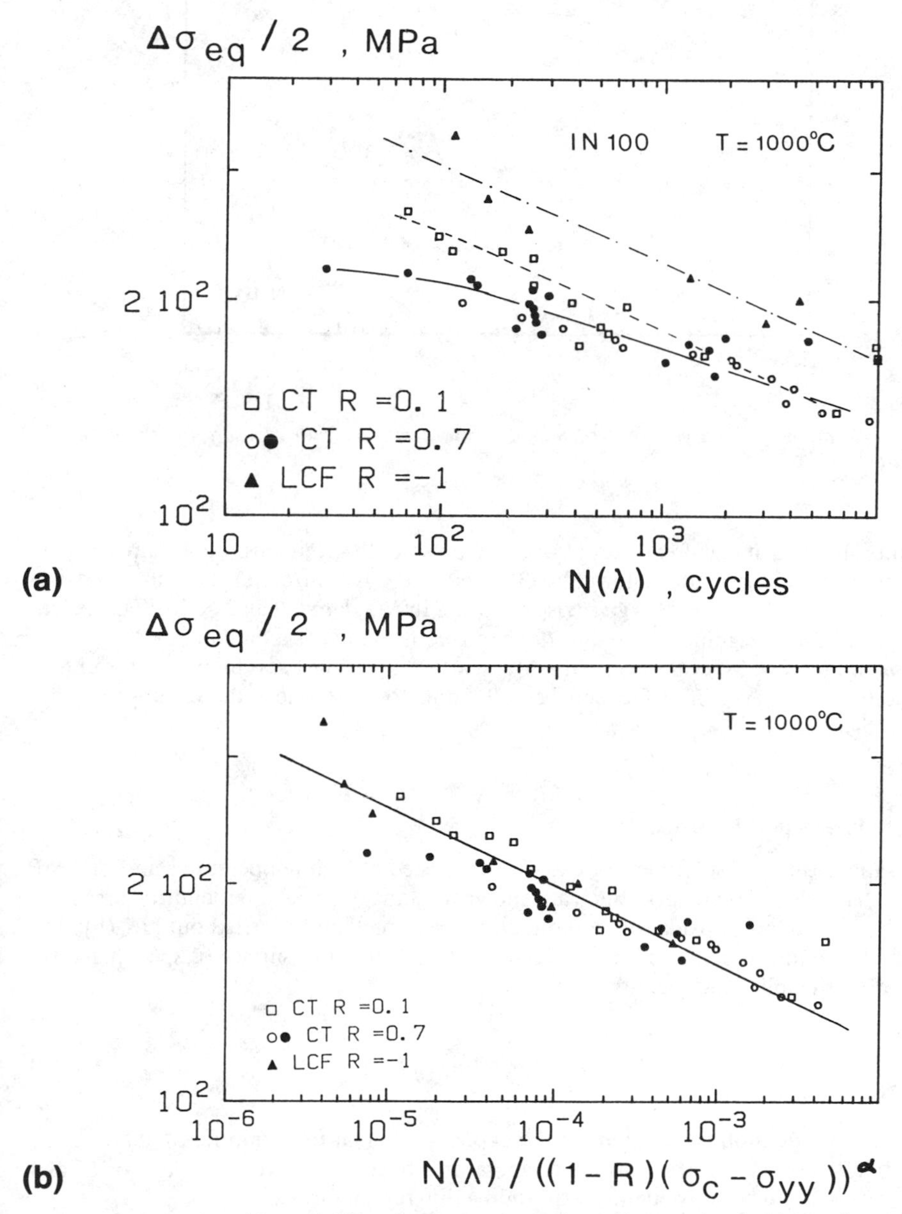

FIG. 3—*Variation of the equivalent stress range in a volume element ahead of the crack tip* ($\Delta\sigma_{eq}$) *as a function of* (a) *the number of cycles to break it* $N(\lambda)$ *at 1000°C and* (b) *the ratio* $(N(\lambda))[(1 - R)(\sigma_c - \sigma_{yy})]^{\alpha}$. *Data are from CT specimens (load ratio* R = *0.1 or 0.7) and LCF specimens (*R = *−1) tested at high frequency.*

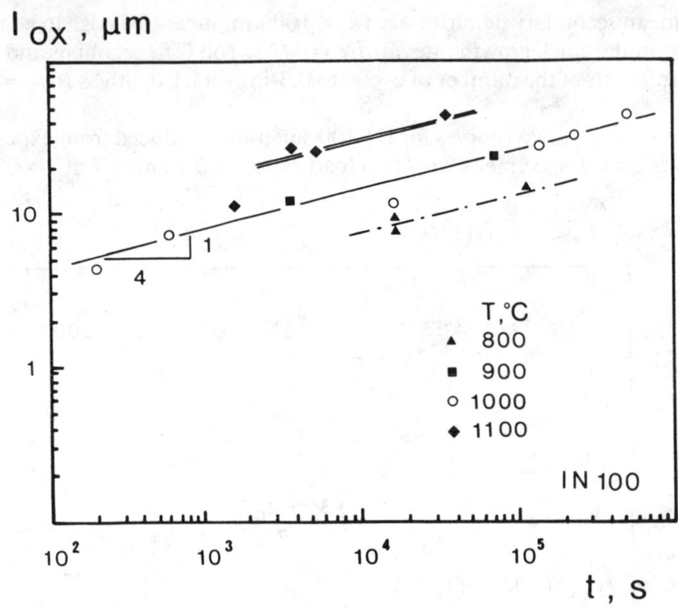

FIG. 4—*Variation of the average interdendritic oxide depth* lox *with exposure time* t *at various temperatures.*

LCF results obtained at high frequency (1 or 2 Hz) on smooth specimens. Data from CT specimens show a pretty large scatter at lower FCGR (below 5×10^{-8} m/cycle) as usually in coarse-grained nickel-base alloys, and this gives rise to scatter in $N(\lambda)$ larger than 2×10^3. The dependence on load ratio is significantly reduced when $\Delta\sigma_{eq}$ is plotted as a function of $N(\lambda)/[(1 - R)(\sigma_c - \sigma_{yy})]^\alpha$, see Fig. 3*b*, within experimental scatter. Equation 1 was found to account for the load ratio dependence of FCGR between 10^{-8} and 10^{-5} m/cycle at the various temperatures investigated.

Kinetics of Interdendritic Oxidation

Observations on LCF specimens of cast IN-100 tested at high temperatures and of TMF specimens have shown that cracks nucleate and grow along oxidized interdendritic areas.

Quantitative studies of oxidation without stress were previously carried out [*18,19*]. The depth of interdendritic oxide spikes l_{ox} measured from the outer surface of specimens was found to obey the following equation

$$l_{ox} = \alpha_{ox}(T)t^{1/4} \tag{3}$$

where α_{ox} is an oxidation constant and t the exposure time at temperature. Metallographic measurements on specimens at various temperatures have confirmed this behavior (Fig. 4 [*19*]). Equation 3 can be conveniently written in a differential form as

$$d(l^4_{ox}) = \alpha^4_{ox}(T)dt \tag{4}$$

where α_{ox} varies as a function of temperature according to an Arrhenius law

$$\alpha_{ox}(T) = \alpha_{ox}^{0} \exp(-Q/RT) \tag{5}$$

where T is temperature in Kelvin, $R = 8.315 \text{ J} \cdot \text{K}^{-1}$ and Q an activation energy.

Fatigue Crack Growth in Preoxidized CT Specimens

Critical experiments were carried out on CT specimens that were first precracked to $a/w \approx 0.4$, then oxidized in a furnace at high temperature and finally tested at a given temperature. A typical FCGR curve is shown at 650°C for a precracked IN-100 specimen which was oxidized 70 h at 1000°C together with that of a virgin precracked specimen (Fig. 5). The loading

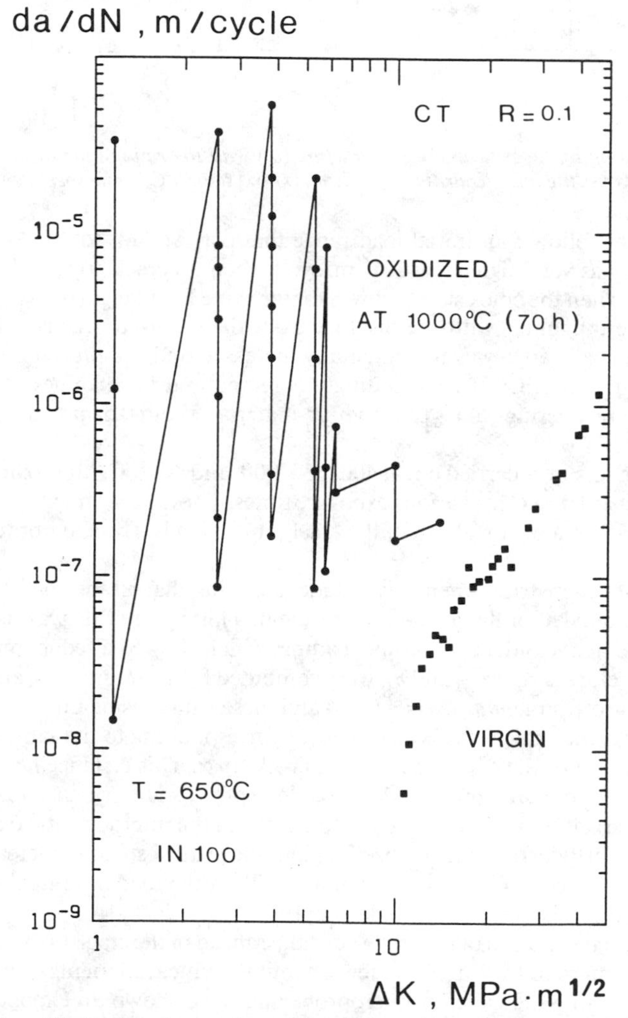

FIG. 5—*Variation of fatigue crack growth rate* (da/dN) *as a function of stress intensity range (ΔK) at 650°C for virgin specimens and a specimen that has been oxidized at 1000°C (load ratio* R = *0.1).*

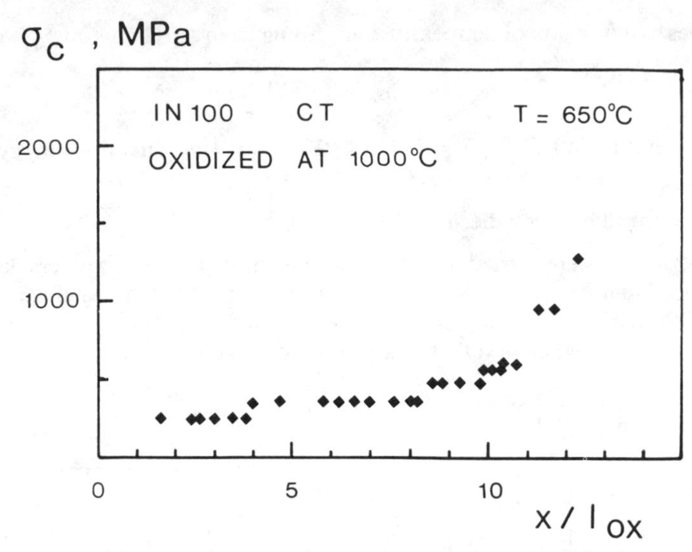

FIG. 6—*Variation of the critical stress to fracture* (σ_c) *with the ratio of the distance to the tip of the oxidized precrack over the interdendritic oxide depth* (x/lox) *at 650°C (same specimen as in Fig. 8).*

procedure was as follows: an initial load range that corresponds to $\Delta K \approx 1$ MPa·m$^{1/2}$ was applied. FCGR was very high a few 10^{-5} m/cycle when no crack growth should occur in the virgin material. Then the crack slowed down for the same load range (this is due to the increase in local fracture toughness with the distance from the oxidized precrack, as will be shown later). Then the load range was incrementally increased and the procedure was repeated. A saw-tooth variation in FCGR is accordingly observed and large values of peak FCGR are observed before recovering values of the virgin material, about 0.5-mm ahead of the oxidized crack front.

Such experiments were carried out at 400, 650, 900, and 1000°C. Most oxidation treatments were carried out at 1000°C for various exposure times. These experiments [*20*] showed a large increase in FCGR in a region ahead of the crack tip, which has been embrittled by the oxidation treatment.

The analysis of these experiments was made assuming that all the coefficients except σ_c in Eq 1 were not altered after the oxidation treatment. This assumption was supported by crack growth measurements under monotonic loading, which showed a reduction of local fracture toughness. Therefore $\Delta\sigma_{eq}$, $\Delta\sigma_{yy}$, and σ_{yy} were computed from ΔK and K_{max} and σ_c was deduced from the crack growth rate $da/dN = 1/N(\lambda)$ and these values using Eq 1.

Figure 6 shows the critical stress σ_c, at 650°C, corresponding to the same experiment as in Fig. 5, as a function of the distance from the crack tip x. This distance has been normalized using the ratio of x over the interdendritic oxide depth l_{ox}. The critical stress to fracture σ_c is very low at the crack tip up to about eight to ten times the interdendritic oxide depth. When the crack grows farther from the oxidized region, the critical stress σ_c increases more rapidly and approaches values typical of the virgin alloy. Thus the oxidation treatment embrittles a zone about ten times larger than the oxide depth.

This exponential variation of σ_c with the distance ahead of the crack tip is linked with oxygen diffusion along interdendritic areas. A few quantitative measurements of oxygen concentration were made using the electron microprobe and have shown an exponential decrease of oxygen concentration with the distance along interdendritic areas of the oxidized precrack. This behavior can be described using Fisher's model for intergranular diffusion [*21*]. Thus the diffusion distance at a given concentration varies as $t^{1/4}$ which gives a physical basis to the $t^{1/4}$ kinetics of interdendritic oxidation.

The exponential variation of the critical stress with the distance from the oxidized crack front reflects its variation with oxygen concentration ahead of the crack tip. The variation of σ_c with the distance was described through an empirical equation

$$\sigma_c = \sigma_{co}(T)f(x/l_{ox}) \tag{6}$$

where σ_{co} is a constant at a given temperature.

FCGR in a preoxidized cracked specimen can thus be described by Eqs 1 and 6 together.

Damage Equations during Oxidation-Fatigue Interactions

Let us consider a volume element of size λ ahead of the crack tip. This volume element will endure an incremental damage dD during dN cycles under oxidizing and cyclic loading conditions. This damage increment is given by

$$dD = \{(\Delta\sigma_{eq}/2S_0)^M/[(1 - R)(\sigma_c - \sigma_{yy})/S_0]^\alpha\}dN \tag{7}$$

$$\text{with} \quad R = 1 - \Delta\sigma_{yy}/\sigma_{yy}, \quad \text{when} \quad \Delta\sigma_{yy} \leq \sigma_{yy}$$
$$R = 0 \quad \Delta\sigma_{yy} > \sigma_{yy}$$

and

$$\sigma_c = \min\{\sigma_{cv}; \sigma_{cox}(t, T)\} \tag{8}$$

where σ_{cv} is the critical fracture stress of virgin alloy and σ_{cox} that of the alloy embrittled by oxidation which is defined from Eq 5 as

$$\sigma_{cox}(t, T) = \sigma_{co}(T)f(\lambda/l_{ox}) \tag{9}$$

where l_{ox} is given from Eq 4.

Using Eqs 8 and 9 σ_c is computed at every cycle and the number of cycles to break the volume element will be given by the condition

$$D = \int_0^{N(\lambda)} dD = 1 \tag{10}$$

$N(\lambda)$ is thus computed through the set of Eqs 7 to 9 and 4 using an iterative procedure, cycle by cycle, until the condition of Eq 10 is fulfilled.

The application of this procedure to isothermal LCF is straightforward. All the relevant coefficients at a given temperature are easily deduced by interpolation between identified values.

The application to TMF loading is slightly more complex. As temperature varies, the right side of Eq 3 has to be averaged over the cycle, and an average oxidation constant α_{ox} is computed as

$$\overline{\alpha}_{ox} = \left[\int_0^{\Delta t} \alpha_{ox}^4[T(t)]\, dt/\Delta t\right]^{1/4} \tag{11}$$

where Δt is the cycle period.

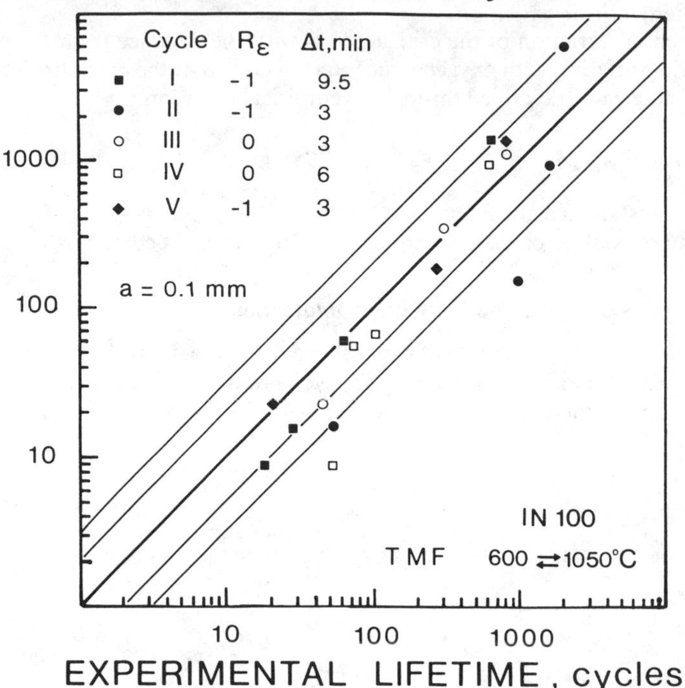

FIG. 7—*Comparison between calculated life to 0.1-mm-deep crack in thermal-mechanical fatigue of bare IN-100 specimens and experimental data.*

An equivalent temperature has to be defined in order to apply Eq 7 since its parameters are temperature-dependent. This temperature was taken as the temperature of the maximum stress in the TMF cycle or as the temperature of the minimum stress. $N(\lambda)$ was therefore defined as the geometric mean of the values that are computed in each hypothesis. The model does not give any difference between in-phase and out-of-phase cycles per se, but of course differences in stress levels induced by cycle shape changes will give rise to differences in life prediction (The model gives for instance good predictions of life for an in-phase cycle as for a diamond-shaped cycle, see Fig. 7).

Identification of damage equations in virgin and oxidized CT specimens has been made accounting for the local stress redistributions due to the existence of a long crack. The application of these equations to a long crack in a real component would require use of local stresses ahead of the crack tip and thus a finite element computation of the cracked structure. However, if the prediction is limited to short crack lengths (and thus to engineering crack initiation) the stresses computed for the uncracked structure could be used in a first approximation.

Similarly, the stress applied to the bulk LCF or TMF specimens can be used for short crack length. This assumption is borne out by the fact that surface cracks in smooth specimens of various cast superalloys including IN-100 propagate at constant FCGR provided the crack depth is small enough (less than about 0.5 mm in solid specimens and 0.3 mm in hollow specimens) as recalled in the introduction [*4,5,8,9*]. FCGR is crack length dependent only for deeper cracks.

Comparison with Experimental Data

Equations 7 to 10 were used to compute the number of cycles to 0.1-mm crack depth in isothermal LCF. A large set of data under push-pull loading using axial strain control was available for IN-100 at 1000°C. Continuous strain cycling tests at high frequency (1 to 2 Hz), 5×10^{-2} Hz and 5×10^{-3} Hz as well as strain hold tests in tension were carried out [*4*]. Experimental data to 0.3-mm crack depth were available from potential drop measurements and N(0.1 mm) was taken as N(0.3 mm)/3 since surface cracks are known to propagate at a constant rate up to this depth in IN-100. Most calculated values are within a factor of three of experimental data (Fig. 8). (Note that Eq 1 was fitted to data from CT at 1000°C. Thus LCF life at high frequency is underestimated as evidenced by Fig. 8 and shown by the discrepancy with the best fit line of Fig. 3*b*.)

Equations 7 to 11 were used to compute the life under thermal-mechanical loading of bare IN-100 hollow specimens using the various TMF cycles described in the introduction. Here the number of cycles to 0.1-mm crack depth was available from plastic replicas on interrupted tests. Predictions are very good since most computed values are within a factor of three of experimental data (Fig. 8).

The same set of equations was applied to the TMF life of aluminized IN-100 [*6,7*]. Standard cycle II was used with a period of 3 min and in one case with a period of 1 h. Other cycles were also used as described in [*7*]. Equations for the bare alloy were used and the substrate was simply assumed to carry the whole load applied to the specimen. Predictions of the model to

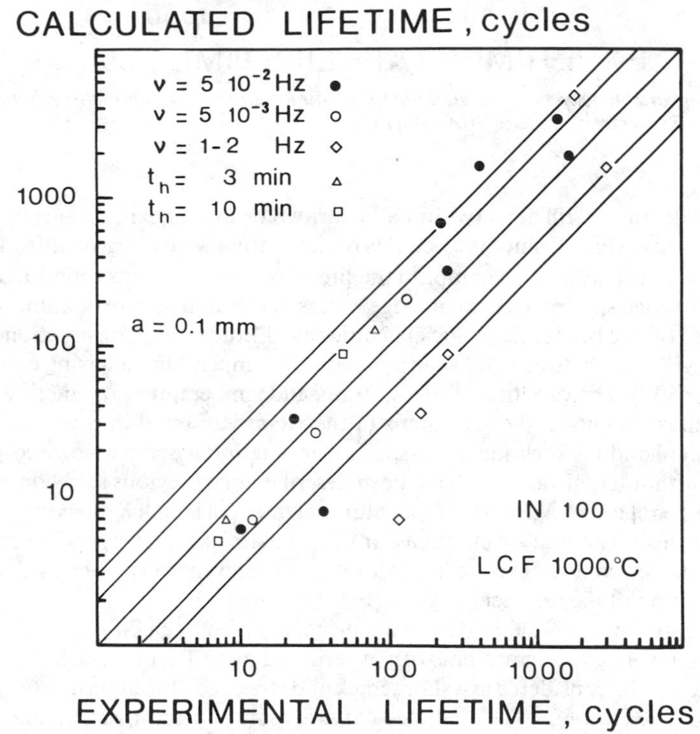

FIG. 8—*Comparison between calculated life to 0.1-mm-deep crack in low cycle fatigue at 1000°C and experimental data.*

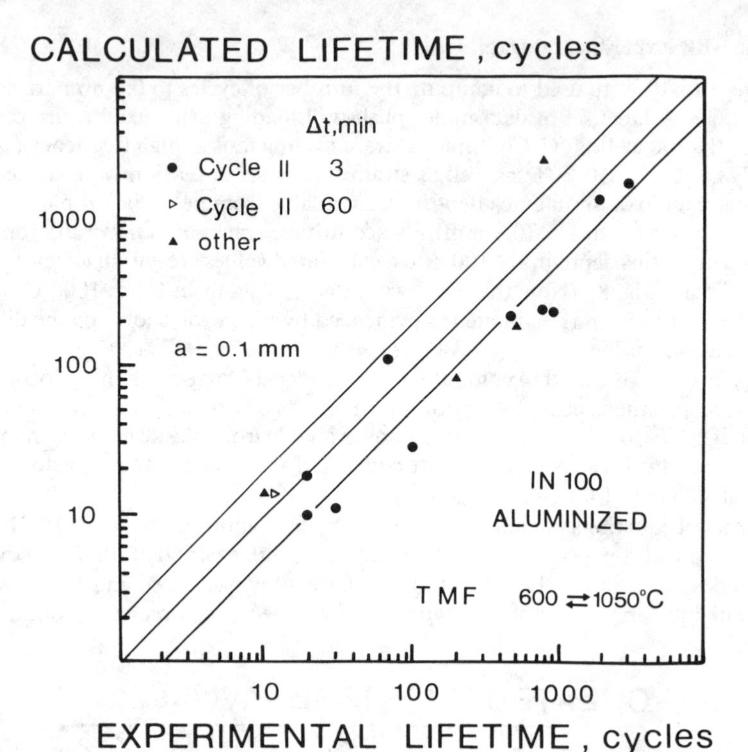

FIG. 9.—*Comparison between calculated life to 0.1-mm-deep crack in thermal-mechanical fatigue of aluminized IN-100 specimens and experimental data.*

0.1-mm crack depth are still good (within a factor two or four of experiment) but tend to be slightly conservative (Fig. 9) since the benefits of the coating were not accounted for.

The model was also applied to compute the life of wedge specimens submitted to thermal shock. IN-100 wedge specimens with an edge radius of 1-mm in a bare or aluminized condition were tested on the burner rig of Société Nationale d'Etude et de Construction de Moteurs d'Aviation (SNECMA) between 200°C and maximum temperature (allowing cooling for 20 s and heating for 60 s). The variation of stress, strain, and temperature as a function of time was available for the elements of the specimens in the vicinity of the thin edge [*7*]. The experimental procedure and the inelastic stress-strain computation based on the visco-plastic Chaboche model with internal variables have been described in a previous study on MAR-M509 [*22*]. The number of cycles $N(\lambda)$ to break a volume element of length λ was computed for each element of the mesh. The crack front was assumed to be straight and perpendicular to the midplan of the wedge specimens. The mean of all values for a given abcissa was used to plot crack length as a function of the number of cycles (Fig. 10).

For a coated specimen [*7*] submitted to a thermal cycle between 200 and 1050°C, the predicted curve is in very good agreement with experiment up to 1.5-mm crack depth. A deeper crack can no longer be considered as a short crack and stress redistribution due to crack growth should be taken into account. The predicted curve is more conservative for a maximum temperature of 1000°C, nevertheless life to 1-mm crack is within a factor of two of experiment. Further crack growth rate is well accounted in both cases (below 1.5-mm crack depth where the short crack approximation holds).

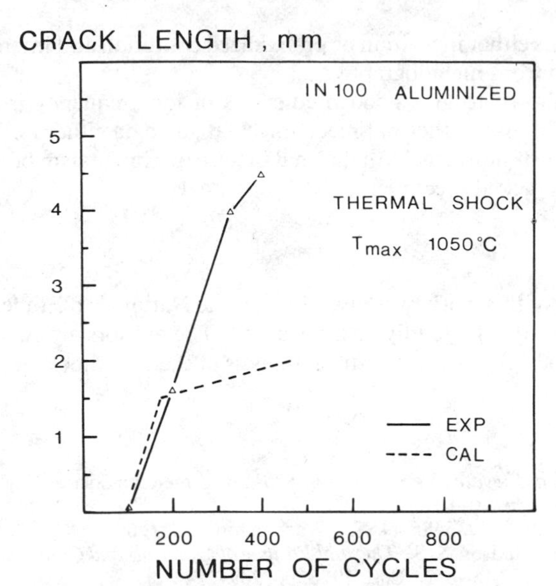

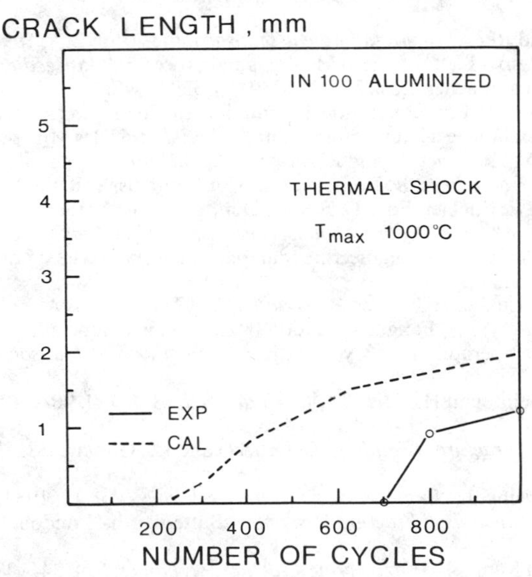

FIG. 10—*Variation of crack length versus the number of cycles in thermal shock experiments on aluminized IN-100. Comparison between experiment (solid line) and calculation (dashed line) :* (a) *for a maximum temperature of 1050°C and* (b) *for a maximum temperature of 1000°C.*

Conclusions

A model was proposed to compute the engineering life to crack initiation in IN-100 cast superalloy under thermal-mechanical fatigue loading.

This model describes the microcrack growth phase and takes into account oxidation fatigue interactions. The parameters of damage equations were identified using fatigue crack growth data on CT specimens either in a virgin or preoxidized condition and interdendritic oxidation kinetic data deduced from metallography.

The model was shown to give good predictions of low-frequency isothermal low cycle fatigue data at 1000°C and of thermal-mechanical fatigue data either for bare or aluminized alloy. The growth law of short cracks in thermal shock experiments can be accounted for using stress analyses of uncracked specimens.

Acknowledgments

Financial support of this work by SNECMA (Société Nationale d'Etude et de Construction de Moteurs d'Aviation) is gratefully acknowledged. The authors are indebted to SNECMA engineers for the experiments and the stress analysis of thermal shock wedge specimens.

References

[*1*] Taira, S., *Fatigue at Elevated Temperatures, STP 520,* American Society for Testing and Materials, Philadelphia, 1973, pp. 80–101.
[*2*] Spera, D. A., NASA-TND,-5485, NASA, Washington, DC, 1969.
[*3*] Halford, G. R. and Manson, S. S., *Thermal Fatigue of Materials and Components, ASTM STP 612,* D. A. Spera and D. F. Mowbray, Eds., 1976, pp. 239–254.
[*4*] Reger, M. and Rémy, L., *Materials Science and Engineering A,* Vol. 101, 1988, pp. 47–54 and 533–63.
[*5*] Malpertu, J. L. and Rémy, L., *Metallurgical Transactions A,* Vol. 21A, 1990, pp. 389–399.
[*6*] Bernard, H. and Rémy, L., "Advanced Materials and Processes," *Proceedings of EUROMAT 89,* H. E. Exner and V. Schumacher, Eds., Vol. 1, 1989, pp. 529–534.
[*7*] Bernard, H., "Influence d'une Protection d'aluminiure sur l'endommagement du superalliage à base de nickel IN 100 en fatigue à haute température," Thesis, Ecole des Mines de Paris, 1990.
[*8*] Rémy, L., Reger, M., Reuchet, J., and Rezai-Aria, F., in *High Temperature Alloys for Gas Turbines 1982, Conference Proceedings,* R. Brunetaud, D. Coutsouradis, T. B. Gibbons, Y. Lindblom, D. B. Meadowcroft, and R. Stickler, Eds., D. Reidel, Dordrecht, The Netherlands, 1982, pp. 619–632.
[*9*] Rémy, L., in *Fatigue 84, Proceedings of the Second International Conference on Fatigue and Fatigue Thresholds,* C. J. Beevers, Ed., Engineering Materials Advisory Services, London, United Kingdom, 1984, Vol. 1, pp. 15–30.
[*10*] Antolovich, S. D., Liu, S., and Bauer, R. *Metallurgical Transactions* A, Vol. 12A, 1981, pp. 73–81.
[*11*] Reuchet, J. and Rémy, L., "Fatigue-oxidation Interaction in a Superalloy, Application to Life Prediction in High Temperature Low Cycle Fatigue," *Metallurgical Transactions A,* Vol. 14A, 1983, pp. 141–149.
[*12*] Neu, R. W. and Sehitoglu, H., *Metallurgical Transactions A,* Vol. 20A, 1989, pp. 1755–1767 and 1769–1783.
[*13*] McClintock, F. A., *Fracture of Solids,* D. C. Drucker and J. J. Gilman, Eds., Interscience, New York, 1963, pp. 65–102.
[*14*] Chalant, G. and Rémy, L., *Engineering Fracture Mechanics,* Vol. 18, 1983, pp. 939–952.
[*15*] Knott, J. F., *Fundamentals of Fracture Mechanics,* Butterworths, London, United Kingdom, 1973, pp. 251–256.
[*16*] Rézai-Aria, F. and Rémy, L., *Engineering Fracture Mechanics,* Vol. 34, 1989, pp. 283–294.
[*17*] Tracey, D. M., *Journal of Engineering Materials Technology,* Vol. 98, 1976, pp. 146–151.
[*18*] Tracey, D. M., *Journal of Engineering Materials Technology,* Vol. 99, 1977, pp. 187–188.
[*19*] Rice, J. R., *Fatigue Crack Propagation, ASTM STP 415,* American Society for Testing and Materials, Philadelphia, 1967, pp. 247–311.
[*20*] Reger, M. and Rémy, L., *Metallurgical Transactions A,* Vol. 19A, 1988, pp 2259–2268.
[*21*] François, M. and Rémy, L., unpublished results, Centre des Matériaux, 1986.
[*22*] Malpertu, J. L., Thesis, Ecole des Mines de Paris, 1987.
[*23*] Fisher, J. C., *J. Applied Physics,* Vol. 22, 1951, p. 74.
[*24*] Rézai-Aria, F., François, M., and Rémy, L., *Fatigue Fracture Engineering Material Structures,* Vol. 11, 1988, pp. 277–289.

Yavuz Kadioglu[1] *and Huseyin Sehitoglu*[1]

Modeling of Thermomechanical Fatigue Damage in Coated Alloys

REFERENCE: Kadioglu, Y. and Sehitoglu, H., **"Modeling of Thermomechanical Fatigue Damage in Coated Alloys,"** *Thermomechanical Fatigue Behavior of Materials, ASTM STP 1186,* H. Sehitoglu, Ed., American Society for Testing and Materials, Philadelphia, 1993, pp. 17–34.

ABSTRACT: A life prediction model that determines the contribution of fatigue, creep and environmental damage to failure was developed for an aluminide coated nickel-based superalloy, Mar-M247. In the first phase of the study, isothermal (IF) and thermomechanical fatigue (TMF) experiments were conducted to investigate the experimental damage mechanisms. In the second phase, an analytical technique was advanced to compute the stress fields due to a surface inclusion in a half-space where the inclusion simulates the oxide spike. The technique is based on Eshelby's equivalent inclusion method and elucidates the mismatch in elastic moduli and thermal expansion coefficients of the matrix and the oxide spike on local strain fields. The fatigue life results of several experiments, along with the local stress-strain field in the vicinity of an oxide spike, were employed to define the model constants. Life prediction bounds are established corresponding to short-time coating protection, where the coating provides inconsiderable protection to substrate, and long-time coating protection, where the coating provides appreciable protection to the substrate. For in-phase loading, since the failure is governed by creep damage, the nature of coating protection did not influence the fatigue lives. The out-of-phase predictions corresponding to short-time coating protection and the experimental data coincided as the maximum temperature increased in the experiments, confirming that the coating provides unsubstantial protection at higher temperatures.

KEYWORDS: thermomechanical fatigue, high temperature, oxidation, nickel-based superalloy, coating, life-prediction, surface inclusion

Oxidation damage mechanism restricts the use of many advanced materials at elevated temperatures for exacting applications. Bare nickel based superalloys undergo copious oxidation at temperatures exceeding 700°C. Diffusion or overlay coatings are the most common among those developed [*1–2*] to circumvent the deleterious oxidation effects. These coatings fulfill their protective role against oxidation under stress-free conditions. However, if the material experiences combined thermal and mechanical loading, the integrity of the oxide and the surrounding material could be severely hindered. The oxide properties could differ substantially from the substrate and the coating, resulting in complex local strain fields. The evaluation of these strain fields is imperative in advancing models of fatigue failure at high temperatures.

Considerable research has focussed on monotonic tension [*3–5*], high cycle fatigue [*6–15*], creep [*16–19*], and low cycle isothermal fatigue [*20–25*] behavior of coated alloys. Few studies have considered the thermomechanical fatigue (TMF) behavior of coated superalloys [*26–28*]. In this material system, damage in out-of-phase loading, in which the material experiences tensile mechanical strain at the minimum temperature, has been found to exceed that

[1] Research associate and professor (temporarily, director, mechanics and materials program, National Science Foundation, Washington, DC), respectively, Department of Mechanical Engineering, University of Illinois, Urbana, IL 61801.

in the in-phase loading condition [*26–27*]. Consequently, the fatigue life of the coated alloy may be lower by 7.5 times of the uncoated alloy [*27*].

Although numerous life prediction methods have been forwarded for uncoated superalloys, few studies were directly concerned with the fatigue life prediction of coated superalloys. For example, coating cracking lives have been linked to total strain, expressed as the summation of the thermal expansion mismatch strain and the mechanical strain [*29–30*], although this approach is considered rudimentary. A fatigue crack-growth model has also been proposed [*31*] in which the penetration of a coating crack into the base metal has been analyzed using fracture mechanics concepts. In recent work [*27*], the mechanical damages for the coating and the substrate have been calculated separately and then combined to produce an optimum prediction damage parameter. Hysteresis energy has been propounded for estimating coating cracking lives in Ref *27*. The shortcomings of hysteresis energy approach are well-recognized. None of these studies explicitly considered the stresses and strains associated with mismatches arising from oxidation. Since oxidation damage plays a discernible role in the damage evolution in coated and uncoated alloys, the need for physically based models is crucial.

Under thermomechanical loading conditions, additional strains on the base metal, the oxide, and the coating may arise primarily due to thermal expansion mismatch, elastic moduli mismatch, and with diffusion between coating and substrate, phase transformation, and chemical reaction with the environment also playing a role. These additional strains and stresses may promote the early formation of the cracks, which forms an easy path for the oxidation environment to reach the substrate. The local oxidation at the coating metal interface is modeled as a semispherical inhomogeneity on the surface of half space. A substrate oxidation model based on the premature cracking of the coating is proposed. Then, a life prediction methodology which accommodates the thermal expansion and elastic modulus mismatches between the oxide and the matrix is employed to estimate the thermomechanical fatigue lives of the superalloy investigated in this study.

In summary, the objectives of this paper are (1) to present new results on stress distribution in the surface oxide and the matrix under thermal and mechanical loads and (2) to forward a life prediction model to handle the coated alloys based on the reported rigorous results of mismatch strains on oxide spikes.

Experimental Procedure

Material

The material studied is a Ni-based superalloy, Mar-M247 which was coated with Alpak-S1 aluminide coating. The coating consists of a manganese and aluminum powder slurry that is applied by painting, or dipping, or spraying with an air brush. The parts are heated in a protective atmosphere at 1093°C for two hours to form a nickel-aluminum coating containing manganese. This coating resulted in higher ductility compared to simple aluminide coatings.[2] The test bars were initially sectioned from cast turbine rotors and then coated by using the technique explained above. The microstructure of the substrate consists of a γ matrix strengthened by cubic γ' nominally 0.75 μm in size. Also present were intermittent grain boundary carbides and script-type MC carbides. The typical microstructure of the substrate is shown in Fig. 1, and the nominal chemical composition of Mar-M247 is listed in Table 1.

[2] Michael Barber, General Motors Corp., P. O. Box 420, Indianapolis, IN 46206-0420. , private communication, 1990.

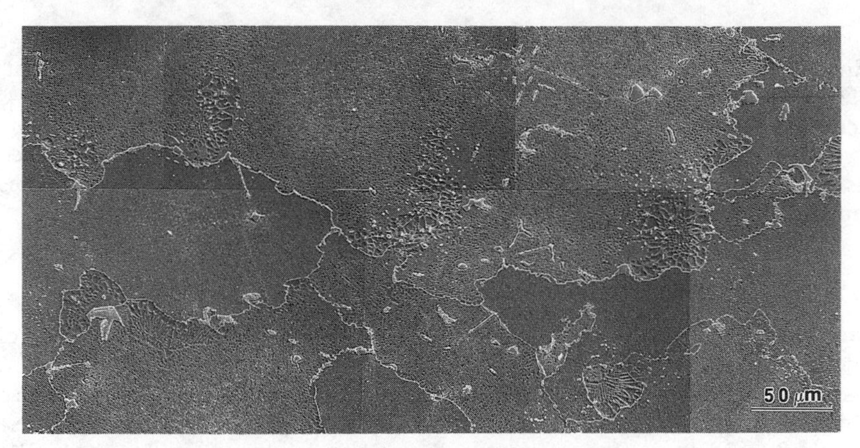

FIG. 1—*Microstructure of Mar-M247.*

Test Equipment

Tests were performed on a computer controlled 100-kN closed loop servohydraulic test system under strain control utilizing a SOMAT 1002 automation system. RF induction heating (Lepel 2.5 kW capacity) allowed rapid heating of the specimen. Thermocouples were attached to the specimen's gage section to measure the instantaneous temperature. In some of the low strain range tests thermocouples were attached to the shoulder of the specimen to prevent premature crack initiation.

The cooling portion of temperature cycling occurred naturally (that is, no forced cooling was used). Total strain was measured over the gage length with a 12.7-mm axial extensometer utilizing either quartz or ceramics rods. Strain, temperature, and load data were recorded by the computer. The thermal strain profile for the specific alloy was first recorded under zero load. Then, the total strain, which is controlled, was obtained by adding the thermal strain and the desired mechanical strain using the following equation

$$\varepsilon_{net} = \varepsilon_{mech} + \varepsilon_{th} \tag{2.1}$$

where ε_{net} is net strain, ε_{mech} and ε_{th} are mechanical strain and thermal strain, respectively.

Test Conditions

The isothermal fatigue (IF) and thermomechanical fatigue (TMF) experiments were performed under total strain control with $R_\varepsilon = -1$ (that is, R_ε = minimum mechanical strain/maximum mechanical strain). Strain ranges of 0.01, 0.0045, and 0.003 35 were considered. In the experiments, the temperature was cycled between T_{min} = 500°C and T_{max} = 871°C. The tests were initiated at the mean temperature of T_0 = 685°C. Additionally, thermomechanical

TABLE 1—*Nominal chemical composition of Mar-M247 by weight %.*

C	Ni	Cr	Co	W	Al	Ta	Ti	Mo	Hf	B	Zr	Fe
0.13	bal	8.4	10.0	10.0	5.5	3.05	1.05	0.65	1.4	0.015	0.055	0.25

fatigue experiments were also conducted with T_{min} = 500°C and T_{max} = 1038°C. In these experiments the mean temperature was T_0 = 769°C. The strain rate was maintained at $\dot{\varepsilon} = 5.0 \times 10^{-5}$ 1/s. All tests were concluded at a 15% load drop, although several specimens fractured completely before 15% load drop was attained. Upon conclusion of the experiments, the specimens were cut longitudinally and prepared for examination under scanning electron microscopy (SEM). Details of these experiments along with microstructural observations can be found in Ref *32*.

Stress State in the Vicinity of Oxide Spikes

Under elevated temperature conditions, a protective oxide scale forms on the surface of the specimen, which separates the substrate from the environment. However, spalling and cracking of the protective oxide scale occurs due to stresses developed in the scale [*33,34*]. The principal sources of stress in the oxide scale are the thermal stresses [*35*] due to the difference between the thermal expansion coefficients of the oxide and the matrix. Although there is zero thermal stress at the oxide formation temperature, upon cooling by ΔT, a stress is generated in the oxide layer and is given by [*36*]

$$\sigma^{ox} = \frac{E_{ox}\Delta T(\alpha^{ox} - \alpha^{m})}{1 + 2\left(\dfrac{E_{ox}t_{ox}}{E_{m}t_{m}}\right)} \tag{3.1}$$

where E and α are the elastic modulus and the thermal expansion coefficient, respectively, and t is the thickness. The subscripts *ox* and *m* denote oxide and metal, respectively. It should be noted that this equation is applicable only to a uniform surface scale. The stress fields due to thermal expansion differential between oxide spikes and the metal is complex, exhibiting spatial variation, and cannot be predicted by this equation. Oxide spikes penetrate from the surface towards the inside of the substrate. The oxide spike morphology could form at the surface or at the coating/substrate interface upon failure of the coating.

In this study, we model the oxide spike as a surface inhomogeneity subjected to thermal and mechanical loading as shown in Fig. 2. Although the technique can handle different aspect ratios (a,b,c) of the oxide spike [*32*] and different remote stress states, we will consider the case of a spherical spike (a = b = c) under uniaxial remote loading. In the literature, the solutions have been reported for the loading types which result in axi-symmetric eigenstrains [*37*] or certain types of eigenstrains in the surface inclusions [*38*]. However, the problem of a surface inhomogeneity sustaining nonsymmetric eigenstrains has not been addressed in the literature and will be studied here.

Stress Field Due to a Semi-Ellipsoidal Surface Inhomogeneity

The stress field due to an inhomogeneity in an infinite medium can be evaluated following Eshelby's solutions [*39,49*]. The transformation problem solved by Eshelby considers an inhomogeneity embedded in an infinite medium. Upon loading, the inhomogeneity is assumed to undergo a stress-free transformation under a fictitious equivalent eigenstrain ε^{*}_{kl}. However, since its deformation is constrained by the surrounding matrix, a perturbed strain field results. The problem, then, is to find how a remotely applied loading, ε^{0}_{kl}, is disturbed by the existence of the inhomogeneity. Eshelby proved that the stress field in a single ellipsoidal inhomogeneity sustaining an eigenstrain, ε^{p}_{kl} (that is, $\varepsilon^{p}_{kl} = (\alpha^{ox} - \alpha^{m})\,\Delta T\,\delta_{kl}$ in this work) is constant and given by

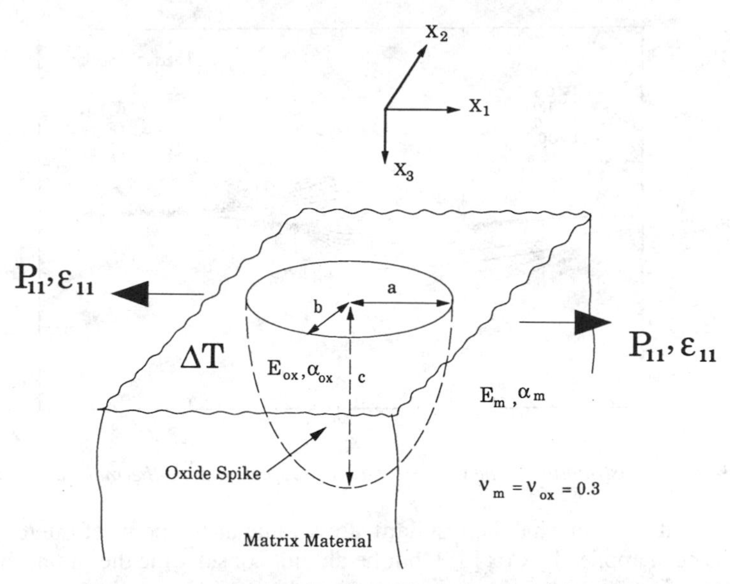

FIG. 2—*An oxide spike in a semi-infinite plane subjected to thermal and mechanical loads.*

$$\sigma_{ij}^{ox} = \sigma_{ij}^{0} + \sigma_{ij} = C_{ijkl}^{ox}(\varepsilon_{kl}^{0} + \varepsilon_{kl}^{cn} - \varepsilon_{kl}^{p}) = C_{ijkl}^{m}(\varepsilon_{kl}^{0} + \varepsilon_{kl}^{cn} - \varepsilon_{kl}^{*} - \varepsilon_{kl}^{p}) \quad (3.2)$$

where $\varepsilon_{ij}^{0}(\sigma_{ij}^{0})$ represents the undisturbed remote strain (stress) field. C_{ijkl}^{ox} and C_{ijkl}^{m} denote the inhomogeneity (that is, oxide) and the matrix elastic moduli, respectively. Constrained strain is ε_{kl}^{cn} given by

$$\varepsilon_{kl}^{cn} = S_{ijkl}(\varepsilon_{kl}^{p} + \varepsilon_{kl}^{*}) \quad (3.3)$$

where S_{ijkl} is the Eshelby's tensor.

Equation 3.2 is solved along with Eq 3.3 for ε_{kl}^{*}. Then, the uniform stress field inside the inhomogeneity can be found from Eq 3.2. The stress field outside the inhomogeneity is not uniform and is given by

$$\sigma_{ij}^{m}(\mathbf{x}) = C_{ijkl}^{m}(\varepsilon_{kl}^{0} + \varepsilon_{kl}^{cn}) = C_{ijkl}^{m}(\varepsilon_{kl}^{0} + D_{ijkl}(\mathbf{x})(\varepsilon_{kl}^{p} + \varepsilon_{kl}^{*})) \quad (3.4)$$

Analytical determination of $D_{ijkl}(\mathbf{x})$ for ellipsoids is given in Ref *41*. Note that at the interior points of the ellipsoids $D_{ijkl}(\mathbf{x})$ is equal to S_{ijkl}.

The coordinate system and a schematic of oxide spike and the surrounding material is shown in Fig. 2. The dimensions of the ellipsoid are indicated by a, b, and c. The remote stress is denoted as P_{11} and the component of remote strain ε_{kl}^{0} in 11 direction is marked as ε_{11}. The semi-infinite plane could also undergo a temperature change of ΔT. In this study, we take the elastic and thermal properties to be isotropic and $\upsilon_{ox} = \upsilon_{m}$. In the first step, we compute the stress field associated with an inhomogeneity in an infinite medium using Eqs 3.2 through 3.4. We then introduce the free surface by applying a surface traction, $f_i(x_1,x_2)$, to negate the stresses calculated in the first step on the X_1-X_2 plane. Due to the nature of the loading considered in this study, shear stresses on X_1-X_2 plane vanish, and the only non-zero stresses are the normal stresses. Finally, the stresses calculated in the first step and the stresses due to application of f_i

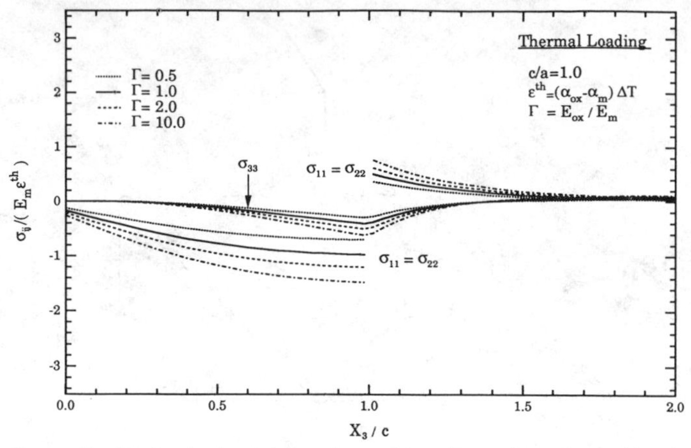

FIG. 3—*Stress distribution in the vicinity of an oxide spike under a thermal loading of* ΔT.

are superimposed to determine the resultant stress value at the point of interest. A similar approach has been applied by Cox [*38*], but he did not consider the thermomechanical mismatch problem. Details of the approach can be found in Ref *32*.

In Fig. 3, the stress field in the vicinity of a semispherical oxide spike is presented under a thermal loading of ΔT for several elastic modulus ratios. The normalized stress components are $\sigma_{ij}/E_m\varepsilon^{th}$ which is the stress tensor normalized by the product of matrix modulus and thermal mismatch strain. Thermal mismatch strain, assuming isotropic properties, is defined as $\Delta T(\alpha_{ox} - \alpha_m)$. We denote Γ as the ratio of oxide elastic modulus to matrix elastic modulus. The region $0 < X_3/c < 1$ represents the oxide spike, $X_3/c > 1$ represents the matrix region. The stresses at the oxide metal interface are three-dimensional and increase with increasing Γ. In Fig. 4, the stress field under mechanical loading is presented for the elastic modulus ratio of 2 which is approximately the ratio for the oxide/substrate system considered in our study. We note the nearly one-dimensional local stress fields due to mechanical loading. This is in constrast to the previous figure where the stress fields were highly multiaxial.

The effects of elastic modulus ratio, E_{ox}/E_m, on the strain field at the tip of the oxide spike are shown in Fig. 5 for mechanical loading and in Fig. 6 for thermal loading. In Fig. 5, vertical axis represents the ratio of the mechanical strains at the oxide tip to the remotely applied strain.

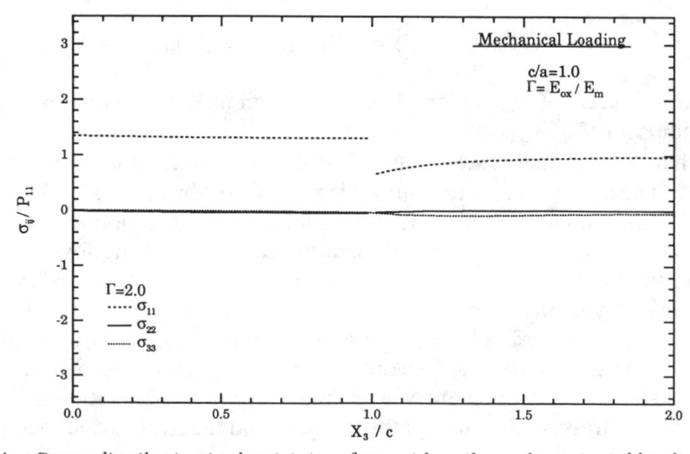

FIG. 4—*Stress distribution in the vicinity of an oxide spike under uniaxial loading.*

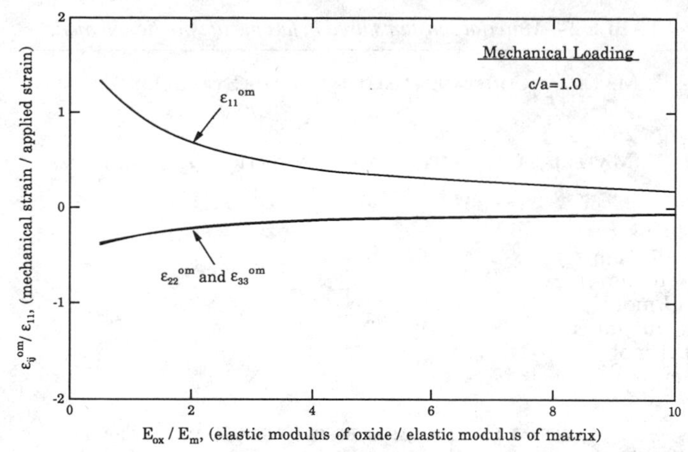

FIG. 5—*Strain distribution at the tip of the oxide due to mechanical loading.*

The horizontal axis is E_{ox}/E_m and denotes the oxide modulus to matrix modulus ratio. The superscript on the strains, *om*, designates oxide strains due to mechanical loading. The vertical axis in Fig. 6 is the ratio of mechanical strain at the oxide tip to the thermal mismatch strain. The superscript on the strains, *ot*, designates oxide strains due to thermal loading. The results illustrate that the magnitude of local strains in the oxide increases as its modulus falls below that of matrix. For the case $E_{ox}/E_m < 3$, the combined thermal and mechanical strains ($= \varepsilon_{11}^{om} + \varepsilon_{11}^{ot}$) could readily exceed the remote strain. To characterize the advance of oxide tip, we will later use ($= \varepsilon_{11}^{om} + \varepsilon_{11}^{ot}$) at $X_3/c = 1^-$ where the minus sign corresponds to the location infinitesimally less than 1 (that is, oxide tip).

Life Prediction Methodology for Coated Mar-M247

In this study, the life prediction methodology proposed by Neu and Sehitoglu [*42,43*] is revised to estimate the fatigue lives of coated superalloys. Mechanical strain range, strain rate, temperature and phasing of temperature, and the mechanical strain-range effects reside in this model. The model has been successfully employed in the life prediction of 1070 Steel [*43*], a

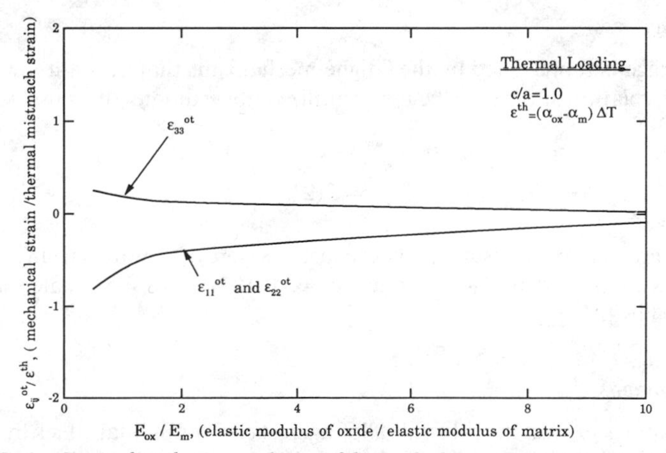

FIG. 6—*Strain distribution at the tip of the oxide due to thermal loading* ΔT.

TABLE 2—*Material constants used in fatigue life prediction model.*

MATERIAL CONSTANTS USED IN FATIGUE STRAIN LIFE TERM

$C = 0.014$
$d = 0.186$

MATERIAL CONSTANTS USED IN OXIDATION DAMAGE TERM

$a' = 0.75$
$\beta = 1.5$
$B = 6.93 \times 10^{-3}\ s^{-0.5}$
$\delta o = 2.16 \times 10^{-10}\ \mu m/s^{-0.75}$
$D^{ox} = 1.54 \times 10^{4}\ \mu m^2/s$
$Q^{ox} = 175.9$ kJ/mol
$D\gamma' = 8.57 \times 10^{3}\ \mu m^2/s$
$Q\gamma' = 163.3$ kJ/mol
$h_{cr} = 461.4\ \mu m$
$\xi^{ev} = .462$

MATERIAL CONSTANTS USED IN CREEP DAMAGE TERM

$A = 5.88 \times 10^{25}\ s^{-1}$
$m = 11.6$
$\Delta H = 536.4$ kJ/mol
$\xi^{creep} = 0.34$

nickel-based superalloy, Mar-M247 [*32,44*] and a metal matrix composite, particulate silicon carbide reinforced aluminum [*45*].

In the proposed model, total damage per cycle is considered as the summation of fatigue, creep and oxidation damage terms

$$D^{total} = D^{fatigue} + D^{oxidation} + D^{creep} \tag{4.1}$$

This equation can also be written in terms of the life, N_f, and when linear damage is equal to unity failure occurs

$$\frac{1}{N_f} = \frac{1}{N_f^{fatigue}} + \frac{1}{N_f^{oxidation}} + \frac{1}{N_f^{creep}} \tag{4.2}$$

We note that although the damage terms are expressed separately in Eq 4.2 they are coupled through strain, temperature, and mismatch strains.

Fatigue Damage

Fatigue damage is represented by the fatigue mechanisms that occur at low temperatures. The strain-life relationship given below is utilized to estimate the pure fatigue damage component

$$\frac{\Delta\varepsilon_{mech}}{2} = C(2N_f^{fatigue})^d \tag{4.3}$$

where C and d are material constants. These constants were determined from low temperature isothermal tests conducted at $T = 500°C$ and $\dot{\varepsilon} = 5.0 \times 10^{-5}\ sec^{-1}$ on coated superalloys and are given in Table 2.

Oxidation Damage

Since the coatings protect the substrate alloy against environmental attack by forming a protective oxide scale, the oxidation damage in the substrate is activated only after complete coat-

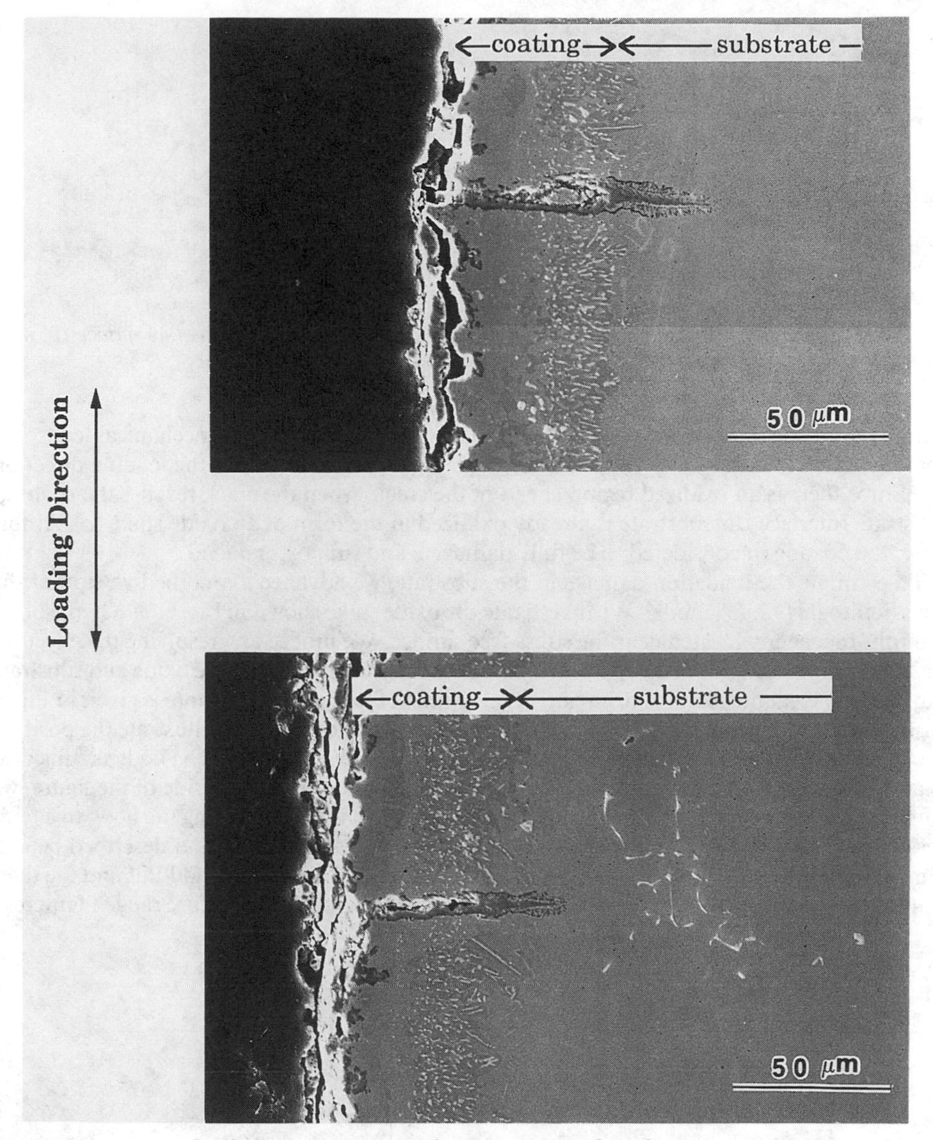

FIG. 7—*Micrographs showing a coating crack penetrating to the substrate on TMF OP tested Mar-M247/Alpak-S1.* $\Delta\varepsilon_m$ = *1%,* T_{max} = *871°C,* T_{min} = *500°C.*

ing cracking (that is, neglecting internal oxidation). The formation of the protective oxide scale on the coating surface is followed by spalling and microfracture of this scale. This exposes fresh material surfaces to the oxidizing environment, resulting in the formation of a new oxide scale, which again ruptures when the oxide reaches a critical thickness. The repeated oxide fracture provides the conduit for crack growth into the substrate. If the coating cracks, this provides an easy path for the oxidizing environment to reach the substrate alloy and result in local oxidation in the substrate at the tip of coating crack. In Fig. 7, a complete coating crack followed by oxidation and cracking of substrate is shown. The proposed model for the formation of an oxide spike on the substrate is shown schematically in Fig. 8. In the model it is proposed that

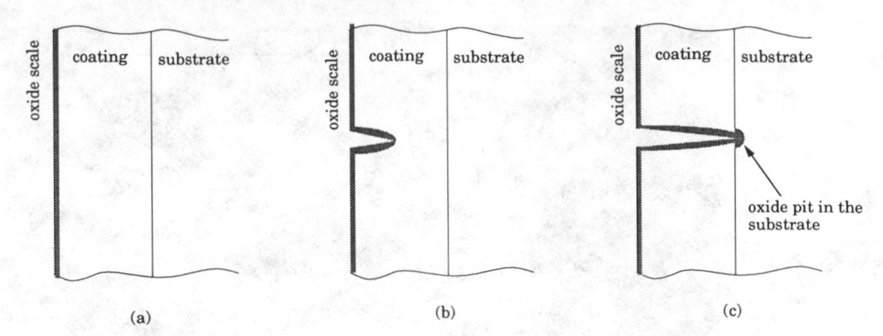

FIG. 8—*Schematic of oxide formation on the substrate:* (a) *Oxidation of the coating surface,* (b) *initiation of a coating crack, and* (c) *oxidation of the substrate.*

initially an oxide scale forms on the surface of the coating (*a*). Under mechanical loads, the coating cracks and this crack propagates along a path perpendicular to the loading direction (*b*). Since there is an oxidized region ahead of the crack, when the crack reaches the coating/substrate interface the substrate is already oxidized in the form of an oxide spike (*c*). At this time, the coating in considered to be fully damaged, and supports no load.

To estimate the oxidation damage in the substrate, we advance a modified version of the Neu-Sehitoglu [*42,43*] model. We investigate the oxide spike shown in Fig. 8*c* as a hemispherical inhomogeneity located near the stress-free surface. We proceed by resolving the effects of elastic modulus and thermal expansion coefficient mismatch between the oxide and substrate alloy on the local mechanical behavior. A schematic of the local strain state is given in Fig. 9 for thermal and mechanical loadings. On the left hand side of Fig. 9, we illustrate the position of ε_{11}^{om} at the tip of oxide with a period · that corresponds to $X_3/c = 1^-$. The body is subjected to a remote strain of ε_{ij} under isothermal conditions. On the right hand side of the figure, we illustrate the ε_{11}^{ot} component at the oxide tip (denoted as ·) upon subjecting the body to a temperature change of ΔT. The stress field is calculated by using the method described before. Throughout the analysis, an average value was employed for the elastic modulus and the thermal expansion coefficient of the superalloy Mar-M247 over the temperature range of interest.

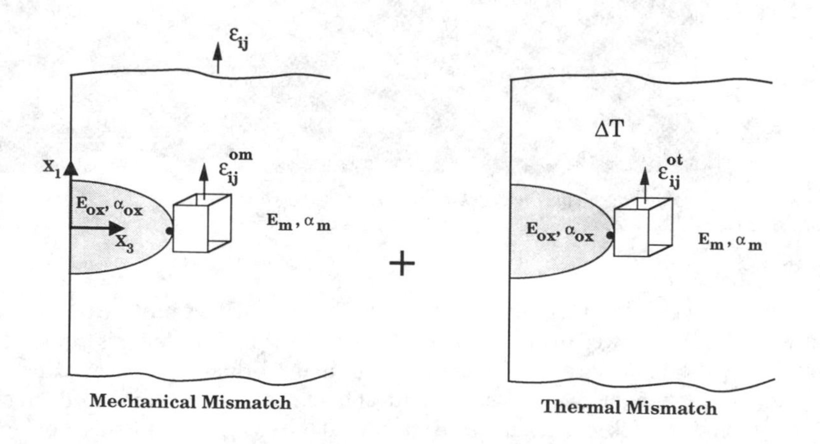

FIG. 9—*Strain state at the tip of the oxide spike under mechanical and thermal loading.*

TABLE 3—*Average mechanical and thermal properties.*

Material	CTE, α (mm/mm/°C)	Elastic Modulus, E (MPa)	Reference
Mar-M247	16.0×10^{-6}	180 000	*44*
Al_2O_3	8.0×10^{-6}	340 000	*47*

In the analysis of stresses and strains the temperature change ΔT is imposed with initial temperature as equal to T_{max}. Furthermore, the physical and mechanical properties of the oxide spike were taken as that of aluminum oxide (Al_2O_3). The average values used in the analysis are listed in Table 3.

The oxidation damage in the substrate alloy after the complete coating failure is given as

$$\frac{1}{N_f^{ox}} = \left[\frac{h_{cr}\,\delta_0}{B\Phi^{ox}(K_{peff}^{0} + K_{peff}^{\gamma'})}\right]^{-1/\beta} \frac{2(\Delta\varepsilon_{mech}^{ox})^{2/\beta+1}}{\dot{\varepsilon}^{(1-a'/\beta)}} \tag{4.4}$$

where h_{cr} is a critical oxide length where environmental attack trails behind the crack tip advance, δ_0 is the ductility of the environmentally effected material, Φ^{ox} is the phasing factor for environmental damage, K_{peff}^{0} is the effective parabolic oxidation constant and $K_{peff}^{\gamma'}$ is the effective parabolic γ' depletion constant, B, a', and β are constants, and the values suggested in Refs *42* and *43* were applied in this analysis. The mechanical strain range $\Delta\varepsilon_{mech}^{ox}$ is obtained from results shown in Figs. 3 and 4. It represents the total strain component ($= \varepsilon_{11}^{om} + \varepsilon_{11}^{ot}$) at $X_3/c = 1^-$ where the minus sign corresponds to the location infinitesimally less than 1 (oxide tip). The effective parabolic constants, K_{peff}^{0} and $K_{peff}^{\gamma'}$, are defined as

$$K_p^{eff} = \frac{1}{t_c}\int_0^{t_c} D_0 \exp\left[-\frac{Q}{RT(t)}\right] dt \ (\text{cm}^4/\text{s}) \tag{4.5}$$

where t_c is the period of the cycle, D_0 is the diffusion coefficient, Q is the activation energy for oxidation, R is the universal gas constant, and $T(t)$ is temperature as a function of time. The phasing factor, Φ^{ox}, was introduced to quantify relative oxidation damage between phasings. The phasing factor is defined as follows

$$\Phi^{ox} = \frac{1}{t_c}\int_0^{t_c} \phi^{ox}\, dt \tag{4.6}$$

$$\phi^{ox} = \exp\left[-\frac{1}{2}\left(\frac{\dot{\varepsilon}_{th}/\dot{\varepsilon}_{mech} + 1}{\zeta_{ox}}\right)^2\right] \tag{4.7}$$

The form of ϕ^{ox} was chosen to represent the observed severity of oxide cracking for different phasing conditions. The parameter ζ_{ox} is a measure of the relative amount of oxidation damage for different thermal strain to mechanical strain ratios and is extracted from the experiments [*42,43*]. Since the substrate damage is considered here, all of the constants were taken as that of uncoated Mar-M247 [*32,44*] and are given in Table 2.

Employing Eq 4.4 implies that there is only short-term coating protection (that is, coating fractures prematurely early in life). On the other hand, if D_{ox} is taken as zero, this signifies that the coating provides long-term protection, hence the failure of the coated component would be governed by the creep and fatigue damage mechanisms only. The actual number of cycles for a coating crack to initiate, advance within the coating, and reach the substrate is not explicitly evaluated. In the current work, the oxidation damage equation is revised by incorporating the function $(1/\Psi)$ to reflect the protective role of coating against pure environmental attack. The oxidation damage equation for the coating/substrate system reads

$$\frac{1}{N_f^{ox}} = \left[\frac{h_{cr}\,\delta_0}{B\Phi^{ox}(K^0_{peff} + K^{\gamma'}_{peff})}\right]^{-1/\beta} \frac{2(\Delta\varepsilon^{ox}_{mech})^{2/\beta+1}}{\dot{\varepsilon}^{(1-a'/\beta)1/\Psi}} \tag{4.8}$$

where $\Psi \rightarrow 1$ for short-term coating protection and $\Psi \rightarrow \infty$ for long-term coating protection. Although Ψ is expected to be a function of experimental variables such as strain range, strain rate, temperature and coating character, metallurgical variables, for simplicity, in this work, it is taken as

$$\Psi = \Psi_1(\Delta\varepsilon_{mech}, T_{max}) = r_0/\Delta\varepsilon_{mech} \tag{4.9}$$

where r_0 is a constant. Then using one TMF out-of-phase experiment r_0 was found to be 0.012 for the experiments corresponding to T_{max} = 871°C and 0.010 for the T_{max} = 1038°C TMF OP experiments.

Creep Damage

The proposed creep damage term is a function of temperature, effective stress, and hydrostatic stress components and takes into account the creep damage mechanisms which may operate under compression [*42,43*]. The total creep damage is obtained by integrating the creep damage in each cycle throughout the fatigue life of the material

$$D^{creep} = \int_0^{t_c} A\Phi^{creep} \exp(-\Delta H/RT(t))(\alpha_1\bar{\sigma} + \alpha_2\sigma_H)/K)^m\,dt \tag{4.10}$$

where Φ^{creep} is the phasing factor for creep, ΔH is the activation energy for the rate-controlled creep mechanism, R is the gas constant, $T(t)$ is temperature as a function of time, $\bar{\sigma}$ is the effective stress, σ_H is the hydrostatic stress, and K is the drag stress. The constants α_1 and α_2 account for the degree of damage occurring under tension and compression. A and m are material constants. The constant m and activation energy ΔH are taken as those of the uncoated alloy [*44*]. The constant A is calculated from a high strain range TMF IP test. As is evident from Fig. 10 creep damage is dominant under in phase TMF loading. Many cracks are started at the coating/substrate interface and therefore environment has a small effect on

FIG. 10—*Micrograph showing a crack initiated from coating/material interface running along the grain boundary along with other multiple cracks propagating through inside of TMF IP tested Mar-M247/Alpak-S1.* $\Delta\varepsilon_m$ = *1%,* T_{min}... T_{max} = *871°C,* T_{min} = *500°C.*

TABLE 4—*Simple constitutive law to predict stresses.*[a]

$$\dot{\varepsilon}^{in} = A_c \exp\left[-\Delta H_c/R(T + 273)\right] f(\bar{\sigma}/K)$$

$$f(\bar{\sigma}/K) = \begin{cases} \exp\left[(\bar{\sigma}/K)^{17.5} - 1\right] & \bar{\sigma}/K \geq 1 \\ (\bar{\sigma}/K)^{11.6} & \bar{\sigma}/K < 1 \end{cases}$$

[a] $A_c = 1.33 \times 10^{23}\ s^{-1}$,
$\Delta H_c = 536.4$ kJ/mol,
$E = 253\,900 - 107.8$ T (MPa),
$K = K_{sat}$ (in this study), and
$K_{sat} = 886.1 - 0.376$ T (MPa).

the operating damage mechanisms under TMF in phase condition. The phasting factor, Φ^{creep}, was introduced to account for the effect of mechanical strain-temperature phasing. The form of the creep phasing factor is the same as the oxidation phasing factor

$$\Phi^{creep} = \frac{1}{t_c} \int_0^{t_c} \phi^{creep}\, dt \tag{4.11}$$

$$\phi^{creep} = \exp\left[-\frac{1}{2}\left(\frac{\dot{\varepsilon}_{th}/\dot{\varepsilon}_{mech}^{-1}}{\zeta^{creep}}\right)^2\right] \tag{4.12}$$

Stress-strain response during fatigue loading is required to calculate the creep damage experienced by the material. A unified constitutive equation proposed by Sehitoglu-Slavik [46] was used to calculate the hysteresis loops. The form of the constitutive equation and the material constants for the Mar-M247 are given in Table 4. The model predicts the isothermal, thermomechanical, and strain rate sensitivity effects successfully.

Life Prediction Results

Life prediction results for 871°C IF, 500 to 871°C TMF OP and 500 to 871°C TMF IP prediction are given in Figs. 11 through 13. The 500 to 1038°C TMF OP experimental results and

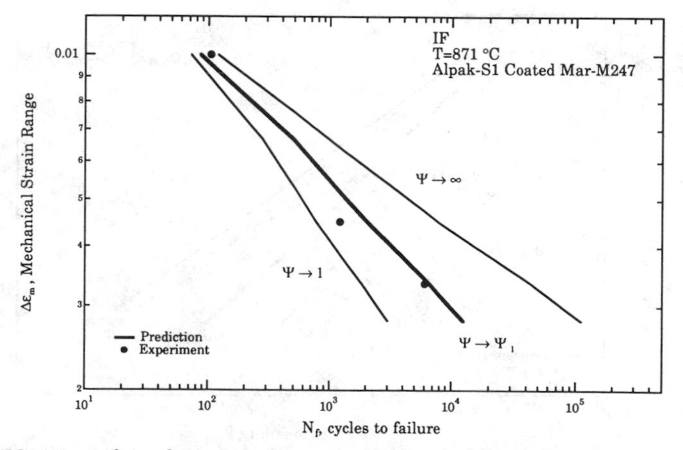

FIG. 11—*Life prediction results on Alpak-S1 coated Mar-M247; IF loading.*

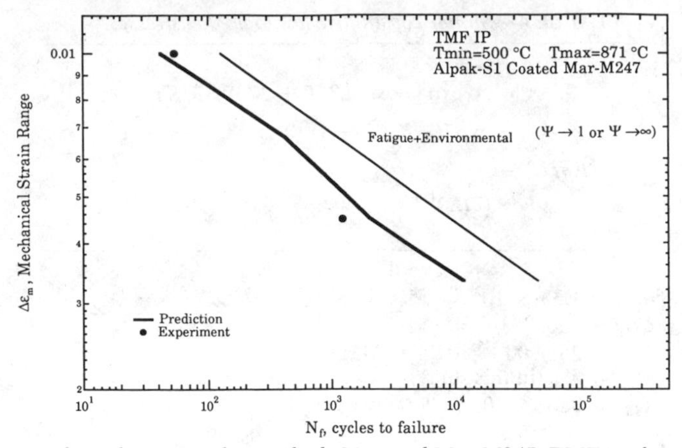

FIG. 12—*Life prediction results on Alpak-S1 coated Mar-M247; TMF in-phase loading.*

predictions are shown in Fig. 14. In Figs. 11 through 14, the lower bound (that is, $\Psi \rightarrow 1$) represents the predicted life based on the proposed oxidation mechanism, assuming that the coating fails prematurely, and results in exposure of the substrate to the environment. The upper bound ($\Psi \rightarrow \infty$) represents the predicted life, if the coating provided full protection against environmental attack. This means that $D_{ox} = 0$, and only fatigue and creep damage mechanisms are the dominant damage mechanisms and will cause the failure of the material. In the TMF IP case creep damage is dominant as is evident from Fig. 10. Therefore, the protection provided by the coating against oxidation does not have a significant effect on the fatigue lives of the coated alloy.

Discussion of the Model and the Results

In this work an oxide spike was modeled as a semispherical surface inhomogeneity. The stress field in the vicinity of the oxide spike was calculated employing a technique based on Eshelby's method. Then the calculated strain at the tip of oxide spike permitted estimation of

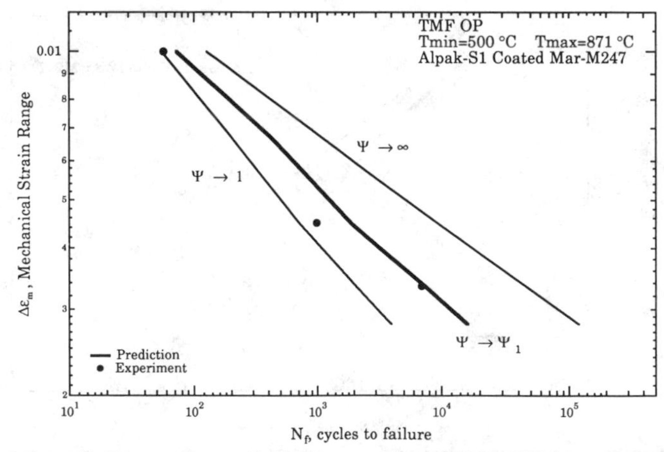

FIG. 13—*Life prediction results on Alpak-S1 coated Mar-M247; TMF out-of-phase loading.*

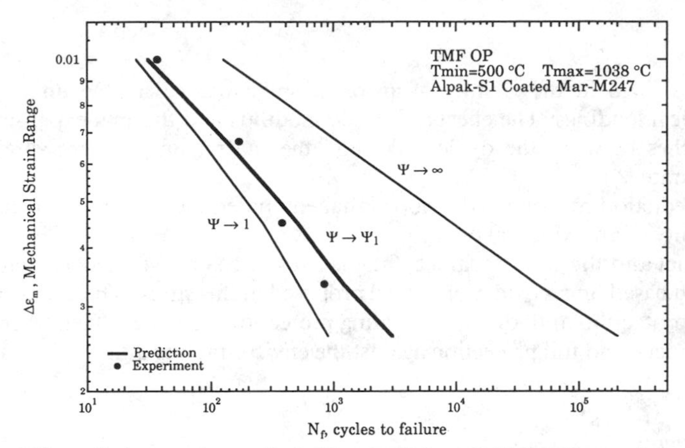

FIG. 14—*Life prediction results on Alpak-S1 coated Mar-M247; ITMF out-of-phase loading.*

the useful life of coated alloys under isothermal and thermomechanical loading conditions. The model proposed can be readily extended to other systems with different oxide and metal modulus and expansion coefficients.

The model presented here is a significant improvement over previously proposed models since it considers the local stress-strain behavior near the oxide and surrounding matrix. It has been known that local mechanical behaviors are dependent on the differences in the mechanical and thermal properties of the oxide spikes and the surrounding matrix, but the numerical values for the strains (stresses) have not been reported in the past, except for the uniform oxide scales. It should be noted that the model calculates elastic strain and stress concentrations. For systems where the matrix plasticity is notable, modifications in the model are necessary.

The results, as noted in Figs. 5 and 6, confirm that strains at the oxide tips increase considerably as the oxide elastic modulus to metal elastic modulus ratio decreases. We note that the aspect ratio of the oxide spike (that is, c/a in Fig. 2) did not have appreciable influence on the results provided that oxide elastic modulus to metal elastic modulus ratio remained above 1. For example, when oxide elastic modulus to metal elastic modulus ratio was equal to 1, the increase of c/a from 1 to 5 changed the stress component in X_1-direction of the oxide tip by only 15% under thermal loading. This led to the choice of the semispherical oxide shape in our model calculations.

In the life prediction model, the performance of the coating is reflected through the term Ψ. The performance of a coating is not only a function of the coating properties but also related to the properties of the substrate (matrix) alloy itself. Therefore, the term Ψ is specific to the coating/substrate system under investigation.

In calculating the damage due to environmental attack, the strains in X_1-direction were used. Due to the geometry and assumed isotropic thermal and mechanical properties for the oxide spike and the matrix, the resulting stresses and strains in X_1 and X_2-directions are same under thermal loading. Under mechanical loading considered in this study the generated strain field in X_1-direction is significantly greater than those of the other principle directions.

Finally, for uncoated alloys the model is anticipated to hold, but the Ψ term in Eq 4.8 should not be used. Although oxide spikes could form on the surface of uncoated alloys, one should recognize the differences in the oxide spike formation sequence on the substrate in the coated and uncoated alloys; therefore, some of the environmental damage constants should be re-evaluated accordingly.

Conclusions

1. The stress field in the vicinity of an oxide spike was calculated under thermal and mechanical loadings. The effects of elastic modulus and thermal expansion coefficient mismatches between the oxide spike and the matrix on the stress-strain field were demonstrated.
2. A life prediction model was developed that embodied the local mechanical behavior in the vicinity of an oxide spike.
3. Isothermal and thermomechanical fatigue lives of coated Mar-M247 were successfully predicted based on an oxidation model proposed in this study. The lives with short-term coating protection and long-term coating protection were identified, where the coating provides zero and full protection against the environmental attack respectively.

Acknowledgments

Initial portions of the work were supported by General Motors Allison Gas Turbine Division, Indianapolis. The work on modeling surface inhomogeneities was supported by Manufacturing Research Center, College of Engineering, University of Illinois.

References

[*1*] "Materials and Coatings to Resist High Temperature Corrosion," D. R. Holmes and A. Rahmel, Eds., Applied Science Publisher, London, 1978.
[*2*] "Gas Turbine Applications in Coatings for High Temperature Applications," E. Lang, Ed., Applied Science Publishers, London, U.K., 1983.
[*3*] Strang, A., "Effects of Corrosion Resistant Coatings on the Structure and Properties of Advanced Gas Turbine Blading Alloys," Published by CIMAC, Br Natl Comm, London, U.K., 1979, Vol. 3, pp. GT33.1–GT33.32.
[*4*] Shankar, S., Koenig, D. E., and Dardi, L. E., "Vacuum Plasma Sprayed Metallic Coatings," *Journal of Metals,* Vol. 33, No. 10, Oct. 1981, pp. 13–20.
[*5*] Hsu, L. and Stetson, A. R. "Evaluation of the Mechanical Properties and Environmental Resistance of Rene' 125 and X-40 Superalloys Coated with Controlled Composition Reaction-Sintered Co-Ni-Cr-Al-Y," *International Conference on Metallurgical Coatings,* San Diego, CA, April 1980, pp. 419–428.
[*6*] Liewelyn, G., "Protection of Nickel-Base Alloys Against Sulfur Corrosion by Pack Aluminizing," *Hot Corrosion Problems Associated with Gas Turbines, ASTM STP 421,* American Society for Testing and Materials, Philadelphia, 1967, pp. 3–20.
[*7*] Felix, P. C. and Villat, M., "High Temperature Corrosion Protective Coating for Gas Turbines," *Sulzer Technical Review,* Vol. 58, No. 3, 1976, pp. 97–104.
[*8*] Grunling, A. W., Schneider, K., and Arnim, H. V., COST 50 RII D2, BBC Final Report, Feb. 1981.
[*9*] Bartocci, R. S., "Behavior of High-Temperature Coatings for Gas Turbine Engines," *Hot Corrosion Problems Associated with Gas Turbines, ASTM STP 421,* American Society for Testing and Materials, Philadelphia, 1967, pp. 169–187.
[*10*] Betts, R. K., "Selection of Coatings for Hot-Section Components of the LM1500 Marine-Environment Engine," GE Report R67FPD357, Marine and Industrial Department, Lynn, Massachusetts/Cincinnati, OH, Oct. 2, 1967. (Ref 16 in Myers and Geyer, *Sampe Quart,* Vol. 1, 1970, pp. 18–28).
[*11*] Schneider, K. and Gruling, H. W., "Mechanical Aspects of High Temperature Coatings," *Thin Film Solids,* Vol. 107, 1983, pp. 395–416.
[*12*] Belyaev, M. S., Zhukov, N. D., Krivenko, M. P., and Terekhova, V. V., "Effect of Aluminum Coatings on the Fatigue of Alloy ZhS6U," *Problemy Prochnosti,* Vol. 9, No. 11, Nov. 1977, pp. 34–39 (pp. 1316–1320 English).
[*13*] Puyear, R. B., "High Temperature Metallic Coatings," *Machine Design,* July 1962, pp. 177–183.
[*14*] Lane, R. and Geyer, N. M., "Superalloy Coatings for Gas Turbine Components," *Journal of Metals,* Vol. 18, Feb. 1966, pp. 186–191.
[*15*] Paskeit, G. F., Boone, D. H., and Sullivan, C. P., "Effect of Aluminide Coating on the High-Cycle

Fatigue Behavior of a Nickel Base High Temperature Alloy," *Journal of the Institute of Metals,* Vol. 100, Feb. 1972, pp. 58–62.
[*16*] Strang, A. and Lang, E., "Effect of Coatings on the Mechanical Properties of Superalloys," in *Behavior of Superalloys in Aggressive Environments,* Petten, Netherlands, 1982, pp. 469–506.
[*17*] Castillo, R. and Willet, K. P., "The Effect of Protective Coatings on the High Temperature Properties of a Gamma Prime-Strengthened Ni-Based Superalloy," *Metallurgical Transactions A,* Vol. 15, 1984, pp. 229–236.
[*18*] Whitlow, G. A., Beck, C. G., Viswanathan, R., and Crombie, E. A., "The Effects of a Liquid Sulfate/Chloride Environment on Superalloy Stress Rupture Properties at 1300°F (704°C)," *Metallurgical Transactions A.,* Vol. 15, 1984, pp 23–28.
[*19*] Kolkman, H. J., "Creep, Fatigue, and Their Interaction in Coated and Uncoated Rene 80," *Materials Science and Engineering,* Vol. 89, 1987, pp. 81–91.
[*20*] Wells, C. H. and Sullivan, C. P., "Low Cycle Fatigue of Udimet 700 at 1700°F," *Transactions of the American Society for Metals,* Vol. 61, March 1968, pp. 149–155.
[*21*] Wright, P. K., "Oxidation Fatigue Interactions in a Single-Crystal Superalloy," *Low Cycle Fatigue, ASTM STP 942,* American Society for Testing and Materials, Philadelphia, 1988, pp. 558–575.
[*22*] Wood, M. I., "Mechanical Interactions between Coatings and Superalloys under Condition of Fatigue," *Surface and Coating Technology,* Vols. 39–40, 1989, pp. 29–42.
[*23*] Kortovich, C. S. and Sheinker, A. A., "Strainrange Partitioning Analysis of Low Cycle Fatigue of Coated and Uncoated Rene' 80," *AGARD Conference Proceeding,* No. 243, April 1978, pp. 1-1, 1-23.
[*24*] Halford, G. R. and Nachtigall, A. J., "Strainrange Partitioning Behavior of the Nickel-Base Superalloys, Rene' 80 and IN100," *AGARD Conference Proceeding,* No. 243, April 1978, pp. 2-1, 2-14.
[*25*] Au, P. and Patnaik, P. C., "Isothermal Low Cycle Fatigue Properties of Diffusion Aluminide Coated Nickel and Cobalt Based Superalloys," *Surface Modification Technologies III,* T. S. Sudarshan and D. G. Bhat, Eds., The Minerals, Metals and Materials Society, 1990, pp. 729–748.
[*26*] Swanson, G. A., Linask, I., Nissley, D. M., Norris, P. P., Meyer, T. G., and Walker, K. P., "Life Prediction and Constitutive Models for Engine Hot Section Anisotropic Materials Program," NASA Contractor Report 179594, National Aeronautics and Space Administration, Washington, DC, 1987.
[*27*] Heine, J. E., Warren, J. R., and Cowles, B. A., "Thermomechanical Fatigue of Coated Blade Materials," Final Report, WRDC-TR-89-4027, Wright Research and Development Center, 27 June 1989.
[*28*] Bain, K. R., "The Effects of Coatings on the Thermomechanical Fatigue Life of a Single Crystal Turbine Blade Material," *AIAA/SAE/ASME/ASEE 21st Joint Propulsion Conference,* 1985, pp. 1–6.
[*29*] Strangman, T. E., "Thermal Fatigue of Oxidation Resistant Overlay Coatings for Superalloys," Ph.D. Thesis, University of Connecticut, Storrs, CT, 1978.
[*30*] Leverant, G. R., Strangman, T. E., and Langer, B. S., "Parameters Controlling the Thermal Fatigue Properties of Conventionally-Cast and Directionally-Solidified Turbine Alloys," *Superalloys: Metallurgy and Manufacturing Proceedings of the Third International Symposium,* Claitor's Publishing Division, 1976, pp. 285–295.
[*31*] Strangman, T. E. and Hopkins, S. W. "Thermal Fatigue of Coated Superalloys," *Ceramic Bulletin,* Vol. 55, No. 3, 1976, pp. 304–307.
[*32*] Kadioglu, Y., "Modelling of Thermo-Mechanical Fatigue Behavior in Superalloys," Ph.D. Thesis, University of Illinois at Urbana-Champaign, IL, 1992.
[*33*] Stringer, J., "Stress Generation and Relief in Growing Oxide Films," *Corrosion Science,* Vol. 10, 1970, pp. 513–543.
[*34*] Hancock, P. and Hurst, R. C., "The Mechanical Properties and Breakdown of Surface Oxide Films at Elevated Temperatures," *Advances in Corrosion Science and Technology,* R. W. Staehle and M. G. Fontana, Eds., Plenum Press, New York, 1974.
[*35*] Birks, N. and Meier, G. H., *Introduction to High Temperature Oxidation of Metals,* Edward Arnold Ltd., London, U.K., 1983.
[*36*] Oxx, G. D., "Which Coating at High Temperature," *Product Engineering,* Vol. 29, No. 3, 1958, pp. 61–63.
[*37*] Kouris, D., Tsuchida, E., and Mura, T., "The Hemispheroidal Inhomogeneity at the Free Surface of an Elastic Half Space," *ASME Journal of Applied Mechanics,* Vol. 56, 1989, pp. 70–76.
[*38*] Cox, B. N., "Surface Displacement and Stress Field Generated by a Semi-Ellipsoidal Surface Inclusion," *ASME Journal of Applied Mechanics,* Vol. 56, 1989, pp. 564–570.
[*39*] Eshelby, J. D., "The Determination of the Elastic Field of an Ellipsoidal Inclusion, and Related Problems," *Proceedings of the Royal Society,* London, U.K., Vol. 241A, pp. 376–396.

[40] Eshelby, J. D., "The Elastic Field Outside an Ellipsoidal Inclusion," *Proceedings of the Royal Society*, London, U.K., Vol. 252A, pp. 561–569.
[41] Moschovidis, Z. S., "Two Ellipsoidal Inhomogeneities and Related Problems Treated by the Equivalent Inclusion Method," Ph.D. Thesis, Northwestern University, Evanston, IL, August 1975.
[42] Neu, R. and Sehitoglu, H., "Thermo-Mechanical Fatigue, Oxidation and Creep: Part I-Experiments," *Metallurgical Transactions*, Vol. 20a, 1989, pp. 1755–1767.
[43] Neu, R. and Sehitoglu, H., "Thermo-Mechanical Fatigue, Oxidation and Creep: Part II-Life Prediction," *Metallurgical Transactions*, Vol. 20A, 1989, pp. 1769–1783.
[44] Sehitoglu, H. and Boismier, D. A., "Thermo-Mechanical Fatigue of Mar-M247: Part 2-Life Prediction," *ASME Journal of Engineering Materials Technology*, Vol. 112, 1990, pp. 80–89.
[45] Sehitoglu, H. and Karayaka M., "Prediction of Thermomechanical Fatigue Lives in Metal Matrix Composites," *Metallurgical Transactions*, Vol. 23A, July 1992, pp. 2029–2038.
[46] Slavik, D. and Sehitoglu, H., "A Constitutive Model for High Temperature Loading Part I-Experimentally Based Forms of the Equations," *Thermal Stress, Material Deformation, and Thermomechanical Fatigue, Proceedings of the Pressure Vessels and Piping Conference*, ASME PVP, Vol. 123, 1987, pp. 65–73.
[47] Samsonov, G. V., *The Oxide Handbook*, Plenum Press, New York, 1973, p. 183.

DISCUSSION

R. J. DiMelfi[1] *(written discussion)*—The speaker has presented a very clever way of modeling the contribution to total damage from oxidation damage resulting from the penetration of surface oxide "spikes". It occurred to me that one might extend such an analysis to the case where ongoing oxidation at a crack tip is associated with the crack propagation process itself. This perhaps could be accomplished via modification of the free surface and oxide spike geometry. Such an approach might capture the synergism associated with oxidation damage and fatigue crack damage.

Y. Kadioglu and H. Sehitoglu (authors' response)—We thank Dr. DiMelfi for raising an interesting point. In the present work, we modeled the oxide spike as a surface inclusion on a semi-infinite plane. It is possible to model a surface crack with the oxide emanating from the crack tip, and revise our oxide failure model, or develop new fracture mechanics parameters to represent the crack tip oxide stress interactions. Admittedly, this would be a substantial analytical and computational effort, and is set aside for future research.

[1] Argonne National Lab., RE-207, Argonne, IL 60439.

M. P. Miller,[1] *D. L. McDowell,*[2] *R. L. T. Oehmke,*[3] *and S. D. Antolovich*[4]

A Life Prediction Model for Thermomechanical Fatigue Based on Microcrack Propagation

REFERENCE: Miller, M. P., McDowell, D. L., Oehmke, R. L. T., and Antolovich, S. D., **"A Life Prediction Model for Thermomechanical Fatigue Based on Microcrack Propagation,"** *Thermomechanical Fatigue Behavior of Materials, ASTM STP 1186*, H. Sehitoglu, Ed., American Society for Testing and Materials, Philadelphia, 1993, pp. 35–49.

ABSTRACT: A thermomechanical fatigue (TMF) life prediction model is developed based on the concept of microcrack propagation. The model is used to correlate high temperature fatigue lives of two nickel-base superalloys.

Due to its complexity, a well-accepted framework for correlation of TMF life has been elusive. Various approaches have been taken, ostensibly nonisothermal generalizations of isothermally derived models. The proposed model explicitly accounts for damage from all three high temperature fatigue (HTF) damage mechanisms prevalent in metals, i.e. fatigue, oxidation, and creep. The general form of the microcrack propagation equation is

$$\frac{da}{dN} = \left.\frac{da}{dN}\right|_{\text{fatigue}} + \left.\frac{da}{dN}\right|_{\text{creep}} + \left.\frac{da}{dN}\right|_{\text{ox}}$$

No crack growth is actually monitored in the present analysis; rather the equation is integrated between an appropriate initial and final crack size. The constants and exponents in this integrated form of the equation are then fit from test data. The fatigue and oxidation components are correlated using the ΔJ parameter with an additional time and temperature dependence included in the oxidation term. The ΔJ parameter has demonstrated excellent versatility in the correlation of fatigue macrocrack growth and also implicitly contains the elastic deformation information that is so important in the life of high-strength Ni-base superalloys. The creep component of microcrack propagation is correlated using $\hat{C}$, which is similar to the creep crack growth parameter C_t. Isothermal as well as TMF test data for MAR-M247 are analyzed. The mechanical strain versus temperature relationships for the TMF tests include in-phase (IP), out-of-phase (OP), and counterclockwise diamond history (DH). Lives of the isothermal, IP, and OP tests are correlated within a factor of ± 2 of the median life. Lives of the DH tests are predicted within a factor ± 2 as well. Isothermal, hold time, and bithermal test data for as-cast MAR-M246 are also correlated within ± 2 of the median life using the approach.

KEYWORDS: thermomechanical fatigue, microcrack propagation, ΔJ-integral, oxidation, creep, creep-fatigue interaction

[1] Graduate student, The George W. Woodruff School of Mechanical Engineering, Georgia Institute of Technology, Atlanta, GA 30332-0405.

[2] Associate professor, The George W. Woodruff School of Mechanical Engineering, Georgia Institute of Technology, Atlanta, GA 30332-0405.

[3] Software engineer, Materials Testing Division, MTS Systems Corporation, 14000 Technology Drive, Eden Prairie, MN 55344.

[4] Professor, Department of Mechanical and Materials Engineering, Washington State University, Pullman, WA 94551-0969.

Introduction: Thermomechanical Fatigue Damage Mechanisms

In metals subjected to cyclic thermal and mechanical loading, creep and oxidation can occur in addition to classical fatigue processes. These mechanisms compete and interact depending on the temperature and load ranges. The manner in which current life prediction models for TMF account for the different types of damage and interaction varies significantly. Some models explicitly include separate creep and oxidation terms, while others simply imply time dependence through a modification of strain rate or cycle time.

In this paper, fatigue damage is the cyclic plasticity-driven, time and temperature independent damage that exists whenever cyclic loading occurs. Traditionally, creep refers to a material undergoing viscous deformation at a constant stress level. This type of deformation leads to intergranular creep cavity growth and rupture. Under TMF loading, however, creep deformation contributes to the formation and propagation of microcracks. In general, creep crack growth must be distinguished from bulk creep damage. Metals exposed to environment at high temperature are subject to corrosion by oxidation. This corrosion is accelerated by a tensile stress [*1*]. During TMF, brittle oxides can enhance the nucleation and propagation of fatigue microcracks and impede the rewelding of crack surfaces during unloading.

Damage Interaction

At elevated temperatures, under cyclic loading, the inelastic deformation of metals is comprised of both plastic and viscous components. As stated above, plasticity is associated with fatigue damage and viscous deformation is associated with creep microcrack propagation. Since both plastic and viscous deformation mechanisms can serve as driving forces for microcrack propagation, fatigue and creep damage mechanisms are intimately coupled under TMF loading. In other words, neither pure fatigue nor pure creep tests would exhibit this type of damage. For this reason, it is very important to experimentally approximate the type of creep-fatigue interaction that is expected in the application of interest.

Oxidation damage, defined within the context of TMF loading, is implicitly coupled with fatigue damage. While corrosion cracking under a constant stress can be a very significant process, it is not relevant to the present topic.

Life Prediction Models for Thermomechanical Fatigue

Life prediction models for thermomechanical fatigue (TMF) generally take the form of: (1) frequency-modified strain-life or stress-life approaches [*2–4*]; (2) parametric damage approaches [*5–9*]; (3) continuum damage approaches [*10–15*]; or (4) damage rate approaches [*16–17*] including microcrack propagation models [*18*]. Of these, only the damage rate approaches use a physically measurable quantity such as crack length or creep cavity radius as specific definitions of damage. The others rely on cycle fraction, time fraction, stiffness degradation, or other related parameters as measures of damage. While the use of such damage measures may facilitate life prediction for a particular dataset, they are not necessarily versatile enough for more complex applications.

Microcrack propagation models monitor the evolution of microcrack length. The problems associated with using different approaches for macrocrack initiation (microcrack propagation) and macrocrack propagation are avoided using microcrack propagation equations if they are philosophically compatible with their macrocrack counterparts.

The macrocrack correlation parameter, ΔJ (cyclic J-integral [*19,20*]), has been used to correlate fatigue microcrack growth in a number of creep- fatigue studies including TMF [*21–24*]. There are size scale limitations on the use of ΔJ as a macrocrack growth correlation parameter, however. For cracks on the order of a grain size, the J-integral should not be considered as representative of the strength of the stress and strain fields. Instead, the J-integral should be regarded as a *local* parameter which reflects the crack tip opening behavior. Due to its local nature, the path independence of ΔJ is irrelevant, and micromechanical issues such as slip band configuration and dislocation and grain boundary interactions become pertinent.

Proposed Model

We propose a microcrack propagation model which explicitly accounts for damage due to fatigue, creep, and oxidation. The general form of each equation is presented along with the particular form used for life correlation of uniaxial smooth specimens.

The general form of the equation is

$$\frac{da}{dN} = \left.\frac{da}{dN}\right|_{\text{fatigue}} + \left.\frac{da}{dN}\right|_{\text{creep}} + \left.\frac{da}{dN}\right|_{\text{ox}} \quad (1)$$

with

$$\left.\frac{da}{dN}\right|_{\text{fatigue}} = C_f \, \Delta J^{m_f} \quad (2)$$

$$\left.\frac{da}{dN}\right|_{\text{creep}} = C_c \hat{C}^{m_c} \quad (3)$$

$$\left.\frac{da}{dN}\right|_{\text{ox}} = C_o \, \Delta J^{m_o} \, \Delta t^{\Psi} \quad (4)$$

where a is the crack length, and N is the cycle number.

The additive form of Eq 1 implies an explicit decoupling of the damage mechanisms. However, the deformation mechanisms related to the individual microcrack propagation components may be coupled. As discussed earlier, this type of coupling between creep and fatigue damage may be the most physically relevant for propagation of fatigue cracks. There is precedence for the additive form in the literature associated with oxidation-assisted fatigue crack growth [*25*] and creep-fatigue crack growth [*26*]. It is probable that direct damage couplings do exist which render Eq 1 approximate, but such couplings are difficult to quantify.

The fatigue component of microcrack propagation, Eq 2, is correlated using the ΔJ parameter. As mentioned earlier, ΔJ has been used with success in the correlation of microcracks under TMF loading conditions. This parameter has also shown great versatility in isothermal macrocrack growth correlation of both long and short cracks under uniaxial [*27–28*], as well as biaxial fatigue [*29*].

As discussed previously, ΔJ has a different interpretation when used to correlate growth of a crack with length on the order of a grain size. The derivation of a local form for a microcrack will ultimately involve micromechanical analyses. Such analyses were not performed here; instead, forms of the J-integral that were developed for macrocracks were used.

It should also be noted that the influence of a temperature gradient for nonisothermal problems is not addressed in the present model since structural temperature gradients are typically small over a length scale on the order of microcrack length. As pointed out earlier, the approach is local, rather than global, so the inclusion of thermal gradients in the J-integral expressions to ensure path independence [*30*] are unnecessary.

Dowling [*27*] extended a plane stress solution for J derived by Shih and Hutchinson [*31*] for a crack in an infinite plate to an expression for ΔJ for smooth specimens, i.e.

$$\begin{aligned}\Delta J &= \Delta J_e + \Delta J_p \\ &= 2\pi Y^2 \left[\frac{\Delta\sigma\Delta\epsilon_e}{2} + \frac{f(1/n')\,\Delta\sigma\,\Delta\epsilon_p}{2\pi}\right] a\end{aligned} \tag{5}$$

where Y is a geometric correction factor, n' is the cyclic hardening exponent, $\Delta\sigma$ is the stress range, and $\Delta\epsilon_e$ and $\Delta\epsilon_p$ are the elastic and plastic strain ranges, respectively. The function $f(1/n')$ is given by

$$f(1/n') = 3.85 \sqrt{\frac{1}{n'}}(1 - n') + \pi n' \tag{6}$$

We can combine C_f with $2\pi Y^2$ to produce the coefficient C'_f. Equation 2 can now be written as

$$\frac{da}{dN} = C'_f\,(\alpha a)^{mf} \tag{7}$$

with α defined as

$$\begin{aligned}\alpha &= \alpha_e + \alpha_p \\ &= \frac{\Delta\sigma\,\Delta\epsilon_e}{2} + \frac{f(1/n')\,\Delta\sigma\,\Delta\epsilon_p}{2\pi}\end{aligned} \tag{8}$$

For pure fatigue, Eq 7 becomes the microcrack propagation equation and can be integrated from an initial to final crack size to obtain the life of the specimen. Note that the two components in Eq 8 are analogous to cyclic elastic and plastic work terms, but not uniquely identifiable with such terms, in general. Upon integrating Eq 7, we obtain

$$\frac{Z_f}{C'_f} = \alpha^{mf} N_f \tag{9}$$

$$\begin{aligned} Z_f &= \frac{a_f^{(1-m_f)} - a_o^{(1-m_f)}}{(1 - m_f)} \quad &\text{if } m_f \neq 1 \\ Z_f &= \ln(a_f) - \ln(a_o) \quad &\text{if } m_f = 1 \end{aligned} \tag{10}$$

where a_o and a_f are the initial and final crack sizes, respectively. From Eq 10, it is obvious that the initial and final crack sizes chosen are irrelevant if the integration proceeds over the fatigue life of the material, provided a_o and a_f are the same for all cases considered and the loading conditions are invariant from cycle to cycle for each test considered. Altering the values of a_o and a_f would only serve to change the value of Z_f. Values chosen for the analyses in this paper were $a_o = 5 \times 10^{-5}$ in. (1.27 μm) and $a_f = 5 \times 10^{-3}$ in. (127 μm).

Examining Eq 9, it is apparent that the parameter α assumes the role of the driving force for pure fatigue microcrack propagation. A plot of α versus N_f for high temperature fatigue tests performed at different strain rates should then disclose any frequency dependence of damage, since enhanced oxidation will shorten the lives of tests at lower frequencies compared to those at higher frequencies. In many fatigue-life theories the plastic strain amplitude $\Delta\epsilon_p/2$ is the assumed driving force for fatigue damage. For high-strength materials with limited ductility like nickel-base superalloys, the elastic strain amplitude becomes very significant, and cer-

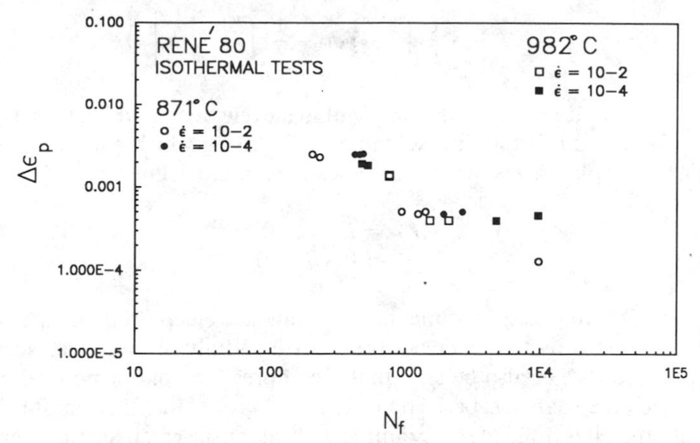

FIG. 1—*Plastic strain range* $\Delta\epsilon_p$ *versus number of cycles to failure* N_f *for René 80; isothermal fatigue tests.*

tainly contributes to microcrack propagation. The parameter α contains the elastic strain amplitude information explicitly, in contrast to $\Delta\epsilon_p/2$.

A striking example of the advantages of α over $\Delta\epsilon_p/2$ as a fatigue parameter is found in the analysis of isothermal fatigue tests of the superalloy René 80 [*32*]. Data from tests at two strain rates and two temperatures were analyzed. Figures 1 and 2 are plots of $\Delta\epsilon_p/2$ versus N_f and α versus N_f, respectively, for tests conducted on René 80 at 871 and 982°C. Comparing these two figures, it is obvious that α more successfully separates the frequency dependence of the lives at both temperatures, especially at the lower strain amplitudes. The expected strain rate dependence is observed using α, and is very consistently separated. Use of $\Delta\epsilon_p/2$ produces an apparent inverse rate sensitivity.

Consistent with the use of ΔJ as the driving force for fatigue crack growth, a parameter resembling either C^* [*33–34*] or C_t [*35–36*] would be a logical choice to correlate the creep component of microcrack growth. Forms for $C^*(t)$ or C_t for TMF loading do not presently exist so the stress power release rate parameter $\hat{C}$ is introduced, i.e.

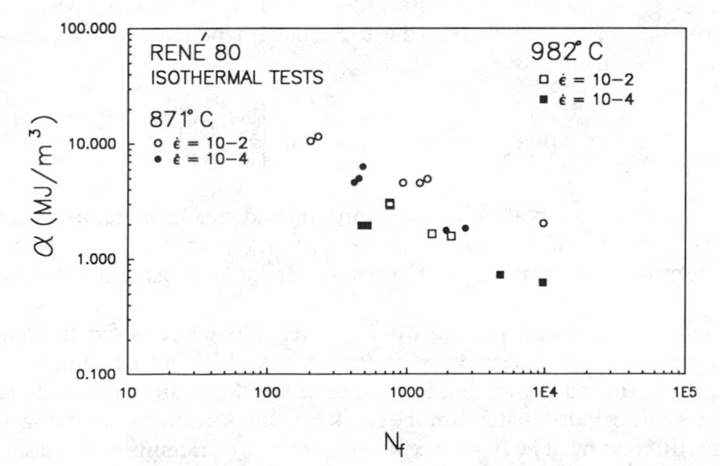

FIG. 2—*Fatigue microcrack propagation parameter* α *versus* N_f *for René 80; isothermal tests.*

$$\hat{C} = \left\langle a \left[\frac{1}{t_t} \int_0^{t_t} \sigma \dot{\epsilon}_c \, dt - \frac{1}{t_c} \int_0^{t_c} \sigma \dot{\epsilon}_c \, dt \right] \right\rangle \tag{11}$$

where $\dot{\epsilon}_c$ is the creep strain rate, t_t is the time within a cycle during which tensile-going, tensile strain accumulates, and t_c is the time within a cycle during which compressive-going, compressive strain accumulates. The Macaulay brackets, $\langle\ \rangle$, are defined as

$$\begin{aligned} \langle f \rangle &= f, \quad f > 0 \\ &= 0, \quad f \leq 0 \end{aligned} \tag{12}$$

This definition of $\hat{C}$ implicitly assumes that compressive creep strain has the effect of negating the damage induced by tensile creep strain during a fully reversed test, such that no net damage is produced. It can also be seen that a compressive hold or no-load hold time test would produce no creep damage, based on $\hat{C}$. It is emphasized that the definition of $\hat{C}$ is material-dependent, since it is difficult, for example, to heal tensile creep damage during compressive creep if the cavity surfaces oxidize.

In Eq 3, C_c and m_c are the experimentally determined creep constant and exponent, respectively. Any test for which creep damage accumulates could conceivably be used to determine these constants although the type of creep damage, the creep deformation or damage experienced within the test is expected to conform to that which is expected in the application. No special form of Eq 3 must be assumed for uniaxial loading, although the determination of the creep strain rate $\dot{\epsilon}_c$ for any loading configuration may require some effort.

In the oxidation component of microcrack propagation, Eq 4, m_o and ψ are experimentally determined constants, Δt is the cycle time, and the coefficient C_o is defined as

$$C_o = C_o' \exp \left[\frac{-(Q_{ox} - B\hat{\sigma}^k)}{RT_{eff}} \right] \tag{13}$$

where

$$\hat{\sigma} = \langle \sigma_{Tmin} \rangle \tag{14}$$

Here, C_o', B, and k are experimentally determined constants, Q_{ox} is the experimentally determined activation energy of the effective crack tip oxidation and growth process, and T_{eff} is an effective temperature which is introduced due to the incremental form of the microcrack propagation equation and defined as

$$\exp \left[\frac{-Q_{ox}}{RT_{eff}} \right] = \frac{1}{\Delta t} \int_{t_{min}}^{t_{max}} \exp \left[\frac{-Q_{ox}}{RT(t)} \right] dt \tag{15}$$

Here t_{min} and t_{max} are the times at which the minimum and maximum temperature occur during a temperature cycle, respectively, and $\Delta t = t_{max} - t_{min}$. The stress term in Eq 14, σ_{Tmin}, is the stress at the minimum temperature. The parameter $\hat{\sigma}$ is taken as zero for isothermal cyclic loading.

The Arrhenius temperature dependence of C_o is altered to reflect the fact that a material will oxidize more readily when it is subjected to tensile stress [1]. During an out-of-phase (OP) TMF loading cycle, the maximum tensile stress occurs at the minimum temperature. For the less ductile, oxygen-degraded material at the crack tip, this is a much more damaging scenario than in-phase (IP) loading. The oxide becomes more brittle as temperature decreases and is therefore more susceptible to microcrack extension, in effect lowering the activation energy of the microcrack extension process.

In general, the values of the exponents on ΔJ in Eqs 2 and 4, m_f and m_o, respectively, will be different. For lack of more detailed information, however, we will assume that $m_f = m_o$ for all analyses presented herein.

Application of the Microcrack Propagation Model

The microcrack propagation model was used to correlate uniaxial test lives of the nickel-base superalloys MAR-M246 and MAR-M247. In both analyses the microcrack propagation equation was employed in its integrated form since the loading cycles did not vary in each test. As stated previously, the initial and final crack sizes of 1.27 and 127 μm, respectively, were chosen for both studies.

MAR-M246

The MAR-M246 tests analyzed include: (1) isothermal fully reversed tests at 650, 800, and 900°C at strain rates of 10^{-2}, 10^{-3}, and 10^{-4} s^{-1}; (2) isothermal tensile and compressive strain hold time and no-load hold tests at 900°C; and (3) in-phase (IP) and out-of-phase (OP) bithermal fatigue tests between temperatures of 650°C and 900°C at a mechanical strain rate of 10^{-4} s^{-1}.

Bithermal tests are similar to TMF tests except that the temperature excursions are discrete and occur under no-load between the mechanical strain excursions [*37*]. These tests were introduced into the present program as a possible step between isothermal and true TMF testing conditions which were unavailable.

The near-net cast MAR-M246 was machined into button-end LCF specimens which were 140 mm long and had gage lengths and gage diameters of 19.1 and 6.4 mm, respectively. The material had a dendritic structure with large radially oblate grains with an average grain size of 2.5 mm along the major axis of the grain. Other details of the material as well as the test data are found in Ref *38*.

The first step in the analysis was to assume that fatigue was the only active damage mechanism. The microcrack propagation equation then reduces to Eq 7 and an α value was computed for each test. As noted, the frequency and temperature dependence can be qualitatively examined from a plot of α versus N_f. Figure 3 is a plot of α versus N_f for the isothermal tests performed on MAR-M246. Some observations from Fig. 3 are as follows:

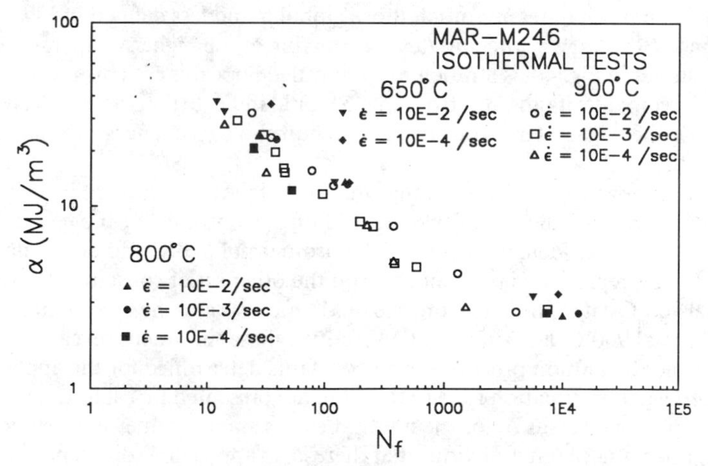

FIG. 3—*Plot of α versus* N_f, *MAR-M246; isothermal tests.*

1. The lives of the 650°C tests show no appreciable strain rate dependence.
2. The lives of the 800 and 900°C tests exhibit strain rate dependence.
3. The tests at 800°C are typically shorter lived than comparable tests at 650°C, but a similar temperature dependence is not observed in the test results at 900°C. These tests appear to overlay the entire 650°C, 800°C test band.

Due to the lack of a continuous temperature dependence at the upper temperature (observations 2 and 3 above), a possible morphology change was investigated. To explore this possibility, coarsening and/or dissolution of γ' precipitates was investigated. At 900°C, the homologous temperature of MAR-M246 is approximately 0.7. In tests conducted on the Ni-base superalloy René 80, Antolovich [*32*] has reported that coarsening of large γ' precipitates takes place under fatigue conditions near this homologous temperature. He also found that the coarsening was accompanied by dissolution of the smaller γ' precipitates. This coarsening and dissolution process resulted in a more ductile material. Antolovich showed that lives at elevated temperatures could be comparable to lives at lower temperatures, provided that the improved fatigue resistance that accompanies an increase in ductility is on the same order as the accelerated material degradation that accompanies the increased oxidation rate at the elevated temperature.

The effect of stress state on the morphology of γ' precipitates has been investigated by a number of investigators for uniaxial loading [*32,39–41*]. It has been demonstrated experimentally that under reversed plastic straining, coarsening rates are extremely rapid [*40*]. There is also some indication that the dislocations generated during such a process enhance coarsening rates by providing high diffusion rate paths connecting large and small precipitates [*40–41*]. In all cases, the reversed loading leads to globular precipitates of indefinite shape.

The γ' precipitate system is the primary strengthening mechanism in MAR- M246; hence, this coarsening and dissolution process may be occurring in the tests conducted within the present study. It is feasible that even though the rate at which this coarsening/dissolution process takes place is a continuous function of temperature, the acceleration rate of this morphology change between 800 and 900°C could be high enough to warrant a seemingly discontinuous change of fatigue constants between these two relatively similar temperatures. Transmission electron microscopy (TEM) was used to assess these possibilities.

Figure 4 is a TEM micrograph of specimen T139-1, a fully reversed test at 650°C with a strain amplitude of 0.003 and a strain rate of 10^{-4} s^{-1}. Figure 5 is a TEM micrograph of specimen G4, a comparable test at 900°C. The large light features in each figure are the large γ' precipitates. These precipitates are much more globular and less defined in Fig. 5, indicating that coarsening of the large γ' precipitates is occurring at the higher temperature. The interprecipitate material in Fig. 4 has a much more mottled appearance. This mottling is actually the smaller γ' precipitate; its absence from Fig. 5 would indicate that the smaller γ' precipitate is dissolving at the higher temperature. These morphology changes are similar to those reported in [*32*].

The precipitate morphology change along with the fatigue life–temperature dependence disparity at 900°C warranted the use of two sets of fatigue constants for the tests conducted on MAR-M246. One set was determined using the isothermal 650°C and 800°C data assuming that the 650°C tests represented pure fatigue and the other set of constants determined using the isothermal 900°C data. The data from the hold time tests and the bithermal tests were used with the isothermal 900°C data since a TEM study of these specimens revealed the same type of coarsening and dissolution processes. The constants determined for the application of the microcrack propagation equation for MAR-M246 are presented in Table 1.

Figure 6 is a plot of σ versus N_f for the hold time tests and the bithermal tests conducted on MAR-M246. From this plot it is obvious that there is no appreciable difference in life between

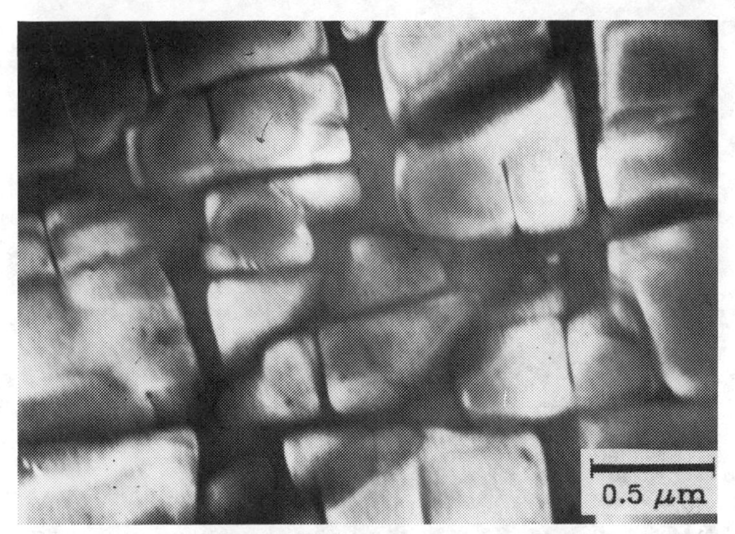

FIG. 4—*Dark field TEM micrograph of specimen T139-1; a MAR-M246 fully reversed test at 650°C at a strain amplitude of 0.003 and a strain rate of* 10^{-4} s^{-1}.

comparable 1000 s tensile, compressive, and no-load hold time tests. This leads to the conclusion that creep deformation has a negligible effect on life for a hold time test. This seems reasonable in view of the rather large grain size of the material and the nature of creep damage in Ni-base superalloys [*42*]. The average grain size of the MAR-M246 castings used in this study is 2500 μm. It seems likely that oxidation would play a much larger role in the damage of such a material than creep.

It is also apparent from Fig. 6 that the bithermal test results resemble hold time tests with the hold time equal to the time it took for the specimen to come to thermal equilibrium during

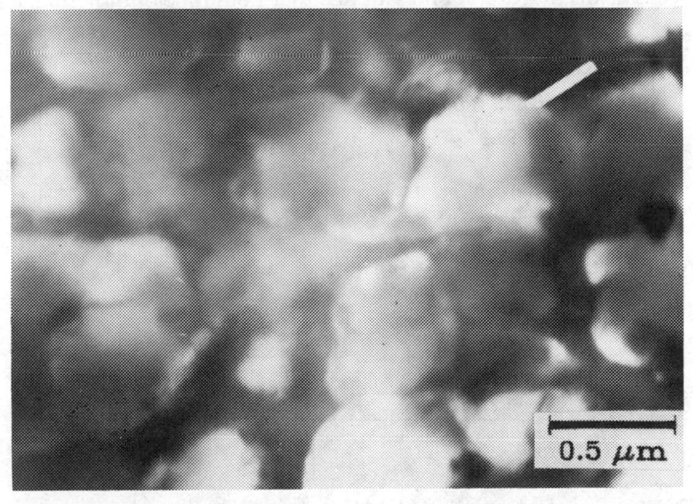

FIG. 5—*Dark field TEM micrograph of specimen G4; a MAR-M246 fully reversed test at 900°C at a strain amplitude of 0.003 and a strain rate of* 10^{-4} s^{-1}.

TABLE 1—*Microcrack propagation equation constants–MAR-M246.*

650°C and 800°C[a]
$m_f = 2.46$
$Z_f = 2.64 \times 10^8$ m$^{-1.46}$
$C_f' = 2595$ m$^{4.92}$ MJ$^{-2.46}$ cycles^{-1}
$\psi = 0.5$
$C_0\|_{800} = 108.8$ m$^{4.92}$ MJ$^{-2.46}$ cycles^{-1} s$^{-0.5}$
900°C
$m_f = 2.20$
$Z_f = 9.96 \times 10^6$ m$^{-1.2}$
$C_f' = 199.2$ m$^{5.40}$ MJ$^{-2.20}$ cycles^{-1}
$\psi = 0.32$
$C_0\|_{900} = 58.13$ m$^{5.40}$ MJ$^{-2.20}$ cycles^{-1} s$^{-0.32}$

[a]Constants B, k, C_c, m_c, and the activation energy of the oxidation process, Q_{ox}, were not determined for MAR-M246.

the temperature excursions. It can be concluded that the bithermal tests yield little additional information beyond that provided by hold time tests in this case.

Since creep did not contribute to damage in the hold time tests, it was further assumed that creep did not contribute to the lives of the bithermal tests. The creep term is not present in the analysis of MAR-M246 for the present set of tests. Since the only damage contributions are associated with fatigue and oxidation, the microcrack propagation equation will reduce to

$$\frac{da}{dN} = \left.\frac{da}{dN}\right|_{\text{fatigue}} + \left.\frac{da}{dN}\right|_{\text{ox}} = C_f'(\alpha a)^{m_f} + C_o\,\Delta t^{\Psi}(\alpha a)^{m_f} \tag{16}$$

It is conceivable, however, that creep could contribute to the damage accumulated during an IP TMF test. The continuous change of temperature as well as mechanical strain may lead to damage mechanisms that are not present in either hold time or bithermal tests.

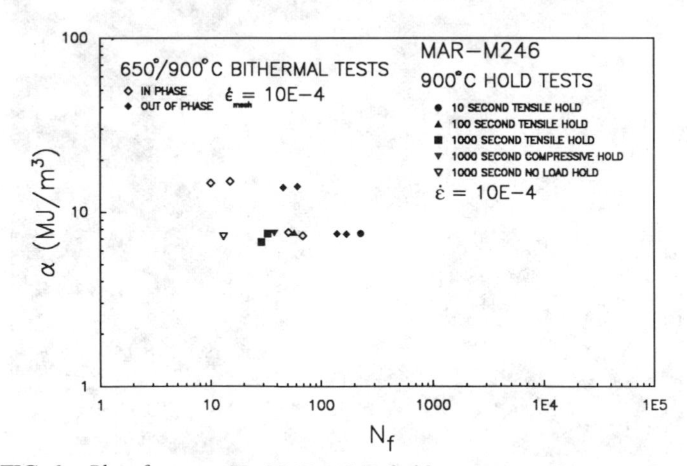

FIG. 6—*Plot of α versus* N_f*, MAR-M246; hold time tests and bithermal tests.*

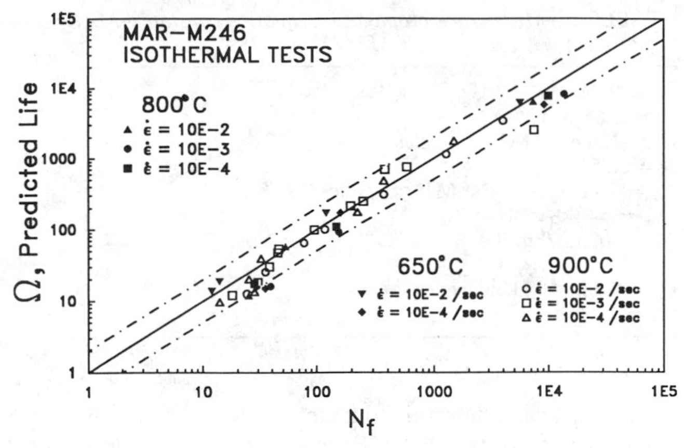

FIG. 7—*Predicted life* Ω *versus* N_f, *MAR-M246; isothermal tests.*

Figures 7 and 8 are plots of the predicted life Ω versus the actual life N_f for the tests conducted on MAR-M246. As can be seen, the correlation is generally within a scatter of ±2 of the median.

MAR-M247

The MAR-M247 tests [*43,44*] analyzed included: (1) isothermal fully reversed tests at 500 and 871°C at strain rates of 10^{-3} and 10^{-5} s^{-1}; and (2) in-phase (IP), out-of-phase (OP), and counter-clockwise diamond history (DH) temperature-strain relation TMF tests between 500°C and 871°C conducted at a mechanical strain rate of 10^{-5} s^{-1}.

The MAR-M247 analysis proceeded in much the same manner as the MAR-M246 analysis. The fatigue constants C_f', and m_f were determined from the 500°C data at a strain rate of 10^{-3} s^{-1}. The oxidation constants C_o', ψ, and the activation energy of the oxidation process Q_{ox} were determined from the 500°C, 10^{-5} s^{-1} tests as well as the 871°C tests at both strain rates. The

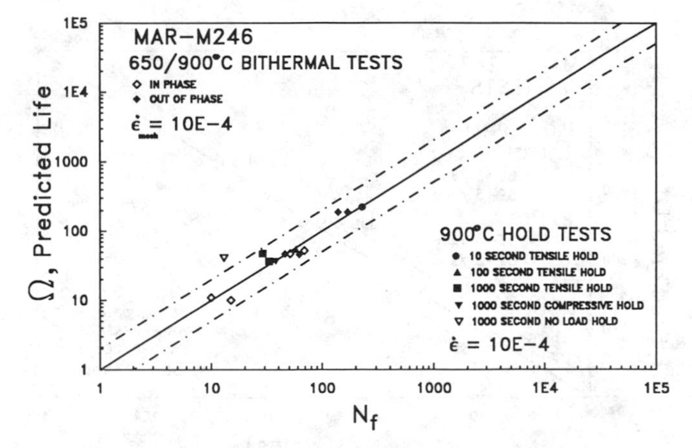

FIG. 8—*Plot of* Ω *versus* N_f, *MAR-M246; hold time tests and bithermal tests.*

TABLE 2—*Microcrack propagation equation constants-MAR-M247.*

$m_f = 2.24$
$Z_f = 1.65 \times 10^7$ m$^{-1.24}$
$C_f' = 120$ m$^{5.48}$ MJ$^{-2.24}$ cycles^{-1}
$\psi = 0.25$
$C_0' = 7.19 \times 10^4$ m$^{5.48}$ MJ$^{-2.24}$ cycles^{-1} s$^{-0.25}$
$Q_{ox} = 54$ KJ/mol
$B = 0.02$ m^3/kmol
$k = 1$
$C_c' = 0.3$ m$^{3.4}$ MJ$^{-1.2}$ cycles^{-1} (= $C_c t_c^{-mc}$)
$m_c = 1.2$

oxidation constants B and k were determined from the OP TMF tests and the creep constants C_c and m_c were determined from the IP TMF tests. The constants determined for the application of the microcrack propagation equation for MAR-M247 are presented in Table 2.

The creep strain rate $\dot{\epsilon}_c$ was determined using a viscoplastic constitutive law for MAR-M247 [*44–46*]. The loading history was run at the prescribed mechanical strain rate, then at a highly accelerated rate. The difference in accumulated strain was attributed to creep strain.

The microcrack propagation equation was then used to predict the lives of the diamond history tests. Figure 9 is a plot of the predicted life Ω versus the actual life N_f for the tests conducted on MAR-M247. Again, in general, the correlation is within a scatter of ± 2 of the median.

Conclusions

A TMF life prediction model based on microcrack propagation is proposed. A microcrack propagation model has a physically appealing definition of damage. The present model contains explicit damage contributions from fatigue, creep, and oxidation. The model uses ΔJ as the parameter correlating fatigue and oxidation microcrack growth, with a frequency and temperature dependence resident in the oxidation term. The oxidation term has an Arrhenius-type temperature dependence with the activation energy modified for TMF loading condi-

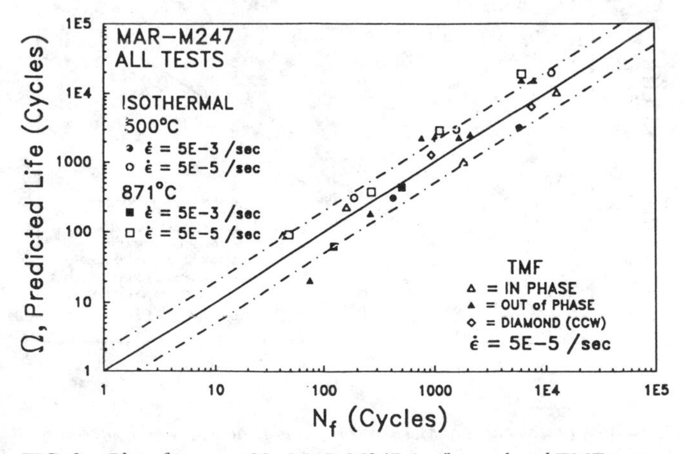

FIG. 9—*Plot of* Ω *versus* N_f, *MAR-M247; isothermal and TMF tests.*

tions. The model uses a stress power release rate-type parameter $\hat{C}$ to correlate creep microcrack growth.

The proposed microcrack propagation model correlates isothermal fully reversed and hold time test lives of MAR-M246 at three temperatures, 650, 800, and 900°C. Bithermal test lives of MAR-M246 between 650 and 900°C were also correlated. The model correlated isothermal and TMF lives of MAR-M247. The TMF tests included in-phase, out-of-phase, and counterclockwise diamond temperature versus mechanical strain histories. The isothermal tests were conducted at 500 and 871°C with the TMF tests run between these two temperatures.

The ΔJ parameter has demonstrated considerable versatility in microcrack and macrocrack correlation. It implicitly contains information regarding the elastic strain range $\Delta\epsilon_e$, which is important for life prediction of Ni-base superalloys. ΔJ provides superior frequency separation for Ni-base superalloy René 80 than offered by $\Delta\epsilon_p$. When used to correlate microcrack growth, ΔJ is a local parameter which reflects crack tip opening behavior.

A viscoplastic constitutive law was beneficial in determining the quantities necessary for the creep component of the microcrack propagation equation for TMF.

Due to complex creep-fatigue couplings, an effort should be made in testing to induce the type of time-dependent damage expected in actual applications. If this is impractical, several types of creep-fatigue interaction tests should be run, with the most conservative constants retained.

From the test results analyzed, the bithermal tests yielded little additional information beyond hold time tests for the Ni-base superalloy MAR-M246. If the application contains continuous temperature cycling, TMF test data, if available, should also be used to determine model constants.

Processes such as dissolution and coarsening of the precipitate microstructure can greatly complicate life analysis. It is necessary to account for morphological changes and to use different sets of constants if the morphology change is great enough, as was the case for MAR-M246 at 900°C.

Acknowledgments

The authors are grateful to the Allison Gas Turbine Engine Division of General Motors Corporation for support of the work on as-cast MAR-M246. Mr. Ritesh Shah is acknowledged for contributions to the TEM study of precipitate coarsening.

References

[*1*] Antolovich, S. D., "Fatigue Mechanisms," *Pressure Vessels and Piping: Design Technology—1982 A Decade of Progress,* S. Y. Zamrik, and D. Dietrich, Eds., The American Society of Mechanical Engineers, N.Y. pp. 533–540.

[*2*] Coffin, L. F., "Fatigue at High Temperatures—Prediction and Interpretation," Proc. Institution of Mechanical Engineers, London, Vol. 188, 1974.

[*3*] Coffin, L. F., "Overview of Temperature and Environmental Effects on Fatigue of Structural Metals," *Fatigue: Environment and Temperature Effects,* J. J. Burke and V. Weiss, Eds., Plenum Press, N.Y., 1983, pp. 1–40.

[*4*] Ostergren, W. J., "A Damage Function and Associated Failure Equations for Predicting Hold Time and Frequency Effects in Elevated Temperature Low Cycle Fatigue," *Journal of Testing and Evaluation,* Vol. 4, 1976, pp. 327–339.

[*5*] Manson, S. S., Halford, G. R., and Hirschberg, M. H., "Creep-Fatigue Analysis by Strain Range Partitioning," NASA TM X-67838, 1971.

[*6*] Halford, G. R. and Saltsman, J. F., "Strainrange Partitioning—A Total Strain Range Version," NASA TM-83023, 1983.

[*7*] Saltsman, J. F. and Halford, G. R., "An Update of the Total-Strain Version of Strainrange Parti-

tioning," *Low Cycle Fatigue, ASTM STP 942,* H. D. Solomon, G. R. Halford, L. R. Kaisand, and B. N. Leis, Eds., Philadelphia, 1988, pp. 329–341.
[*8*] Neu, R. W. and Sehitoglu, H., "Thermomechanical Fatigue, Oxidation and Creep: Part I. Damage Mechanisms," *Metals Transactions A,* Vol. 20A, 1989.
[*9*] Neu, R. W. and Sehitoglu, H., "Thermomechanical Fatigue, Oxidation and Creep: Part II. Life Prediction," *Metals Transactions A,* Vol. 20A, 1989, pp. 1769–1783.
[*10*] Kachanov, L., *Fundamentals of Fracture Mechanics,* Nauka, Moscow, 1974.
[*11*] Rabotnov, Y. N., *Creep Problems in Structural Members,* Amsterdam, North Holland Publishing Co., 1969.
[*12*] Hult, J., "Continuum Damage Mechanics—Capabilities Limitations and Promises," *Mechanisms of Deformation and Fracture,* Pergamon, Oxford, 1979, pp. 233–347.
[*13*] Chaboche, J. L., "Description Thermodynamique et Phenomenologique de la Viscoplasticite Cyclique Avec Endommagement," Ph.D. Thesis, Universite Paris VI, 1978.
[*14*] Chaboche, J. L., "Continuous Damage Mechanics—A Tool to Describe Phenomena Before Crack Initiation," Nuclear Engr. and Design, Vol. 64, 1981, pp. 233–247.
[*15*] Lemaitre, J., "Damage Modelling for Prediction of Plastic or Creep-Fatigue in Structures," Paper L5-1 SMIRT-5 Conference, Berlin, 1979.
[*16*] Majumdar, S. and Maiya, P. S., "A Mechanistic Model for Time Dependent Fatigue," ASME *Journal of Materials and Technology,* Vol. 102, Jan 1980, pp. 159–167.
[*17*] Majumdar, S., "Thermomechanical Fatigue of Type 304 Stainless Steel," *Proceedings,* ASME on Thermal Stress, Material Deformation and Thermo-Mechanical Fatigue, H. Sehitoglu and S. Y. Zamrik, Eds., PVP-Vol. 123, 1987, pp. 31–36.
[*18*] Wareing, J., "Mechanisms of High Temperature Fatigue and Creep-Fatigue Failure," *Fatigue at High Temperature,* R. P. Skelton, Editor, Applied Science, London, 1983, pp. 135–186.
[*19*] Lamba, H. S., "The J-Integral Applied to Cyclic Loading," *Engineering Fracture Mechanics,* Vol. 7, 1975, p. 693.
[*20*] Dowling, N. E. and Begley, J. A., 1976, in *Mechanics of Crack Growth, ASTM STP 590,* Philadelphia, 1976, pp. 82–103.
[*21*] Riedel, H., *Fracture at High Temperatures,* Springer Verlag, New York, 1987.
[*22*] Ohtani, R., Kitamura, T., Nitta, A., and Kuwabara, K., "High-Temperature Low Cycle Fatigue Crack Propagation and Life Laws of Smooth Specimens Derived from the Crack Propagation Laws," *Low Cycle Fatigue, ASTM STP 942,* H. D. Solomon, G. R. Halford, L. R. Kaisand, and B. N. Leis, Eds., Philadelphia, 1988, pp. 1163–1180.
[*23*] Nitta, A. and Kuwabara, K., "Thermal-Mechanical Fatigue Failure and Life Prediction," *Current Japanese Materials Research,* Vol. 3, Elsevier, New York, 1988, pp. 203–222.
[*24*] Ohji, K. and Kubo, S., "Fracture Mechanics Evaluation of Crack Growth Behavior under Creep and Creep-Fatigue Conditions," *High Temperature Creep-Fatigue, Current Japanese Materials Research,* Vol. 3, R. Ohtani, M. Ohnami, and T. Inoue, Eds., Elsevier, Japan, 1988, pp. 91–113.
[*25*] Reuchet, J. and Remy, L., "Fatigue Oxidation Interaction in a Superalloy-Application to Life Prediction in High Temperature Low Cycle Fatigue," *Metallurgical Transactions A,* Vol. 14A, Jan. 1983, pp. 141–149.
[*26*] Saxena, A. and Gieske, B., "Transients in Elevated Temperature Crack Growth," in *Proceedings* of MECAMET—International Seminar on High Temperature Fracture Mechanisms and Mechanics, Vol. III, Dourdan, France, 1987, pp. 19–36.
[*27*] Dowling, N. E., "Crack Growth During Low-Cycle Fatigue of Smooth Axial Specimens," *Cyclic Stress-Strain and Plastic Deformation Aspects of Fatigue Crack Growth, ASTM STP 637,* Philadelphia, 1977, pp. 97–121.
[*28*] Lankford, J., "The Growth of Small Fatigue Cracks in 7075-TG Aluminum," *Fatigue of Engineering Materials and Structures,* No. 5, 1982, pp. 223–227.
[*29*] McDowell, D. L. and Berard, J.-Y., "A ΔJ-Based Approach to Biaxial Low Cycle Fatigue of Shear Damaging Materials," *Proceedings,* Third International Conference on Biaxial/Multiaxial Fatigue, Stuttgart, FRG, April 3–6, 1989.
[*30*] Kishimoto, K., Aoki, S., and Sakata, M., "On the Path Independence of Integral $\hat{J}$," Engineering Fracture Mechanics, Vol. 13, No. 4, 1980, pp. 841–850.
[*31*] Shih, C. F. and Hutchinson, J. W., "Fully Plastic Solutions and Large Scale Yielding Estimates for Plane Stress Crack Problems," *J. of Engineering and Materials Technology,* Vol. 98, 1976, pp. 289–295.
[*32*] Antolovich, S. D., Liu, S., and Baur, R., "Low Cycle Fatigue Behavior of René 80 at Elevated Temperature," *Metallurgical Transactions A,* Vol. 12A, 1981, pp. 473–481.
[*33*] Landes, J. D. and Begley, J. A., "A Fracture Mechanics Approach to Creep Crack Growth," *Mechanics of Crack Growth, ASTM STP 590,* 1976, pp. 128–148.

[34] Goldman, N. L. and Hutchinson, "Fully Plastic Crack Problems: the Center-Cracked Strip Under Plane Strain," *Int. J. Solids Structure,* Vol. 11, 1975, pp. 575–591.
[35] Saxena, A., "Creep Crack Growth Under Non Steady-State Conditions," *Fracture Mechanics: Seventeenth Volume, ASTM STP 905,* J. H. Underwood et al., Eds., Philadelphia, 1986, pp. 185–201.
[36] Saxena A. and Liaw, P. K., "Remaining Life Estimation for Boiler Pressure Parts—Crack Growth Studies," EPRI-CS-4688, Power Research Institute, Palo Alto, July 1986.
[37] Halford, G. R., McGaw, M. A., Bill, R. C., and Fanti, P. D., "Bithermal Fatigue: A Link Between Isothermal and Thermomechanical Fatigue," *Low Cycle Fatigue, ASTM STP 942,* H. D. Solomon, G. R. Halford, L. R. Kaisand, and B. N. Leis, Eds., Philadelphia, 1988, pp. 625–637.
[38] Miller, M. P., "Life Prediction Models for High Temperature Fatigue Based on Microcrack Propagation," M.S. Thesis, Georgia Institute of Technology, Atlanta, 1990.
[39] Antolovich, S. D., Domas, P., and Strudel, J. L., "Low Cycle Fatigue of René 80 as Affected by Prior Exposure," *Metals Transactions A,* Vol. 10A, 1979, pp. 1859–1868.
[40] Antolovich, S. D. and Rosa, E., "Low Cycle Fatigue of René 77 at Elevated Temperatures," *Materials Science and Engineering,* Vol. 47, 1981, pp. 47–57.
[41] Lasalmonie, A. and Strudel, J. L., "Interfacial Dislocation Networks Around γ' Precipitates in Nickel-Base Alloys," *Philosophy Magazine,* Vol. 32, 1975, pp. 937–949.
[42] McDowell, D. L., Antolovich, S. D., and Oehmke, R. L. T., "Mechanistic Considerations for TMF Life Prediction of Nickel-Base Superalloys," Presented at Smirt-10 Post-Conference Seminar on *Inelastic Analysis, Fracture, and Life Prediction,* August 21–22, 1989, University of California, Santa Barbara.
[43] Boismier, D. A. and Sehitoglu, H., "Thermo-Mechanical Fatigue of Mar-M247: Part 2—Experiments," *ASME Journal of Engineering and Materials Technology,* Vol. 112, 1990. pp. 68–79.
[44] Sehitoglu, H. and Boismier, D. A., "Thermo-Mechanical Fatigue of Mar-M247: Part 2—Life Prediction," *ASME Journal of Engineering and Materials Technology,* Vol. 112, 1990, pp. 80–89.
[45] Slavik, D. and Sehitoglu, H., "A Constitutive Model for High Temperature Loading, Part I: Experimentally Based Forms of the Equations," ASME PVP 123, H. Sehitoglu and S. Y. Zamrik, Eds., New York, 1987, pp. 65–73.
[46] Slavik, D. and Sehitoglu, H., "A Constitutive Model for High Temperature Loading, Part II: Comparison of Simulations with Experiments," ASME PVP 123, H. Sehitoglu and S. Y. Zamrik, Eds., New York, 1987, pp. 75–82.

Demirkan Coker,[1] *Noel E. Ashbaugh,*[1] *and Theodore Nicholas*[2]

Analysis of Thermomechanical Cyclic Behavior of Unidirectional Metal Matrix Composites

REFERENCE: Coker, D., Ashbaugh, N. E., and Nicholas, T., **"Analysis of Thermomechanical Cyclic Behavior of Unidirectional Metal Matrix Composites,"** *Thermomechanical Fatigue Behavior of Materials, ASTM STP 1186,* H. Sehitoglu, Ed., American Society for Testing and Materials, Philadelphia, 1993, pp. 50–69.

ABSTRACT: An analytical tool is developed to determine the three-dimensional stress state in a unidirectional composite subjected to axial loading and changes in temperature. A finite difference method is used to analyze a representative volume element of the composite which consists of concentric cylinders. The constituents are assumed to be elastic-plastic materials having temperature dependent properties. An iterative technique using the Prandtl-Reuss flow rule to determine incremental plastic strains is implemented in a computer code capable of predicting the axisymmetric triaxial stresses in a composite under thermomechanical fatigue (TMF) loading conditions. The model is verified with finite element method calculations for the problem of thermal residual stresses resulting from cool-down from the processing temperature. Results for several TMF loading conditions are compared with experimental data and 1-D predictions for a SCS-6 silicon carbide fiber and Ti-24Al-11Nb matrix composite. Significant differences are noted between results based on 1-D and 3-D approximations to the stress state in a composite and are discussed in detail.

KEYWORDS: metal matrix composites, plasticity, micromechanics, concentric cylinder model, thermomechanical fatigue (TMF), analysis, modeling

Interest in metal matrix composites (MMC) for high-temperature aerospace applications has grown in recent years. Because of the high use temperatures envisioned, proposed applications of MMCs almost always involve both thermal and mechanical cyclic loading. Consequently, the accurate prediction of the thermomechanical fatigue (TMF) life of these components is a critical aspect of the design process. Because of the mismatch in coefficient of thermal expansion between fiber and matrix, internal thermal stresses are produced under thermal cycling. These stresses must be accounted for in analysis in addition to those produced by mechanical loading. Life prediction, therefore, requires a knowledge of the individual components of stress or strain in the fiber or matrix, or both, as well as the overall applied stresses.

Parameters used for life prediction in metals subjected to thermomechanical fatigue include stress, strain, energy, or some combination of these. Ellyin et al. [*1*] used total strain energy density to predict TMF life under multiaxial loading. This parameter correlated well with both in-phase (IP) and out-of-phase (OOP) TMF life data for low alloy pressure vessel steel. Spera and Cox [*2*] determined TMF life of components by tracking plastic and creep strains using a computer program called TFLIFE. Halford and Manson [*3*] used strain range partitioning to

[1] Assistant research engineer and senior research engineer, respectively, University of Dayton, Research Institute, Dayton, OH 45469.

[2] Senior scientist, Materials Directorate, Wright Laboratory, Wright-Patterson AFB, OH 45433.

determine life under uniaxial TMF loading. Jaske [*4*] attempted to predict fatigue life using total and plastic strains, stress range, and the product of the maximum stress and total strain range. Taira and Inoue [*5*] used effective strain range to relate multiaxial TMF life to uniaxial life data. The last two investigations concluded that TMF was more damaging than isothermal cycling and OOP cycling was more detrimental than IP cycling.

To determine life in composites the concepts and parameters developed for monolithic metals have been extended to this area. An additional complication which arises in composites is the existence of two or more constituents. To develop life prediction procedures, the work in this area has concentrated on predicting TMF life based on uniaxial fiber or matrix stresses. Dvorak and Tarn [*6*] conducted fatigue experiments on B/Al and other similar composites. They concluded that maximum stress in the matrix determines life and that maximum fiber stress has no effect on fatigue life. They also determined that low cycle fatigue life is associated with cyclic plastic straining. Henry and Stoloff [*7*] conducted TMF life tests on nickel chromium alloy and tantulum carbide fibrous eutectic. They successfully related TMF life to plastic strains in the matrix using the Coffin-Manson law and concluded that fatigue response is matrix-controlled even in the case of such high strains that fiber failure occurs on the first cycle. Johnson et al. [*8*], in their work with SCS-6/Ti-15-3 concluded that fatigue life is controlled by stresses in the 0° fibers. More recently, Russ et al. [*9*] found that a single parameter is inadequate to predict fatigue life from a variety of TMF tests. Rather, they were able to correlate IP and OOP TMF life for SCS-6/Ti-24Al-11Nb using a combination of maximum stress in the fiber and the stress range in the matrix.

To calculate the micromechanical stresses in the fiber and the matrix due to both thermal and mechanical loading, various approaches have been adopted. The most widespread approach involves use of the finite element method to model a representative volume element (RVE) and compute the stress distributions around the fiber. Common RVE geometries that have been used include either a square or a cylindrical fiber in a square matrix and a cylindrical fiber in a cylindrical matrix. Different constitutive behaviors include elastic-plastic behavior [*10–12*], time-dependent behavior using a classical creep law [*13*] and unified viscoplastic theories such as the Bodner-Partom model [*14*], and Bodner-Partom with backstress [*15*]. However, finite element codes are usually run on mainframe computers, are time consuming, allow for solution of only one problem at a time, and require familiarity with finite element analysis and the code. To conduct a parametric study and to support the design of a large number of experiments on different composite systems, a more practical approach is desirable.

One simple approach to micromechanical modeling of composites has been to use average stresses in the constituents. This method has the advantage of being easy to implement into a code and of easily being extended to analyze laminates. Approaches using this method include analysis of square fiber in a square matrix and the vanishing fiber diameter (VFD) model developed by Dvorak and Bahei-El-Din [*16*]. In the VFD model, the presence of the fibers is assumed not to influence the transverse stresses. The transverse properties of a lamina are easily computed using this assumption while the longitudinal properties are calculated from the rule of mixtures. This analysis can be carried out for an elastic fiber surrounded by a matrix material having a variety of constitutive relations to calculate ply properties which, in turn, can be incorporated into lamination theory to analyze a composite. Such a procedure has been implemented into computer codes such as AGLPLY [*17*] which uses thermoplastic matrix behavior and VISCOPLY [*18*] which incorporates the Eisenberg creep model to predict average stresses in the constituents and in a symmetric composite laminate subjected to thermomechanical loading. Aboudi [*19*] modeled a square fiber in a square matrix subcell using first order displacement expansions. The unified theory of Bodner and Partom was used to compute inelastic strains. Hopkins and Chamis [*20*] and Sun [*21*] conducted strength-of-materials type analyses for a square fiber in a square matrix cell to obtain expressions for the constituent microstresses. Chamis and Hopkins [22] modified these expressions based on experimental data for uniaxial lamina and three-dimensional computer simulations of composite behavior. The results were incorporated into a computer program, METCAN [23], which treats material nonlinearity at the constituent level where material behavior is modeled using a time-temperature-stress dependence of a constituent's properties.

The main drawback to all of these simplified material models is the assumption of average uniform stress in the representative volume element. This approach fails to take into account the triaxial stresses that arise due to the mismatch of the Poisson's ratio and the coefficient of thermal expansion of the constituents. A detailed review of these codes and comparisons of the stresses using these methods and using finite element analysis is given by Bigelow et al. [24].

For a more accurate analytical treatment of the triaxial stresses, the concentric cylinder model has been used in the literature due to its simple geometry. Thermoeleastic treatment of the axisymmetric concentric cylinder model with multiple rings is given in Refs 25 and 26. An elastic analysis of a multidirectional coated continuous fiber composite by means of a three-phase concentric cylinder model is given in Ref 27. Ebert et al. [28] and Hecker et al. [29] were the first to introduce elastic-plastic behavior for the constituents in which they modeled two concentric cylinders and verified the model with experiments. Gdoutos et al. used this model to compute thermal expansion coefficients [30] and stress-strain curves and obtained good agreement with experimental results [31]. However, this approach postulated new stress-strain relations in which the strain increments were functions of the stress increment, in contrast to the Prandtl-Reuss relations that relate the strain increments to the current state of stress. This method is a deformation type of theory and could not be applied to nonproportional loadings and ideally plastic material. More recently Lee and Allen [32] obtained an analytical solution for elastic-perfectly plastic fiber and matrix obeying Tresca's yield criterion.

In this paper an analytical tool is developed to compute the three-dimensional stress state in a composite using the concentric cylinder model with multiple rings. The analysis accommodates multiple materials having elastic-plastic behavior with strain hardening. The Prandtl-Reuss flow rule with the von Mises yield criterion is used. In addition, temperature-dependent material properties are taken into account. The analysis was implemented into the computer code FIDEP (Finite Difference Code for Elastic-Plastic Analysis of Composites) which accounts for thermomechanical cyclic loading.

Concentric Cylinder Model

A representative volume element of the composite is modeled as concentric cylinders with the core cylinder representing the fiber and the outer ring representing the matrix (Fig. 1). The fiber and matrix radii are denoted as (a) and (b), respectively. The direction of the z-axis is along the fibers and the cylinders are infinitely long in the axial direction. Cylindrical coordinates are used in the equations.

The following assumptions are made in the analysis. The temperature distribution is uniform and is quasi-static. A perfect bond exists between the constituents of the composite so that there is no slippage or separation of the constituents. The concentric cylinders are in generalized plane strain and are subjected to axisymmetric loadings and displacements so that the shear stresses are zero. The constituent properties are isotropic. The fiber is linearly elastic. The matrix follows a von Mises yield surface and is incompressible in the plastic region, that is, hydrostatic stresses do not cause plastic deformation. The plastic deformation of the matrix is governed by the Prandtl-Reuss flow rule. The following boundary conditions are improved

(1) Radial stress is P_r at $r = b$;
(2) Finite stresses at $r = 0$; and
(3) Continuous radial displacements and radial stresses at the interface.

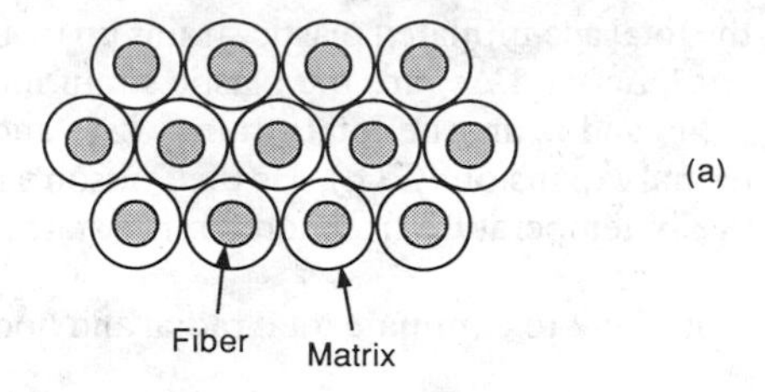

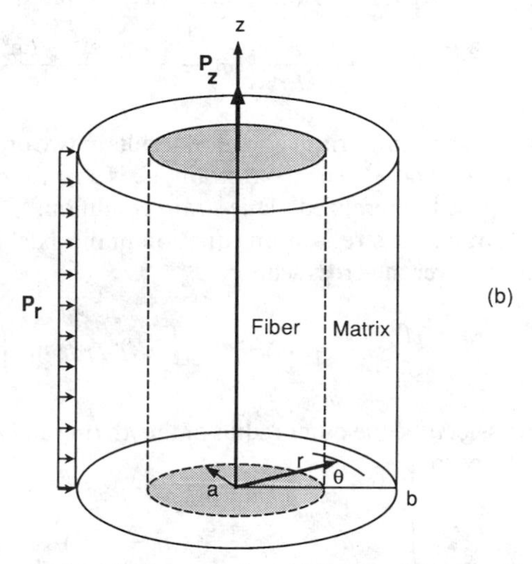

FIG. 1—(a) *Concentric cylinder idealization of a unidirectional composite,* (b) *Concentric cylinder model.*

Governing Equations

Equilibrium and compatibility equations for an axisymmetric case become

$$\frac{d\sigma_r}{dr} + \frac{\sigma_r - \sigma_\theta}{r} = 0 \tag{1}$$

$$\frac{d\varepsilon_\theta}{dr} + \frac{\varepsilon_\theta - \varepsilon_r}{r} = 0 \tag{2}$$

and the stress-strain equations as given in [*33*] are

$$\begin{aligned}
\varepsilon_r &= \frac{1}{E}(\sigma_r - \nu(\sigma_\theta + \sigma_z)) + \alpha T + \varepsilon_r^p + \Delta\varepsilon_r^p \\
\varepsilon_\theta &= \frac{1}{E}(\sigma_\theta - \nu(\sigma_r + \sigma_z)) + \alpha T + \varepsilon_\theta^p + \Delta\varepsilon_\theta^p \\
\varepsilon_z &= \frac{1}{E}(\sigma_z - \nu(\sigma_\theta + \sigma_r)) + \alpha T + \varepsilon_z^p + \Delta\varepsilon_z^p = \text{constant}
\end{aligned} \tag{3}$$

where ε_r^p, ε_θ^p and ε_z^p are the total accumulated plastic strains up to, but not including the current increment of loading, $\Delta\varepsilon_r^p$, $\Delta\varepsilon_\theta^p$ and $\Delta\varepsilon_z^p$ are the plastic strain increments due to the current increment of loading, ε_r, ε_θ and ε_z are the total strains, σ_r, σ_θ and σ_z are the stresses, α is the secant coefficient of thermal expansion (CTE), ν is the Poisson's ratio, E is the Young's modulus, and T is the change in temperature from a reference state in which the stresses and the strains are assumed to be zero.

Substitution of Eq 3 into Eq 2 to eliminate total radial and hoop strains yields

$$\frac{d}{dr}\left[\frac{\sigma_\theta}{E} - \frac{\nu}{E}(\sigma_r + \nu(\sigma_\theta + \sigma_r) - \alpha ET + E(\varepsilon_z - \varepsilon_z^p - \Delta\varepsilon_z^p)) + \alpha T + \varepsilon_\theta^p + \Delta\varepsilon_\theta^p\right] + \frac{(1+\nu)}{Er}(\sigma_\theta - \sigma_r) + \frac{\varepsilon_\theta^p + \Delta\varepsilon_\theta^p - \varepsilon_r^p - \Delta\varepsilon_r^p}{r} = 0 \quad (4)$$

This equation together with the equilibrium equation results in two ordinary differential equations in the two unknowns, σ_r and σ_θ. The axial strain, ε_z, is a constant value across the cross section because of the imposed generalized plane strain condition.

To compute the axial strain, the stress-strain equation in the axial direction (Eq 3) is multiplied by Er and integrated over the cross section

$$\varepsilon_z \int_0^b Erdr = \int_0^b \sigma_z rdr - \int_0^b \nu(\sigma_r + \sigma_\theta)rdr + \int_0^b \alpha ETrdr + \int_0^b E(\varepsilon_z^p + \Delta\varepsilon_z^p)rdr \quad (5)$$

For k concentric cylinders, let a_j be the outer radius of the jth ring and $a_o = 0$, then the integral on the left hand side reduces to

$$\varepsilon_z \int_0^b Erdr = \varepsilon_z \sum_{j=1}^{k} \frac{E_j}{2}(a_j^2 - a_{j-1}^2) \quad (6)$$

The first integral on the right hand side is evaluated using the global equilibrium equation in the axial direction; that is, internal forces are equal to the external applied forces

$$\int_0^b \sigma_z(2\pi r)dr = P_z(\pi r^2) \quad (7)$$

where P_z is the applied axial stress.

Rearranging the terms, the equilibrium equation (Eq 1) can also be written as

$$(\sigma_r + \sigma_\theta)r = \frac{d}{dr}(r^2\sigma_r) \quad (8)$$

Using Eq 8, the second integral on the right hand side of Eq 5 becomes

$$\int_0^b \nu \frac{d}{dr}(r^2\sigma_r)dr = \sum_{j=1}^{k} \nu_j(a_j^2\sigma_r(a_j) - a_{j-1}^2\sigma_r(a_{j-1})) \quad (9)$$

Expanding and recollecting terms in Eq 9

$$\int_0^b \nu \frac{d}{dr}(r^2\sigma_r)dr = \sum_{j=1}^{k-1} (\nu_j - \nu_{j+1})a_j^2\sigma_r(a_j) + \nu_k b^2\sigma_r(b) \quad (10)$$

where $\sigma_r(b)$ is the applied radial stress at $r = b$.

The remaining terms are similarly evaluated assuming constant material properties and temperature distribution in each cylinder. The axial strain for k concentric cylinders then becomes

$$\varepsilon_z = \frac{1}{E_c}\left\{P_z - 2\sum_{j=1}^{k-1} V_j\sigma_r(a_j)(\nu_j - \nu_{j+1}) - 2\nu_k\sigma_r(b) + \alpha_c E_c T + \frac{2}{b^2}\sum_{j=1}^{k} E_j \int_0^b (\varepsilon_z^p + \Delta\varepsilon_z^p) r dr\right\} \tag{11}$$

where, V_j is the volume fraction of the jth cylinder, E_c is the axial composite modulus and α_c is the axial coefficient of thermal expansion of the composite defined by

$$V_j = \frac{a_j^2 - a_{j-1}^2}{b^2}$$

$$E_c = \sum_{j=1}^{k} E_j V_j$$

and

$$\alpha_c = \frac{\sum_{j=1}^{k} \alpha_j E_j V_j}{E_c}$$

In the case of two concentric cylinders representing an elastic fiber and an elastic-plastic matrix, $k = 2$ and $e_z^p = 0$ in the fiber, so that the expression for the axial strain simplifies to

$$\varepsilon_z = \frac{1}{E_c}\left[P_z - 2V_f(\nu_f - \nu_m)\sigma_r(a) - 2\nu_m P_r + \alpha_c E_c T + \frac{2E_m}{b^2}\int_a^b (\varepsilon_z^p + \Delta\varepsilon_z^p) r dr\right] \tag{12}$$

where

$$\alpha_c = (\alpha_f E_f V_f + \alpha_m E_m V_m)/E_c$$

$$E_c = E_f V_f + E_m V_m$$

V_f and V_m are the fiber and matrix volume fractions, respectively, and P_r is the applied radial stress at $r = b$. In Eqs 11 and 12, the axial strain is written in terms of the radial stresses at the interfaces. The total plastic strains are known from the previous loading steps and the new plastic strain increments are related to the stresses through the Prandtl-Reuss flow rule.

Finite Difference Formulation

The method of finite differences is used to solve the two ordinary differential equations. This approach is similar to that used by Mendelson [*33*]. The disk radius is divided into N intervals as shown in Fig. 2. There are thus $N + 1$ stations, the first station being at the center of the disk and the last station at the outer radius. Eqs 1 and 4 are written in finite difference form at midpoints of these intervals in the following way

$$r_{i-1/2} = \frac{r_i + r_{i-1}}{2}, \quad \frac{d}{dr}(\sigma_r)_{i-1/2} = \frac{\sigma_{r,i} - \sigma_{r,i-1}}{r_i - r_{i-1}}$$

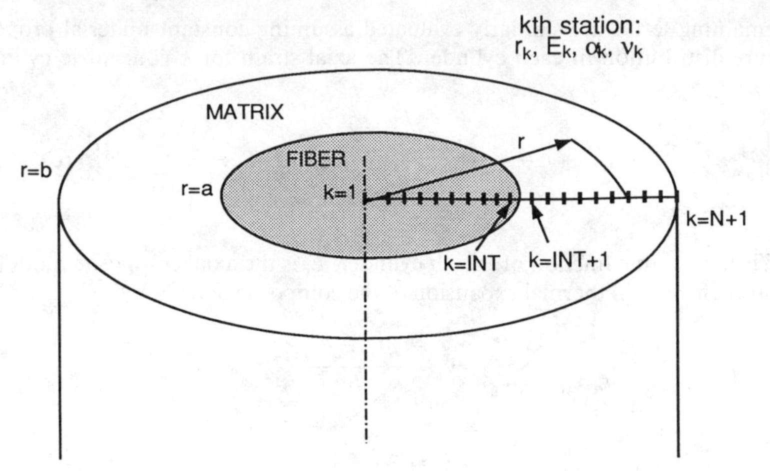

FIG. 2—*Discretization of two concentric cylinders.*

In this manner Eqs 1 and 4 can be written at the midpoints of n stations resulting in $2n$ equations

$$\begin{bmatrix} \dfrac{1}{c_i}\sigma_{r,i-1} - \sigma_{r,i} + a_i\sigma_{\theta,i-1} + a_i\sigma_{\theta,i} = 0 \\ -G_i\sigma_{r,i-1} + D_i\sigma_{r,i} + H_i\sigma_{\theta,i-1} - F_i\sigma_{\theta,i} = Q_i - P_i \end{bmatrix} \quad i = 2, \ldots, n+1 \tag{13}$$

where

$$a_i = \frac{r_i - r_{i-1}}{2r_i},\ b_i = \frac{E_i}{E_{i-1}},\ c_i = \frac{r_i}{r_{i-1}}$$

and

$$D_i = (1 + \nu_i)(\nu_i + a_i),$$

$$F_i = (1 + \nu_i)(1 - \nu_i + a_i),$$

$$G_i = (1 + \nu_{i-1})(\nu_{i-1} - a_ic_i)b_i,$$

$$H_i = (1 + \nu_{i-1})(1 - \nu_{i-1} - a_ic_i)b_i,$$

$$Q_i = E_i\Delta T[\alpha_i(1 + \nu_i) - \alpha_{i-1}(1 + \nu_{i-1})],$$

$$P_i = E_i[P_i^* + \varepsilon_z(\nu_i - \nu_{i-1})],\text{ and}$$

$$P_i^* = a_i\left[(\varepsilon_{r,i}^{pn} + c_i\varepsilon_{r,i-1}^{pn}) - \varepsilon_{\theta,i}^{pn}\left(\frac{1}{a_i} + 1\right) + \varepsilon_{\theta,i-1}^{pn}\left(\frac{1}{a_i} - c_i\right)\right]$$

$$+ \mu_i(\varepsilon_{r,i}^{pn} + \varepsilon_{\theta,i}^{pn}) - \mu_{i-1}(\varepsilon_{r,i-1}^{pn} + \varepsilon_{\theta,i-1}^{pn})$$

where the plastic strains, $\varepsilon_{r,i}^{pn}$, etc., are the updated plastic strains, that is, $\varepsilon_{r,i}^{pn} = \varepsilon_{r,i}^{p} + \Delta\varepsilon_{r,i}^{p}$, etc.

The coefficients at the left hand side are functions of material properties at the ith station. Only the P^* term on the right hand side involves plastic strains. The unknowns are $\sigma_{r,i}$ and $\sigma_{\theta,i}$, $i = 1, \ldots, n + 1$. Using the boundary conditions, $\sigma_{r,n+1} = P_r$ and $\sigma_{\theta,1} = \sigma_{r,1}$, the unknowns reduce to $2n$; $\sigma_{r,i}$, $i = 1, \ldots, n$, and $\sigma_{\theta,i}$, $i = 2, \ldots, n + 1$. The axial stress distribution, $\sigma_{z,i}$, $i = 1, \ldots, n + 1$, is given by Eq 3.

Taking the radial stress terms in Eq 11 to the left hand side of Eq 13 and letting the radial stress at the jth interface correspond to the radial stress at node $i = I_j$, that is, $\sigma_r(a_j) = \sigma_{r,I_j}$ the second expression in Eq 13 becomes

$$-G_i\sigma_{r,i-1} + D_i\sigma_{r,i} + H_i\sigma_{\theta,i-1} - F_i\sigma_{\theta,i} - \frac{2E_i}{E_c}(\nu_i - \nu_{i-1})\sum_{j=1}^{k-1} V_j\sigma_{r,I_j}(\nu_{I_j} - \nu_{I_{j+1}}) = Q_i - E_iP_i^* - E_i(\nu_i - \nu_{i-1})\varepsilon_z^* \quad (14)$$

where

$$\varepsilon_z^* = \frac{1}{E_c}\left\{P_z - 2\nu_kP_r + \alpha_cE_cT + \frac{2}{b^2}\sum_{j=1}^{k} E_j\int_0^b (\varepsilon_z^p + \Delta\varepsilon_z^p)rdr\right\} \quad (15)$$

In the case of two concentric cylinders, $k = 2$ and $j = 1$. Let INT be defined as the index of the highest-numbered node in the fiber and $INT + 1$ be the index of the lowest-numbered node in the matrix. The fiber/matrix interface lies between these two nodes. Then, for all equations in which $i \neq INT + 1$, that is, equations for nodes not adjacent to the interface, $\nu_i - \nu_{i-1}$ is zero and the axial strain vanishes from these equations. For the $INT + 1$st equation, Eq 14 reduces to

$$\left[\frac{2E_mV_f}{E_c}(\nu_f - \nu_m)^2 - G_{INT+1}\right]\sigma_{r,INT} + D_{INT+1}\,\sigma_{r,INT+1} + H_{INT+1}\,\sigma_{\theta,INT} - F_{INT+1}\,\sigma_{\theta,INT+1} = Q_{INT+1} - E_mP_{INT+1}^* - E_m(\nu_m - \nu_f)\varepsilon_z^* \quad (16)$$

Hence Eq 13 is written in matrix form as

$$\mathbf{A}x = \mathbf{B} \quad (17)$$

where $\mathbf{A}$ is a $2n$ by $2n$ matrix of constant coefficients, x is the radial and hoop stress vector of length $2n$ and $\mathbf{B}$ is a vector of length $2n$, as shown in Fig. 3. In matrices $\mathbf{A}$ and $\mathbf{B}$ all the constants are known except the plastic strain increments which are presumed.

Plastic Strain-Total Strain Plasticity Relations

The plastic strain increments are computed using modified Prandtl-Reuss relations derived by Mendelson [33]. In this formulation, modified total strains are used and are defined as follows. Assume a loading path to a given state of stress and total plastic strains ε_{ij}^p. Let the next load step be applied producing additional plastic strains $\Delta\varepsilon_{ij}^p$. Then the modified total strains are

$$\varepsilon_{ij}' = \varepsilon_{ij} - \varepsilon_{ij}^p = \varepsilon_{ij}^e + \varepsilon_{ij}^{th} + \Delta\varepsilon_{ij}^p \quad (18)$$

where ε_{ij}^e is the elastic component of the total strain, ε_{ij}^{th} is the thermal strain, ε_{ij}^p is the accumulated plastic strain up to (but not including) the current increment of load, and $\Delta\varepsilon_{ij}^p$ is the

$$
\left[\begin{array}{ccccccc|ccccccc}
a_2 & -1 & 0 & 0 & \cdots & 0 & 0 & a_2 & 0 & 0 & \cdots & 0 & 0 & 0 \\
0 & 1/c_3 & -1 & 0 & \cdots & 0 & 0 & a_3 & a_3 & 0 & \cdots & 0 & 0 & 0 \\
0 & 0 & 1/c_4 & -1 & \cdots & 0 & 0 & 0 & a_4 & a_4 & \cdots & 0 & 0 & 0 \\
 & & & & & & & & & & & & & \\
0 & 0 & 0 & 0 & \cdots & 1/c_N & -1 & 0 & 0 & 0 & \cdots & a_N & a_N & 0 \\
0 & 0 & 0 & 0 & \cdots & 0 & 1/c_{N+1} & 0 & 0 & 0 & \cdots & 0 & a_{N+1} & a_{N+1} \\
\hline
H_2\text{-}G_2 & D_2 & 0 & 0 & \cdots & 0 & 0 & -F_2 & 0 & 0 & \cdots & 0 & 0 & 0 \\
0 & -G_3 & D_3 & 0 & \cdots & 0 & 0 & H_3 & -F_3 & 0 & \cdots & 0 & 0 & 0 \\
 & & \cdots & \boxed{G^*} & \cdots & & & & & & & & & \\
0 & 0 & 0 & 0 & \cdots & -G_N & D_N & 0 & 0 & 0 & \cdots & H_N & -F_N & 0 \\
0 & 0 & 0 & 0 & \cdots & 0 & -G_{N+1} & 0 & 0 & 0 & \cdots & 0 & H_{N+1} & -F_{N+1}
\end{array}\right]
\left[\begin{array}{c}
\sigma_{r,1} \\ \sigma_{r,2} \\ \vdots \\ \\ \sigma_{r,N-1} \\ \sigma_{r,N} \\ \hline \sigma_{\theta,2} \\ \sigma_{\theta,3} \\ \vdots \\ \sigma_{\theta,N} \\ \sigma_{\theta,N+1}
\end{array}\right]
=
\left[\begin{array}{c}
0 \\ 0 \\ \vdots \\ \\ 0 \\ P_r \\ \hline Q_2\text{-}P_2 \\ Q_3\text{-}P_3 \\ \vdots \\ Q_N\text{-}P_N \\ Q_{N+1} - P_{N+1} \\ -\ddot{D}_{N+1} P_r
\end{array}\right]
$$

$G^* = A(N{+}INT, INT) = 2\, V_f \frac{E_m}{E_c} (\nu_f - \nu_m)^2 - G_{INT+1}$

FIG. 3—*Definition of the A matrix, x vector and B vector in Eq 17.*

increment of plastic strain due to the increment of load. The previous plastic strains ε_{ij}^{p} are presumed to be known, and $\Delta\varepsilon_{ij}^{p}$ is to be calculated. Using this definition the Prandtl-Reuss relations [*33*] become

$$\Delta e_{ij}^{p} = \frac{\Delta e_p}{\varepsilon_{et}} e'_{ij} \tag{19}$$

where e'_{ij} are the modified total deviatoric strains, ε_{et} is the equivalent or effective modified total strain defined by

$$\varepsilon_{et} = (\tfrac{2}{3} e'_{ij} e'_{ij})^{1/2} \tag{20}$$

and Δe_p is the equivalent or effective plastic strain increment defined by

$$\Delta e_p = (\tfrac{2}{3} \Delta e_{ij}^{p} \Delta e_{ij}^{p})^{1/2} \tag{21}$$

The relationship between the effective incremental plastic strain and effective modified total strain is given by [*33*]

$$\Delta e_p = \frac{\varepsilon_{et} - \dfrac{1}{3G}\sigma_{e,i-1}}{1 + \dfrac{1}{3G}\left(\dfrac{d\sigma_e}{de_p}\right)_{i-1}} \tag{22}$$

where G is the shear modulus and $\sigma_{e,i-1}$ is the equivalent or effective stress at the previous step defined by

$$\sigma_e = \sqrt{\tfrac{3}{2} S_{ij} S_{ij}} \tag{23}$$

in which S_{ij} are the deviatoric stresses. Equations 19, 20 and 22 are used together to determine plastic strain increments at each loading step.

Algorithm

The elastic-plastic algorithm for the computer code FIDEP is shown in Fig. 4. The temperature dependent material data and the loading are read from an input file. A load increment and a temperature increment are applied. The finite difference equations (*17*) are solved for the new stress state assuming an elastic deformation, that is, $\Delta\varepsilon_r^p = \Delta\varepsilon_\theta^p = \Delta\varepsilon_z^p = 0$. A new yield surface is computed for the present temperature at all stations. Effective stress (Eq 23) is compared to this new yield surface for each station. If the effective stress is less than the effective stress at all nodes the plasticity subroutine is bypassed to go to the next thermal and mechanical load increments, otherwise a nonzero plastic strain increment is assumed. The finite difference equations (*17*), with the assumed plastic strain increment, are solved for the radial and hoop stresses. The axial stress is computed from these stresses. The modified total strains and the equivalent total strain are computed from Eqs 18 and 20, respectively. The effective incremental plastic strain is computed from Eq 21 and the new incremental plastic strains are computed from the plastic strain-total strain relations (*19*). These strains are then compared with the previous plastic strain increments and if the difference is less than a certain value, the next loading increment is applied. Otherwise the finite difference equations (*17*) are solved with the new incremental plastic strains and the plastic subroutine is repeated until convergence of the incremental plastic strains is obtained.

Results and Discussion

Several problems are solved using the FIDEP code to illustrate the capability for computing micromechanical stresses, to compare with solutions obtained by different techniques, and to analyze the stress states obtained under thermomechanical fatigue conditions. In these com-

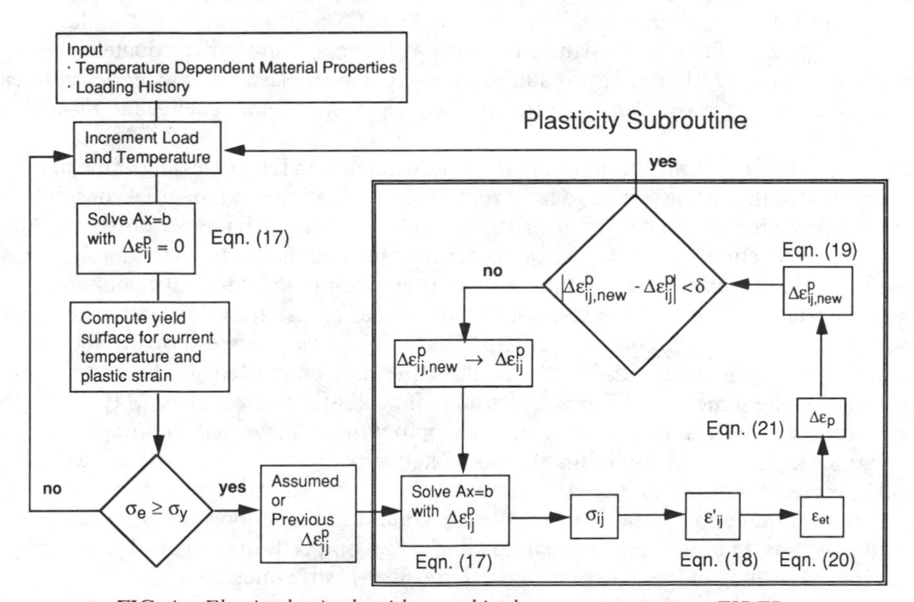

FIG. 4—*Elastic-plastic algorithm used in the computer program FIDEP.*

TABLE 1—*Mechanical properties for SCS-6 fiber and Ti-24Al-11Nb matrix used in the analysis.*

Temperature (°C)	α^a (10^{-6}/°C)	Elastic Modulus, (GPa)	Yield Stress, (MPa)	Plastic Modulus, (GPa)
			SCS-6 Fiber Properties ν = 0.22, E = 414 GPa	
20	4.70			
93	4.81			
204	4.97			
316	5.12			
427	5.26			
538	5.38			
649	5.50			
760	5.60			
871	5.70			
982	5.78			
1010	5.80			
			Ti-24Al-11Nb Matrix Properties ν = 0.22	
20	12.33	94	604	1.3
93	12.47	92	560	0.9
204	12.78	91	498	0.7
316	13.21	89	447	0.7
427	13.75	79	421	0.4
538	14.42	70	381	0.1
649	15.20	50	357	0
760	16.10	25	252	2.3
871	17.11	18	138	2.6
982	18.25	16	38	1.2
1010	18.55	15	30	1.0

[a] α is secant coefficient of thermal expansion with a reference temperature of 1010°C.

putations a two-material concentric cylinder model consisting of SCS-6 silicon carbide fiber and Ti-24Al-11Nb matrix is used. The mechanical properties for SCS-6 and Ti-24Al-11Nb are shown in Table 1 as a function of temperature. The SCS-6 fiber is isotropic and elastic with only the coefficient of thermal expansion varying with temperature. An adequate representation of the Ti-24Al-11Nb matrix was attained using a bilinear elastic-plastic model with temperature dependent elastic and plastic moduli, yield stress and coefficient of thermal expansion.

The first problem was the thermal cool-down associated with the processing of the material. The fiber and matrix are assumed to have zero stresses and strains at the initial consolidation temperature. As the material cools down, the modulus changes with temperature and the difference in coefficient of thermal expansion between fiber and matrix leads to thermal stresses in the fiber and matrix. The three-dimensional stresses computed during the cool-down process are illustrated in Fig. 5. The stresses shown in the matrix material at the fiber-matrix interface are based on zero stress condition at 1010°C. The computed stresses, shown as the symbols in the plot, are compared with results for the same problem solved by the finite element method using the same material properties and same constitutive behavior [*34*]. The agreement with the FEM results is excellent. The computations show that the matrix material reaches the yield surface at approximately 600°C and remains in contact with the yield surface for all lower temperatures. The axial and hoop stresses are in tension at room temperature and are of approximately equal magnitude while the radial stress is compressive and of equivalent magnitudes. Note that subsequent tensile loading of the composite in the axial direction results in additional yielding of the matrix material at the fiber-matrix interface.

To illustrate the numerical differences, the same computations were carried out assuming

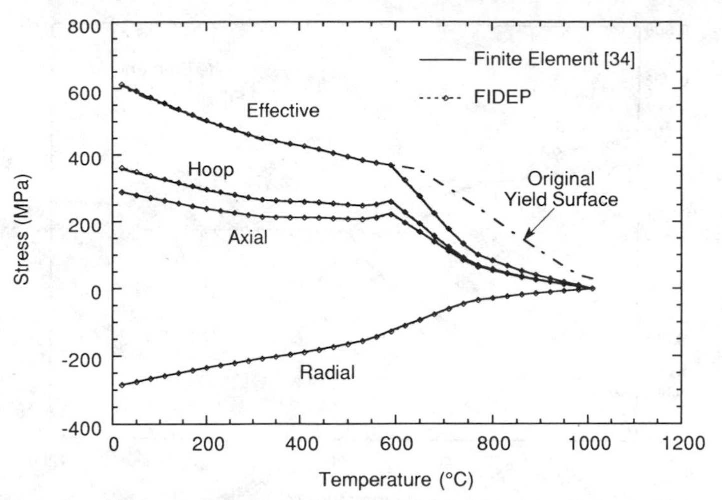

FIG. 5—*Predictions for thermally induced stresses in Ti-24Al-11Nb matrix at the fiber-matrix interface due to cool-down from processing temperature using finite element analysis and FIDEP.*

purely one-dimensional behavior, equivalent to a rule of mixtures formula but taking into account the temperature variation of α amd E and considering elastic-plastic behavior in the matrix material. This computation was carried out for the residual thermal stresses during cool-down using FIDEP with the radial and hoop stresses set equal to zero. The results of that computation are shown in Fig. 6 where the uniaxial results are compared with the 3-D results from the concentric cylinder model in FIDEP. The axial stress after cool-down to room temperature in the matrix (average) is 424 MPa compared to 299 MPa for the 3-D case. The uniaxial solution has an effective stress at room temperature of 424 MPa compared to 610 MPa for the 3-D solution. Of significance in this particular example is the fact that the 3-D solution results in matrix yielding whereas under 1-D conditions no yielding occurs. Consequently, the

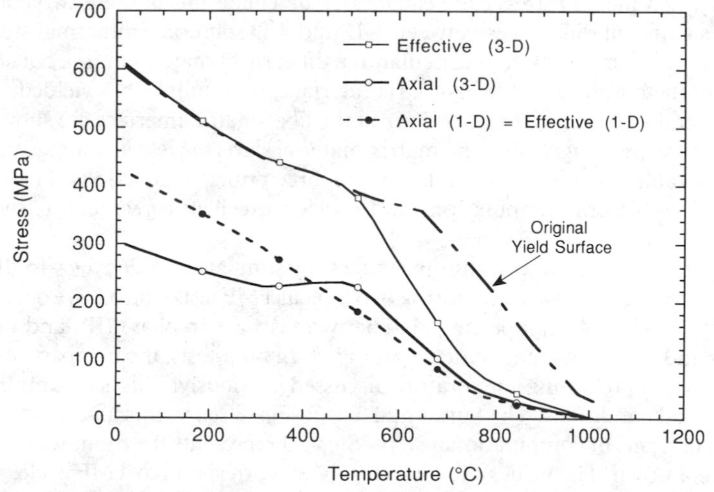

FIG. 6—*Predictions for the axial stresses in Ti-24Al-11Nb matrix during cool-down with the uniaxial model and the concentric cylinder model.*

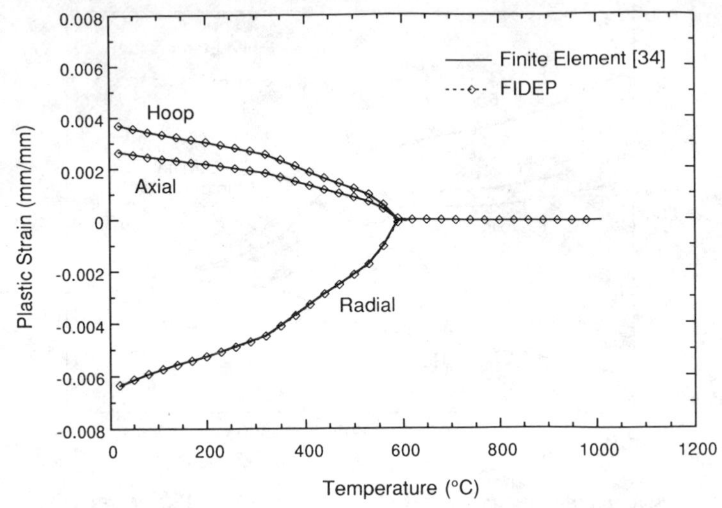

FIG. 7—*Model predictions for plastic strains in Ti-24Al-11Nb matrix at the fiber-matrix interface due to cool-down from processing temperature.*

residual thermal stresses in the axial direction are found to be 42% higher in the 1-D case than in the 3-D case, whereas the effective stress is 30% lower than the 3-D solution.

To compare thermal stresses under purely elastic conditions, the 1-D and 3-D stresses were computed using a single average value of E and α for each material. Choosing $\Delta T = -990°C$ and the following material constants: $E_m = 94$ GPa, $E_f = 414$ GPa, $\alpha_m = 11 \times 10^{-6}/°C$, $\alpha_f = 4.7 \times 10^{-6}/°C$, the following stresses were computed. For the 1-D rule of mixtures solution, the residual thermal stresses in the axial direction at room temperature (20°C) are $\sigma_m = 412$ MPa ($= \sigma_{meff}$) and $\sigma_f = -765$ MPa. The 3-D solution for the same problem using FIDEP provides $\sigma_m = 502$ MPa and $\sigma_{meff} = 872$ MPa. In this case, the axial component of stress in the matrix is 21.8% higher in the 3-D solution while the effective stress in the matrix is a factor of 2.12% higher when the 3-D stress state is considered. The results of these two computations illustrate the significant differences between 1-D and 3-D solutions in thermal stress computations for composite materials. In particular, the differences may result in a conclusion as to whether or not the matrix, at the fiber-matrix interface, for example, has yielded.

The plastic strains developed in the matrix at the fiber-matrix interface are shown in Fig. 7. It can be seen here, as in Fig. 6, that the matrix material starts to yield at approximately 600°C and that measurable plastic strains develop in the three principal directions. These results are also compared with FEM computations and provide excellent agreement as shown in the figure.

The FIDEP code was used to compute stresses for simulated TMF cycles to illustrate the micromechanical stresses developed during two typical TMF tests which have been conducted on a SCS-6/Ti-24Al-11Nb composite [*9*]. The two tests are in-phase (IP) and out-of-phase (OOP) cycles and are shown schematically in Fig. 8. In both tests, the first part of the cycle is the cool-down from processing temperature discussed previously. This is accomplished under zero external load conditions. The horizontal axis in Fig. 8 represents an arbitrary time scale and is shown in terms of computational steps. Step 20 represents the room temperature condition after cool-down. The values shown in the figure correspond to TMF cycles which were evaluated experimentally and involve a maximum stress of 700 MPa and 650 MPa, $R = 0.1$

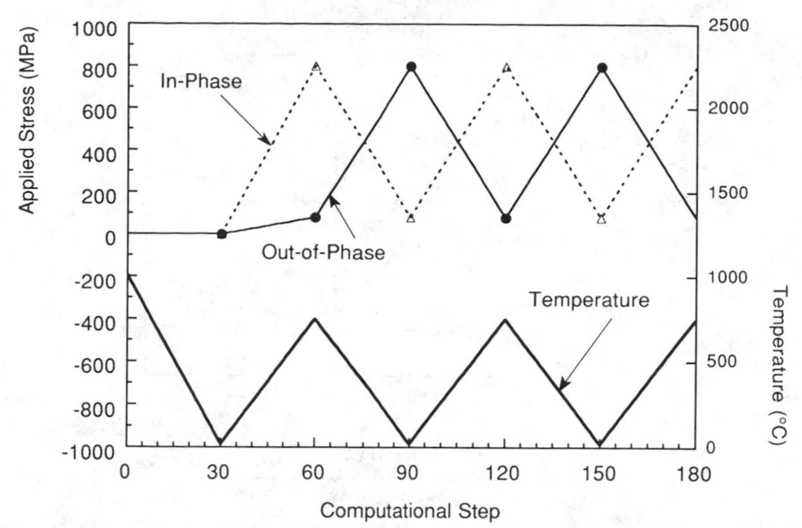

FIG. 8—*Temperature and stress history of a typical in-phase and out-of-phase TMF cycle.*

(ratio of minimum to maximum stress) for the OOP and IP cases, respectively, and a temperature range of 150 to 650°C.

The numerical results for the stress-strain behavior of the composite for stabilized IP and OOP TMF cycles are presented in Fig. 9. The numerical results are obtained after the 2nd or 3rd cycle after which the behavior is purely elastic. The nonlinearity in the stress-strain curves is due to the variation of α and E with temperature. The OOP cycle is nearly vertical since the net expansion during a temperature increase is balanced against the reduction in mechanical strain due to the decrease in load. Conversely, the IP cycle shows a large slope because both mechanical and thermal strain increase during the loading portion of the cycle. In Fig. 9 the

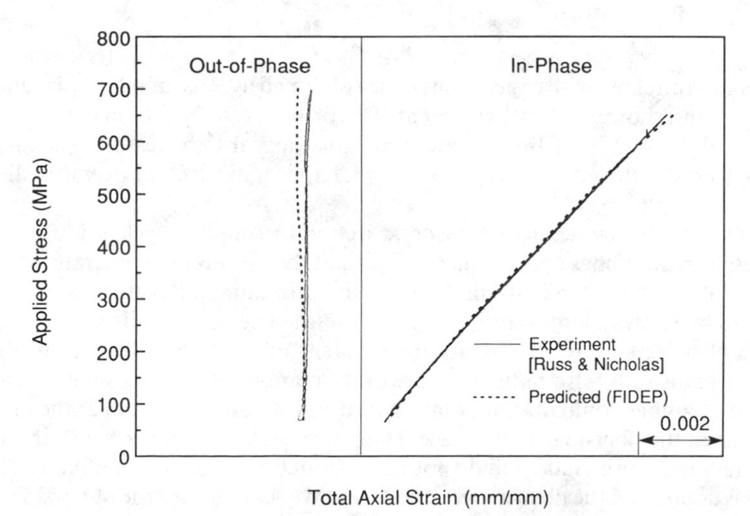

FIG. 9—*Comparison of calculated and experimental stress-strain profiles in in-phase and out-of-phase TMF cycles.*

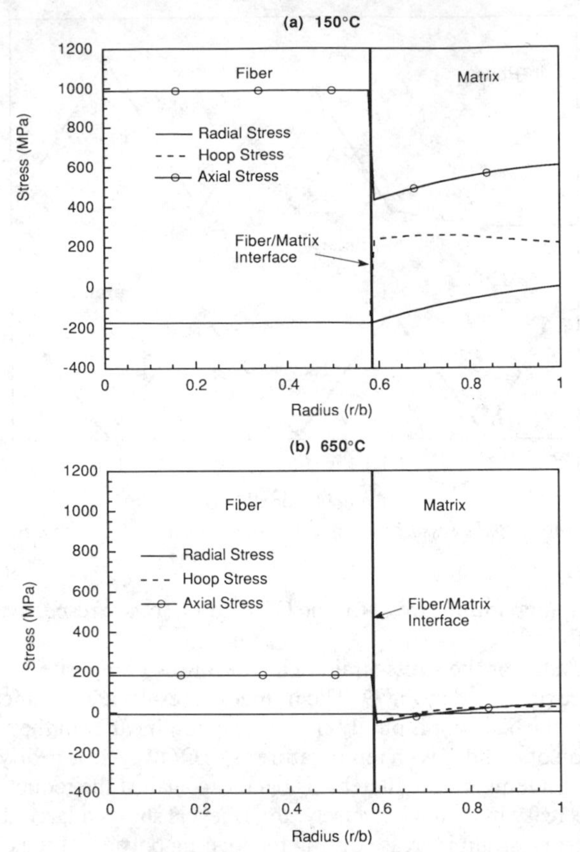

FIG. 10—*Variation of predicted stresses with radius for an out-of-phase TMF cycle;* (a) *150°C,* (b) *650°C.*

calculations are compared with experimental data obtained by Russ and Nicholas after several hundred cycles and show excellent agreement. This provides confidence in the micromechanical stress calculations because the material properties appear to be realistic and any damage which might have developed during cycling does not appear to affect the overall stiffness of the composite.

The variation of the stresses across the cross section for an OOP cycle is illustrated in Fig. 10. The stresses in both fiber and matrix are shown at the minimum temperature (maximum load) in (a) while stresses at maximum temperature (minimum load) are presented in (b). At 150°C, Fig. 10 (a) shows a large tensile axial stress in the fiber and tensile axial stresses in the matrix. The fiber stresses are essentially independent of radial coordinate, but the matrix stresses vary significantly with radius because of the complex 3-D stress state which arises in the concentric cylinder configuration. Matrix axial stresses are maximum at the outer radius and minimum at the fiber-matrix interface. Hoop stresses, on the other hand, are minimum at the outer radius of the cylinder but do not vary as much as the axial component. The matrix radial stress is negative at the fiber-matrix interface and goes to zero at the outer radius because of the imposed stress-free boundary condition. The stresses developed are the net result of the thermal stresses developed at a low temperature, the axial component of which is compression

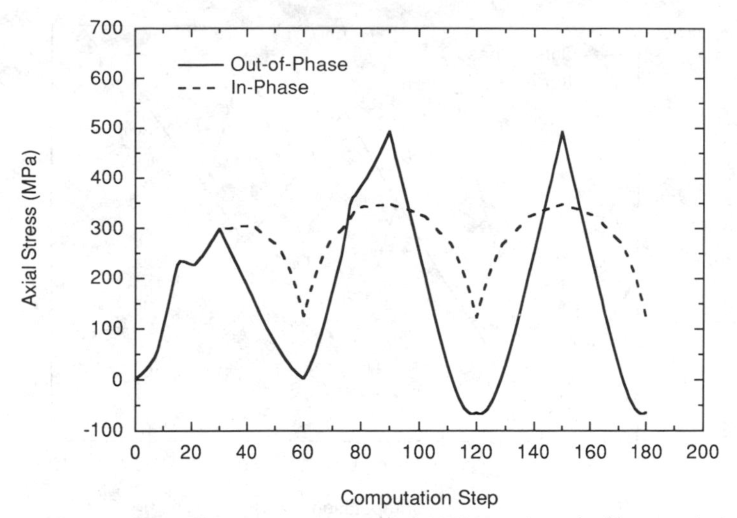

FIG. 11—*Predictions for axial stresses in the Ti-24Al-11Nb matrix at the fiber-matrix interface for in-phase and out-of-phase TMF cycles.*

in the fiber and tension in the matrix, and the applied mechanical stresses which are tension in both fiber and matrix in the axial direction.

At elevated temperature in an OOP TMF cycle, the stress state is generally small as shown in Fig. 10 (b). Under this condition, the thermal stresses are small because at high temperature the stresses are relaxed while the mechanical stresses are small because this is the condition of minimum load.

The axial stresses in the matrix at the fiber-matrix interface are shown in Fig. 11. The steps on the horizontal axis correspond to those of the TMF spectrum schematic in Fig. 8. As noted in the discussion of the previous figure, the OOP cycle has tensile thermal stresses as well as tensile mechanical stresses at minimum temperature which correspond to maximum load. This results in a large stress range in the matrix from approximately 450 MPa at minimum temperature (step 60) to approximately −50 MPa at maximum temperature (step 80). The IP cycle, on the other hand, has a much smaller stress range in the matrix material. At minimum temperature (step 60), there is a thermal residual stress but little contribution from applied (minimum) load. The net stress is approximately 275 MPa. When maximum temperature is reached (step 80), the thermal stress decreases and the mechanical stress increases. The net result is a stress decrease to approximately 200 MPa. In summary, the axial stress range in the matrix is large in an OOP cycle and small in an IP cycle. If failure is dominated by the stress range in the matrix material, the OOP cycle would be the most damaging TMF cycle for this temperature range and applied stress range. This was observed in the experiments of Russ and Nicholas [*9*].

In a similar fashion, the axial stress peaks in the fiber are compared for an IP and OOP TMF cycle in Fig. 12. In the fiber, the thermal stresses at low temperature are compressive because of the low value of α compared to that in the matrix. For an IP cycle, therefore, the axial stresses range from compressive, due to residual thermal stress at low temperature (step 60), to tension due to the applied maximum load at high temperature (step 80). In the OOP cycle, the compressive residual stress at low temperature is offset by the larger mechanical stress due to maximum load, resulting in a net tensile stress of approximately 1000 MPa at minimum

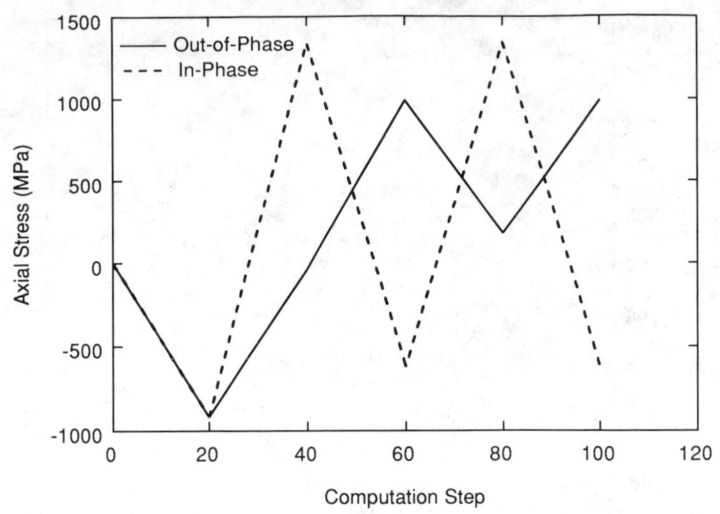

FIG. 12—*Predictions for axial stress peaks in the SCS-6 fiber for in-phase and out-of-phase TMF cycles.*

temperature (step 60). Decreasing the load while increasing the temperature reduces the mechanical stress but also relaxes the compressive residual stress. Since the mechanical stress range exceeds the thermal stress range for this set of conditions, the net effect is a reduction in stress to approximately 200 MPa at maximum temperature (step 80). Thus, an in-phase cycle produces a larger stress range and higher maximum stress in the fiber than an out-of-phase cycle for the conditions evaluated here. This is in contrast to the stress range in the matrix which is maximum in the OOP cycle and smaller in the IP cycle.

For completeness, the radial and hoop stresses in the matrix at the fiber-matrix interface are plotted in Fig. 13. Since the applied load is in the axial direction, its contribution to hoop and

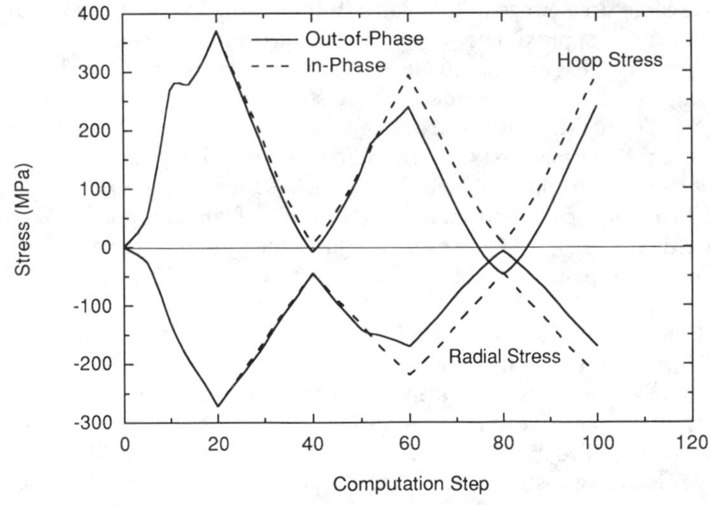

FIG. 13—*Predictions for radial and hoop stresses in Ti-24Al-11Nb matrix at the fiber-matrix interface in TMF cycle.*

radial stresses is minimal and arises solely due to Poisson's ratio and 3-D plasticity effects. Thus, the stresses are mostly due to the thermal cycling, and the IP and OOP cycles produce a similar stress state. At minimum temperature (step 60), the hoop stresses in the matrix are tensile whereas the radial stresses are compressive. At the maximum temperature (step 80), both components relax to nearly zero.

Conclusions

An analytical tool was developed to model a unidirectional composite subjected to thermomechanical cyclic loading and processing conditions. The finite difference method incorporated into a computer code, FIDEP, provides an efficient numerical procedure for analyzing a variety of problems involving thermal and mechanical cycling. The procedure allows the modeling of the constituent materials as elastic-plastic with temperature dependent properties. The concentric cylinder approximation allows the computations to capture the three-dimensional aspects of the stress state in a real composite. Results for a thermal cool-down in a SCS-6/Ti-24 Al-11Nb unidirectional composite compare well with those obtained using the finite element method. Several problems were solved for thermomechanical loading conditions and demonstrate the three-dimensional nature of the stress fields in the matrix material. Predicted stress-strain behavior that stabilized after a few cycles compared favorably with experimental stress-strain curves measured after a few hundred cycles. Comparisons with one-dimensional approximations to the stress field indicate significant differences in the values of the computed axial stresses as well as the effective stresses in the matrix at the fiber-matrix interface due to significant transverse stresses predicted using the 3-D concentric cylinder model. In particular, the one-dimensional solution can show the material to behave purely elastically while the 3-D solution indicates yielding in the matrix for the same material constants and applied loading conditions. This serves to point out the importance of the 3-D nature of the stresses in a composite, particularly when stresses close to yield may be encountered.

Acknowledgment

The work described here was performed at the Materials Directorate, Wright Laboratory (WL/MLLN) under Contract No. F33615-87-C-5243. First author gratefully acknowledges Dr. R. John for his discussions.

References

[1] Ellyin, F., Golos, K., and Xia, Z., "In-Phase and Out-of-Phase Multiaxial Fatigue," *Journal of Engineering Materials and Technology,* Vol. 113, Jan. 1991, pp. 112–118.
[2] Spera, D. A. and Cox, E. C., "Description of Computerized Method for Predicting Thermal Fatigue Life of Metals," *Thermal Fatigue of Materials and Components, STP 612,* D. A. Spera and D. F. Mowbray, Eds., American Society for Testing and Materials, Philadelphia, PA, 1976, pp. 69–85.
[3] Halford, G. R. and Manson, S. S., "Life Prediction of Thermal-Mechanical Fatigue Using Strain-range Partitioning," *Thermal Fatigue of Materials and Components, STP 612,* D. A. Spera and D. F. Mowbray, Eds., American Society for Testing and Materials, Philadelphia, PA, 1976, pp. 239–254.
[4] Jaske, C. E., "Thermal-Mechanical, Low-Cycle Fatigue of AISI 1010 Steel," *Thermal Fatigue of Materials and Components, STP 612,* D. A. Spera and D. F. Mowbray, Eds., American Society for Testing and Materials, Philadelphia, PA, 1976, pp. 170–198.
[5] Taira, S. and Inoue, T., "Thermal Fatigue Under Multiaxial Thermal Stresses," *International Conference on Thermal Stress and Thermal Fatigue, Paper No. 3,* Berkeley, England, 1969.
[6] Dvorak, G. J. and Tarn, J. Q., "Fatigue and Shakedown in Metal Matrix Composites," *Fatigue of Composite Materials, STP 569,* American Society for Testing and Materials, Philadelphia, PA, 1975, pp. 145–168.

[7] Henry, M. F. and Stoloff, N. S., "High Strain Fatigue in a Ni(Cr)-TaC Fibrous Eutectic," *Fatigue of Composite Materials, STP 569,* American Society for Testing and Materials, Philadelphia, PA, 1975, pp. 189–209.
[8] Johnson, W. S., Lubowinski, S. J., and Highsmith, A. L., "Mechanical Characterization of Unnotched SCS_6/Ti-15-3 Metal Matrix Composites at Room Temperature," *Thermal and Mechanical Behavior of Metal Matrix and Ceramic Matrix Composites, STP 1080,* J. M. Kennedy, H. H. Moeller, and W. S. Johnson, Eds., American Society for Testing and Materials, Philadelphia, PA, 1990, pp. 193–218.
[9] Russ, S. M., Nicholas, T., Bates, M., and Mall, S., "Thermomechanical Fatigue of SCS-6/Ti-24Al-11Nb Metal Matrix Composite," *Failure Mechanisms in High Temperature Composite Materials,* AD-Vol. 22/AMD-Vol. 122, G. K. Haritos, G. Newaz, and S. Mall, Eds., American Society of Mechanical Engineers, New York, 1991, pp. 37–43.
[10] Lin, T. H., Salinas, D. and Ito, Y. M., "Elastic-Plastic Analysis of Unidirectional Composites," *Journal of Composite Materials,* Vol. 6, 1972, pp. 48–60.
[11] Adams, D. F., "Inelastic Analysis of a Unidirectional Composite Subjected to Transverse Normal Loading," *Journal of Composite Materials,* Vol. 4, 1970, pp. 310–328.
[12] Foye, R. L., "Theoretical Post-Yielding Behavior of Composite Laminates, Part I—Inelastic Micromechanics," *Journal of Composite Materials,* Vol. 7, April 1973, pp. 178–193.
[13] Foye, R. L., "Inelastic Micromechanics of Curing Stresses in Composites," *Inelastic Behavior of Composite Materials,* Carl T. Herakovich, Ed., 1975 ASME Winter Annual Meeting, Houston, Texas, pp. 177–211.
[14] Kolkailah, F. A. and McPhate, A. J., "Bodner-Partom Constitutive Model and Nonlinear Finite Element Analysis," *Journal of Engineering Materials and Technology,* Vol. 112, July 1990, pp. 287–291.
[15] Sherwood, J. A. and Boyle, M. J., "Investigation of the Thermomechanical Response of a Titanium-Aluminide/Silicon-Carbide Composite using a Unified State Variable Model and the Finite Element Method," *Microcracking-Induced Damage in Composites,* AMD-Vol. 111, MD-Vol. 22, The American Society of Mechanical Engineers, 1990, pp. 151–161.
[16] Dvorak, G. J. and Bahei-El-Din, Y. A., "Plasticity Analysis of Fibrous Composites," *Journal of Applied Mechanics,* Vol. 49, June 1982, pp. 327–335.
[17] Bahei-El-Din, Y. A., "Plastic Analysis of Metal-Matrix Composite Laminates," Ph.D. Dissertation, Duke University, 1979.
[18] Mirdamadi, M., Johnson, W. S., Bahei-El-Din, Y. A., and Castelli, M. G., "Analysis of Thermomechanical Fatigue of Unidirectional Titanium Metal Matrix Composites," *NASA Technical Memorandum 104105,* July 1991.
[19] Aboudi, J., "Damage in Composites-Modeling of Imperfect Bonding," *Composites Science and Technology,* Vol. 28, 1987, pp. 103–128.
[20] Hopkins, D. A. and Chamis, C. C., "A Unique Set of Micromechanics Equations for High-Temperature Metal Matrix Composites," *Testing Technology of Metal Matrix Composites, STP 694,* P. R. DiGiovanni and N. R. Adsit, Eds., American Society for Testing and Materials, Philadelphia, PA, 1988, pp. 159–176.
[21] Sun, C. T., Chen, J. L., Shah, G. T., and Koop, W. E., "Mechanical Characterization of SCS-6/Ti-6-4 Metal Matrix Composite," *Journal of Composite Materials,* Vol. 24, October 1990, pp. 1029–1059.
[22] Chamis, C. C. and Hopkins, D. A., "Thermoviscoplastic Nonlinear Constitutive Relationships for Structural Analysis of High-Temperature Metal Matrix Composites," *Testing Technology of Metal Matrix Composites, STP 964,* P. R. DiGiovanni and N. R. Adsit, Eds., American Society for Testing and Materials, Philadelphia, PA, 1988, pp. 177–196.
[23] Chamis, C. C., Murthy, P. L. N., and Hopkins, D. A., "Computational Simulation of High-Temperature Metal Matrix Composites Cyclic Behavior," *Thermal and Mechanical Behavior of Metal Matrix and Ceramic Matrix Composites, STP 1080,* J. M. Kennedy, H. H. Moeller, and W. S. Johnson, Eds., American Society for Testing and Materials, Philadelphia, PA, 1990, pp. 56–69.
[24] Bigelow, C. A., Johnson, W. S., and Naik, R. A., "A Comparison of Various Micromechanics Models for Metal Matrix Composites," *Mechanics of Composite Materials and Structures,* J. N. Reddy and J. L. Telpy, Eds., The American Society of Mechanical Engineers, 1989, pp. 21–31.
[25] Vedula, M., Pangborn, R. N., and Queeney, R. A., "Modification of Residual Thermal Stress in a Metal-Matrix Composite with the Use of a Tailored Interfacial Region," *Composites,* Vol. 19, No. 2, March 1988, pp. 133–137.
[26] Mikata, Y. and Taya, M., "Stress Field in a Coated Continuous Fiber Composite Subjected to Thermo-Mechanical Loadings," *Journal of Composite Materials,* Vol. 19, November 1985, pp. 554–578.

[27] Pagano, N. J. and Tandon, G. P., "Elastic Response of Multi-Directional Coated-Fiber Composites," *Composites Science and Technology,* Vol. 31, 1988, pp. 273–293.
[28] Ebert, L. J., Fedor, R. J., Hamilton, C. H., Hecker, S. S., and Wright, P. K., Analytical Approach to Composite Behavior, *Technical Report AFML-TR-69-129,* June 1969.
[29] Hecker, S. S., Hamilton, C. H., and Ebert, L. J., "Elastoplastic Analysis of Residual Stresses and Axial Loading in Composite Cylinders," *Journal of Materials, JMLSA,* Vol. 5, No. 4, Dec. 1970, pp. 868–900.
[30] Gdoutos, E. E., Karalekas, D., and Daniel, I. M., "Thermal Stress Analysis of a Silicon Carbide/Aluminum Composite," *Experimental Mechanics,* Vol. 31, No. 3, Sept. 1991, pp. 202–208.
[31] Gdoutos, E. E., Karalekas, D., and Daniel, I. M., "Micromechanical Analysis of Filamentary Metal Matrix Composites Under Longitudinal Loading," *Journal of Composites Technology and Research, JCTRER,* Vol. 13, No. 3, Fall 1991, pp. 168–174.
[32] Lee, J. W. and Allen, D. H., "An Analytical Solution for the Elastoplastic Response of a Continuous Fiber Composite Under Uniaxial Loading," *Research in Structures, Structural Dynamics and Materials 1990, NASA Conference Publication, 3064,* 1990, pp. 55–65.
[33] Mendelson, A., *Plasticity: Theory and Application,* Macmillan, New York, 1968.
[34] Kroupa, J. L., "Thermal Mechanical Fatigue Analysis of Titanium Aluminide MMC," *Fifteenth Annual Mechanics of Composites Review, Materials Laboratory,* Wright-Patterson AFB, OH, pp. 233–242.

R. Zauter,[1] *F. Petry,*[2] *H.-J. Christ,*[1] *and H. Mughrabi*[1]

Thermomechanical Fatigue of the Austenitic Stainless Steel AISI 304L

REFERENCE: Zauter, R., Petry, F., Christ, H.-J., and Maghrabi, H., **"Thermomechanical Fatigue of the Austenitic Stainless Steel AISI 304L,"** *Thermomechanical Fatigue Behavior of Materials, ASTM STP 1186*, H. Sehitoglu, Ed., American Society for Testing and Materials, Philadelphia, 1993, pp. 70–90.

ABSTRACT: The isothermal and thermomechanical fatigue (TMF) behavior of the austenitic stainless steel AISI 304L has been studied in plastic strain control between room temperature and 1073K at a plastic strain amplitude of $\Delta\epsilon_{pl}/2 = 0.5\%$ in vacuum. The isothermal fatigue experiments showed a maximum of the stress amplitude connected with a change of the dislocation slip mode towards planar glide in an intermediate temperature region between about 550K and 800K as a result of dynamic strain ageing processes. The stress amplitudes in TMF tests are strongly influenced by the maximum temperature. When the maximum temperature lies below 873K and the temperature intervals extend over the full range of dynamic strain ageing, the stress amplitudes are very similar to the maximum values under isothermal conditions. With increasing maximum temperature, dynamic recovery by dislocation climb gains importance and the stress amplitudes decrease. When the maximum temperature of the TMF cycling exceeds a limit of about 1000K, dynamic recrystallization is observed. The numbers of cycles to fracture depend strongly on the phase between temperature and plastic strain. In particular, during in-phase cycling with a maximum temperature of 923K and higher, early failure occurs due to creep damage during the tensile half cycle. At maximum temperatures lower than 823K, the absence of creep phenomena results in similar cycle numbers to fracture in out-of-phase and in-phase tests, respectively.

KEYWORDS: austenitic stainless steel, temperature dependence of cyclic stress-strain behavior, thermomechanical fatigue, dynamic strain ageing, microstructural development, dislocation arrangement

Many studies reported in the literature deal with the mechanical behavior of austenitic steels at elevated temperature [*1–6*]. These investigations are of great importance from a technical point of view, since these materials are widely used for applications at intermediate temperature. Nevertheless, there is a lack of systematic studies on the temperature dependence of the deformation behavior, especially under thermomechanical fatigue (TMF) conditions. Only a few studies deal with the TMF behavior of AISI 304 stainless steel from a mechanical as well as a microstructural point of view [*7–14*]. In most cases the temperature dependence has not been taken into account over a wide temperature range and, since the tests have been carried out in air, complications due to the superposition of environmental effects were unavoidable.

In this work we report the results obtained in the framework of an extensive study on the temperature dependence of the cyclic deformation behavior in plastic strain control and on the related microstructural processes of AISI 304L under isothermal and thermomechanical conditions [*15,16*]. The present work focuses primarily on the TMF behavior. A more exten-

[1] Research Assistant, Senior Research Associate and Professor, respectively, Institut für Werkstoffwissenschaften, Lehrstuhl I, Universität Erlangen-Nürnberg, Martensstr. 5, D-W-8520 Erlangen, Germany.
[2] Development Engineer with Siemens-Nixdorf, Augsburg, Germany.

sive report on the isothermal tests is given in Ref *17*. The material's stress response, the dislocation arrangements formed, the damage mechanisms and lifetime are considered as a function of the range and the mean of the temperature interval used during the TMF test. The TMF behavior is discussed with reference to the mechanical behavior and microstructural development under elevated temperature isothermal fatigue conditions.

Experimental Details

The composition of the investigated heat of AISI 304L austenitic stainless steel is given in Table 1. The material was delivered in the form of bars with a diameter of 25 mm. Prior to machining the specimens, a two-step heat treatment was carried out. In the first step, the steel was solution heat-treated by annealing at 1366K for 3 h and subsequently water quenched. In the second step, the temperature was kept constant at 873K for 1000 h. The aim of this pretreatment was to thermally stabilize the microstructure by precipitating almost the total amount of carbon in the form of $M_{23}C_6$ carbides at the grain boundaries to minimize further transformation and precipitation processes during the tests at high temperatures. The grain structure of the material after the heat treatment showed a large portion of twins. The grain size corresponded to a mean linear intercept value of about 50 μm taking these twins into account and to about 115 μm without twins.

Specimens with a cylindrical gauge length of 14 mm and a diameter of 8 mm were used (Fig. 1). The surface of the gauge length was electropolished in a last preparation step to avoid an influence of the surface condition on fatigue life.

The isothermal and thermomechanical fatigue tests were carried out in a servohydraulic test system (of type MTS 880). The load frame was equipped with a self-built vacuum chamber, vacuum-tight hydraulic grips and an induction heating unit. With this equipment the test system permits fatigue tests in high vacuum and in gaseous environments under isothermal or, with the aid of a programmable temperature controller, under anisothermal conditions.

All tests (isothermal and thermomechanical) were carried out in true plastic strain control. For this purpose a modification of the electronic system was necessary to allow the use of the plastic strain signal as feedback signal in the closed loop control system. By means of a trian-

TABLE 1—*Composition of heat of AISI 304L stainless steel under investigation.*

Component	wt %
Cr	18.0
Ni	9.9
C	0.020
Si	0.40
P	0.025
S	0.0007
Mn	1.68
Mo	0.34
Cu	0.26
Co	0.15
Al	0.009
Nb	0.12
Ti	0.006
V	0.096
N	220 ppm
Fe	bal

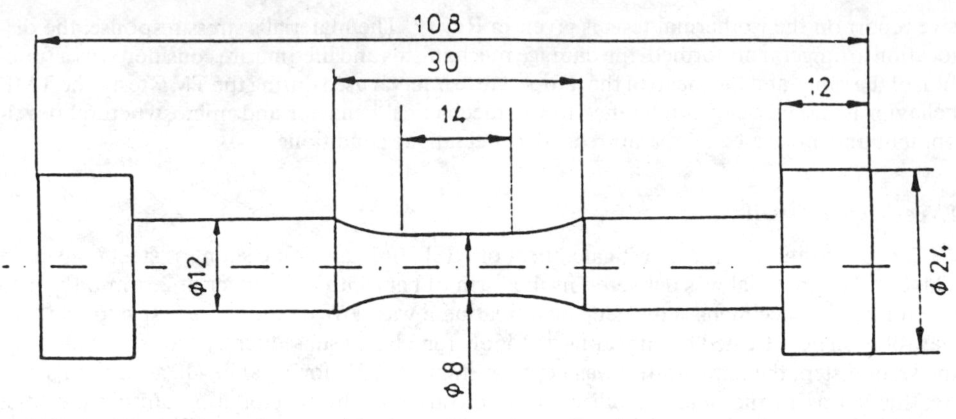

FIG. 1—*The specimen shape used for the isothermal as well as for the TMF tests in this study (dimensions in millimetres).*

gular command signal, this fatigue testing technique permits experiments with constant plastic strain amplitude and with a prescribed constant plastic strain rate throughout each cycle.

In the TMF fatigue tests, the temperature was varied linearly with time and synchronously in-phase or out-of-phase to the plastic strain. The course of the temperature and the plastic strain signal with time and the resulting stress response for these two types of TMF tests are shown schematically in Fig. 2. As a consequence of temperature cycling, asymmetric hysteresis loops resulted showing a mean stress in compression or tension (in-phase and out-of-phase,

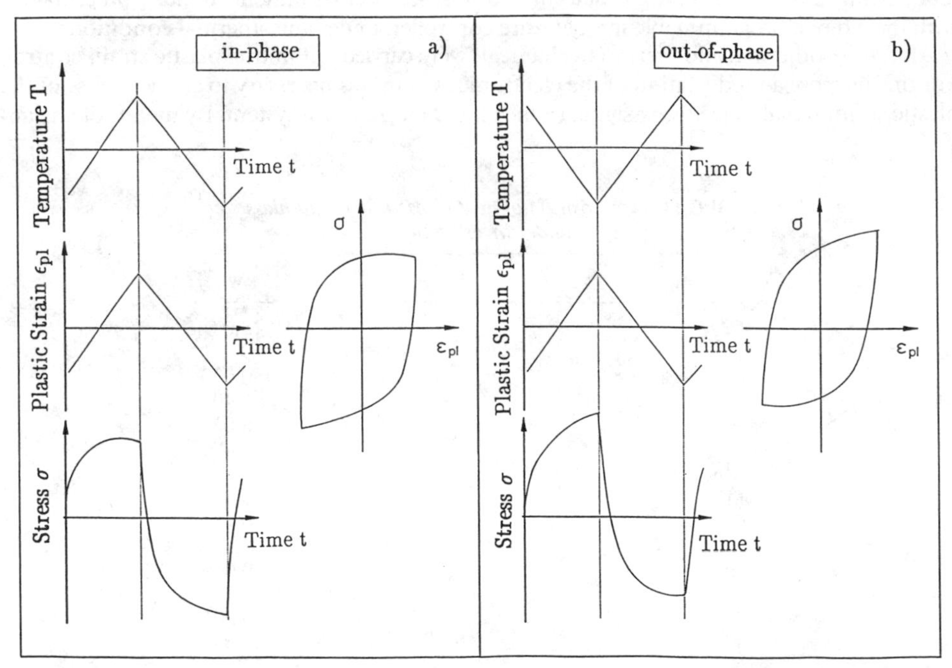

FIG. 2—*Schematic representation of temperature, plastic strain and stress versus time and of the resulting plastic strain–stress hysteresis loop for* (a) *in-phase and* (b) *out-of-phase cycling.*

respectively). The tests were always started at the mean temperature and zero plastic strain value ($\epsilon_{pl} = 0$) with the strain either increasing (in-phase) or decreasing (out-of-phase) with increasing temperature.

It should be noted that the true plastic strain control in the TMF experiments requires a computer control, since the values of the thermal expansion and the Young's modulus depend on temperature. The knowledge of the actually valid values is a prerequisite to calculate the plastic strain from the total strain and the stress, according to

$$\epsilon_{pl} = \epsilon - \epsilon_{th}(T) - \sigma/E(T) \tag{1}$$

where

ϵ = total strain,
$\epsilon_{th}(T)$ = thermal strain (temperature-dependent),
σ = stress, and
$E(T)$ = Young's Modulus (temperature-dependent).

The functions $E(T)$ and $\epsilon_{th}(T)$ were measured independently by a resonance method and dilatometry, respectively. The results of these measurements are stored in the computer in the form of a wave table and are used to provide the actual values corresponding to the existing specimen temperature for an analog strain computer which performs the on-line calculation of ϵ_{pl} (Eq 1).

The striking advantages of the plastic strain control in TMF tests which justify the higher experimental effort are the constant plastic strain amplitude and the constant plastic strain rate independent of the mean temperature and of the width of the temperature interval used. This provides a reasonable basis to compare the results obtained with corresponding isothermal tests.

All the tests reported here were conducted in high vacuum better than 5×10^{-6} mbar and at a plastic strain amplitude of $\Delta\epsilon_{pl}/2 = 0.5\%$ ($\Delta\epsilon_{pl}$: plastic strain range). The cycle time in the TMF experiments was 300 s for a plastic strain rate of $\dot{\epsilon}_{pl}$ of $6.7 \times 10^{-5}\ s^{-1}$ and temperature intervals from 250K up to 400K. This corresponds to cooling/heating rates of 100K/min to 160K/min, respectively. The normally used plastic strain rate $\dot{\epsilon}_{pl}$ of the isothermal tests was $5 \times 10^{-4}\ s^{-1}$. Additional single and multiple step tests with $\dot{\epsilon}_{pl} = 6.7 \times 10^{-5}\ s^{-1}$ were carried out to provide a basis for the comparison of the stress amplitudes under isothermal and anisothermal conditions. In all cases a 20% loss in the stress amplitude in tension was used as a criterion for fracture.

The development of the microstructure (especially the dislocation distribution) was studied, using optical microscopy, transmission electron microscopy (TEM) and scanning electron microscopy (SEM), including also electron-channelling contrast of back-scattered electrons. The last technique requires careful electrolytic polishing of the surface of the specimens to avoid the superposition of topographic contrast due to unevenness.

Results and Discussion

Isothermal High Temperature Fatigue Behavior

Isothermal plastic strain controlled fatigue tests at a plastic strain range of $\Delta\epsilon_{pl} = 1 \times 10^{-2}$ and a plastic strain rate of $\dot{\epsilon}_{pl} = 5 \times 10^{-4}\ s^{-1}$ document a temperature dependence of the cyclic hardening/softening and saturation behavior which is typical of 300 series steels. Figure 3 represents the stress amplitude $\Delta\sigma/2$ as a function of cycle number N (cyclic hardening/softening curves) of AISI 304L measured at various temperatures between 298K and 1048K.

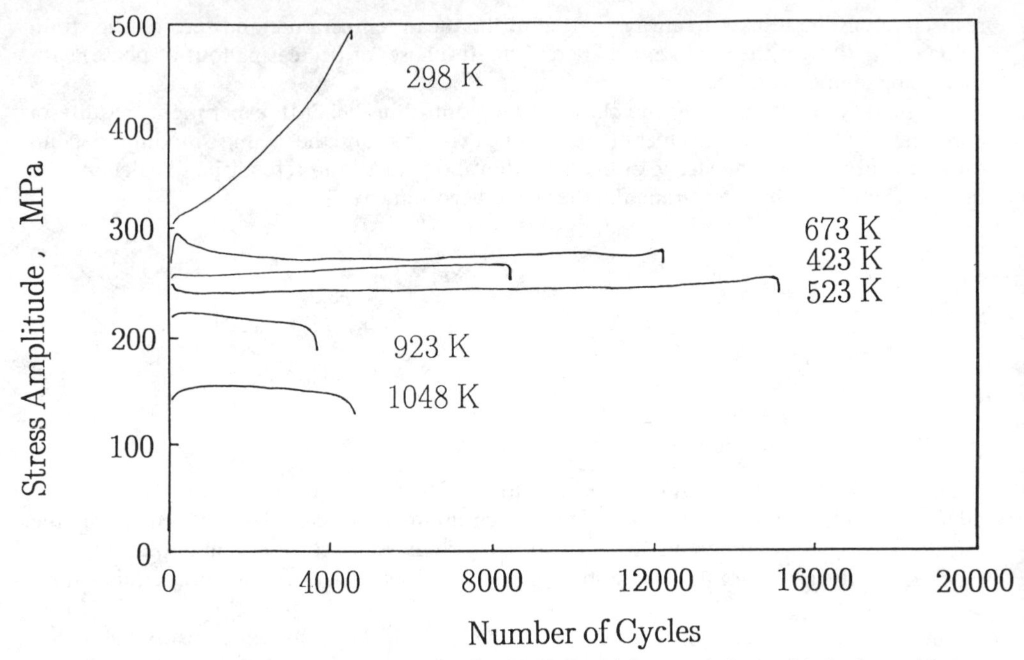

FIG. 3—*Cyclic hardening/softening curves of AISI 304L stainless steel obtained in isothermal fatigue tests at various temperatures.*

Three different characteristic shapes of the cyclic hardening/softening curve appear. At room temperature a regime of cyclic saturation is only indicated at the very beginning of the curve, followed by a drastic increase of $\Delta\sigma/2$ which continues until failure of the specimen. This behavior results from a deformation-induced formation of martensite as could be shown by extensive TEM studies and is reported in detail in Refs *17,18*. The transformation of austenite to martensite is well established for austenitic stainless steels [*19–22*] and depends strongly on the composition of the alloy. During cyclic loading, the amount of martensite increases with the cumulative plastic strain $\epsilon_{pl,cum}$, if a critical value of $\Delta\epsilon_{pl}/2$, accompanied by a threshold stress level, is exceeded (in good agreement with [*22*]), leading to cyclic hardening (see Fig. 3).

In an intermediate temperature range between about 423K and about 823K, the stress amplitude goes through a marked maximum after initial cyclic hardening, followed by cyclic softening, and levels off in a long range of saturation (plateau). At high temperatures ($T >$ 823°K), the maximum of the hardening curve disappears and the initial cyclic hardening curve merges directly into an extended saturation regime.

The maximum stress amplitude and the saturation stress amplitude exhibit the temperature dependence shown in Fig. 4. Since the deformation-induced martensite formation should not be taken into account in this context, the room temperature data plotted in Fig. 4 are taken from the first part of the cyclic hardening/softening curve (Fig. 3). Between room temperature and about 523K both stress amplitudes decrease continuously. $\Delta\sigma/2$ passes a minimum and increases in an intermediate temperature region with increasing temperature. In this interval, a marked difference between the maximum and the saturation value of $\Delta\sigma/2$ exists ($\approx$20 MPa). At temperatures above 823K, $\Delta\sigma/2$ decreases rather rapidly. This type of temperature

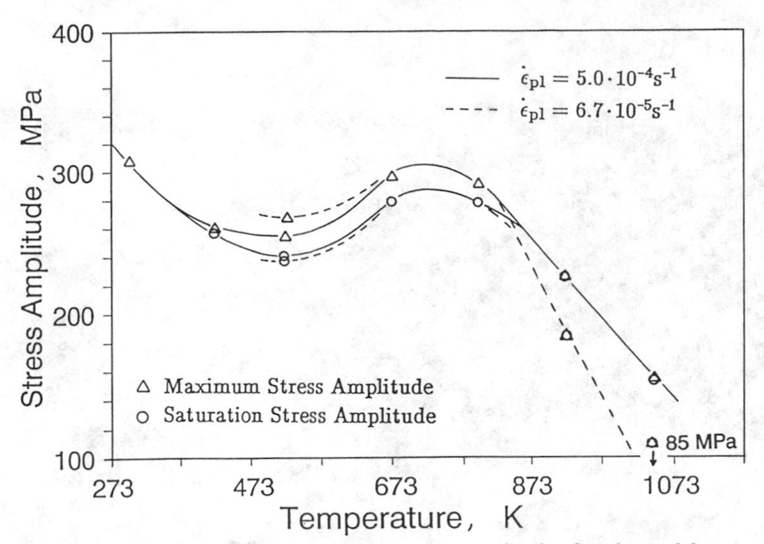

FIG. 4—*Dependence of maximum and saturation stress amplitude of isothermal fatigue tests on temperature (dashed line for reduced* $\dot{\epsilon}_{pl}$*).*

dependence of the mechanical fatigue behavior of austenitic stainless steel is reported in the literature [*2,3,5*] and is considered to be caused by dynamic strain ageing processes. Figure 4 shows additionally the results of fatigue tests with a reduced plastic strain rate of $\dot{\epsilon}_{pl} = 6.7 \times 10^{-5} s^{-1}$ which are used as a basis for the discussion of the TMF behavior of AISI 304L investigated at this $\dot{\epsilon}_{pl}$ value.

The maximum of the stress amplitude at about 723K is connected with a change in the slip mode of the dislocations and can be seen clearly in the dislocation microstructure existing after the fatigue test. In the temperature region from room temperature up to about 523 K, the slip mode at the relatively high plastic strain range applied in this study is wavy. The dislocation arrangements formed during cyclic deformation are characterized by low-temperature cell configurations with relatively broad cell walls in which dislocations do not form a well-defined network. At 673K (near by the maximum in the stress amplitude), a planar slip mode prevails, indicating that the dislocations are more confined to their slip planes due to impeded cross-slip. This leads to a more uniform distribution of the dislocations in the sense that no bifurcation into dislocation-dense and dislocation-poor regions as in the case of cell structures is observable. At higher temperatures (>798K) the slip character becomes wavy again [*6*] and the stress amplitudes decrease. Dynamic recovery by thermally activated dislocation climb gains importance at these high temperatures and gives rise to the formation of cells or subgrains with sharp walls. It could be shown by Ermi and Moteff [*6*] that in this temperature range the stress amplitude is inversely proportional to the mean cell size. In Figs. 5 and 6 typical dislocation arrangements as observed by TEM and SEM, respectively, are shown for comparison.

The increase in $\Delta\sigma/2$ with temperature (Fig. 4) in the intermediate temperature range points to the occurrence of dynamic strain ageing processes, in agreement with earlier work [*3–5,23*], in particular by Tsuzaki et al. [*2*]. This is also supported by the fact that a negative strain rate sensitivity exists ($d\Delta\sigma/d\dot{\epsilon}_{pl} < 0$) as evidenced by additional experiments with different (but during each test constant) values of $\dot{\epsilon}_{pl}$ as well as with $\dot{\epsilon}_{pl}$-changes during cyclic deformation.

Different processes are proposed in the literature to be responsible for dynamic strain ageing

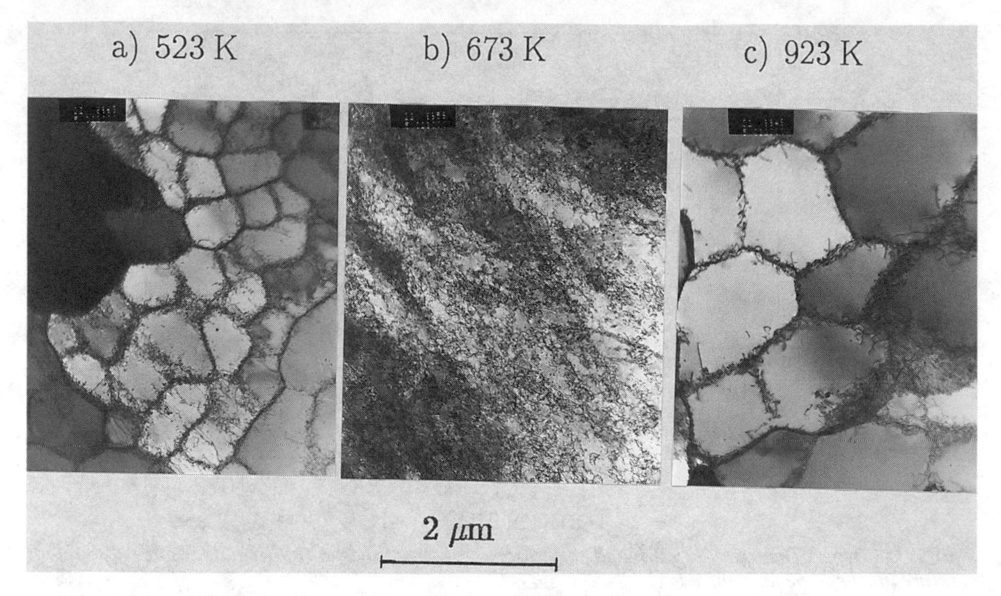

FIG. 5—*Comparison of TEM observations of the dislocation arrangements formed during fatigue testing at the temperatures* (a) *523K,* (b) *673K, and* (c) *923K.*

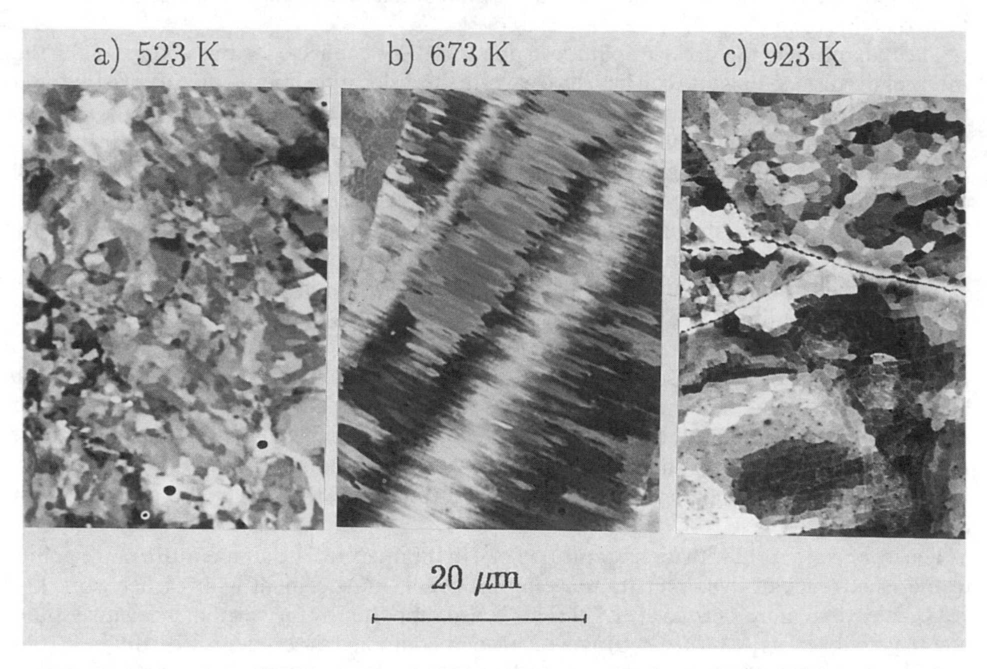

FIG. 6—*Comparison of SEM micrographs (channelling contrast) of typical (global) dislocation arrangements after isothermal fatigue loading at the temperatures* (a) *523K,* (b) *673K, and* (c) *923K.*

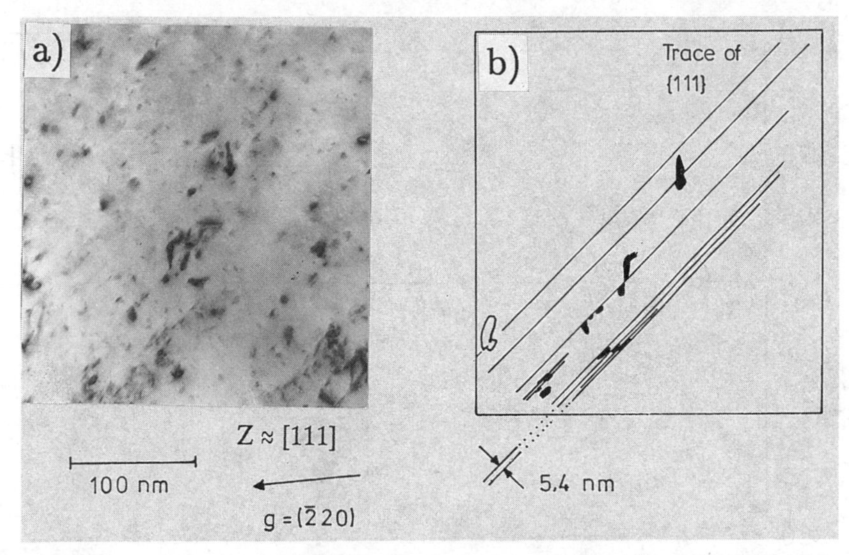

FIG. 7—*Coherent carbide precipitates formed during fatigue at 673K with* $\Delta\epsilon_{pl}/2 = 0.5\%$. *The carbides are cut by gliding dislocations as shown schematically in the right representation.*

in the considered temperature range. At $T < 673$K an interaction of the dislocations with carbon-vacancy pairs and clusters of carbon atoms seems to be the dominant mechanism [*24,25*]. With increasing temperature the mobility of substitutional Cr is raised and can lead to a dynamic interaction with gliding dislocations [*2,24,26*]. Figure 7 documents that in addition also the formation of carbide precipitates must be taken into account. In the intermediate temperature range, during the cyclic loading very small (≈10 nm) coherent particles are formed which have a plate-like morphology and are sheared by gliding dislocations (as shown schematically in Fig. 7*b*). The cutting process is thought to enhance the planarity of the glide behavior and can also explain the softening after the maximum in the cyclic hardening/softening curve at the beginning of the test, which leads to the large difference between maximum and saturation stress amplitude in the considered temperature range. Furthermore, short range ordering processes reported by other authors [*27,28*] must be considered as another possible reason for the planar glide behavior [*29*].

In the isothermal fatigue tests at temperatures below 800K mainly transgranular stage I cracks are formed which then propagate as transgranular stage II cracks. At higher temperatures cavities nucleate at grain boundaries and especially at triple points, forming the well-known wedge cracks. This observation points to an increasing importance of creep damage processes with increasing temperatures. At T ≥ 923K these latter damage contributions dominate obviously over typical fatigue damage effects. This is in agreement with the results reported in [*30*], where, in addition the influence of the strain rate is taken into consideration.

The number of cycles to fracture N_f increases with increasing temperature in the temperature interval between room temperature and 523K. This can easily be understood on the basis of the decreasing stress amplitude. After a maximum of N_f at 523K, N_f is reduced with enhanced temperature in the whole temperature range of dynamic strain ageing up to temperatures in the creep range (923K). With further increasing temperature, N_f rises again. A similar observation is reported in Refs *3* and *31* and explained as a result of the low stress amplitudes which are no longer sufficiently high to cause grain boundary sliding.

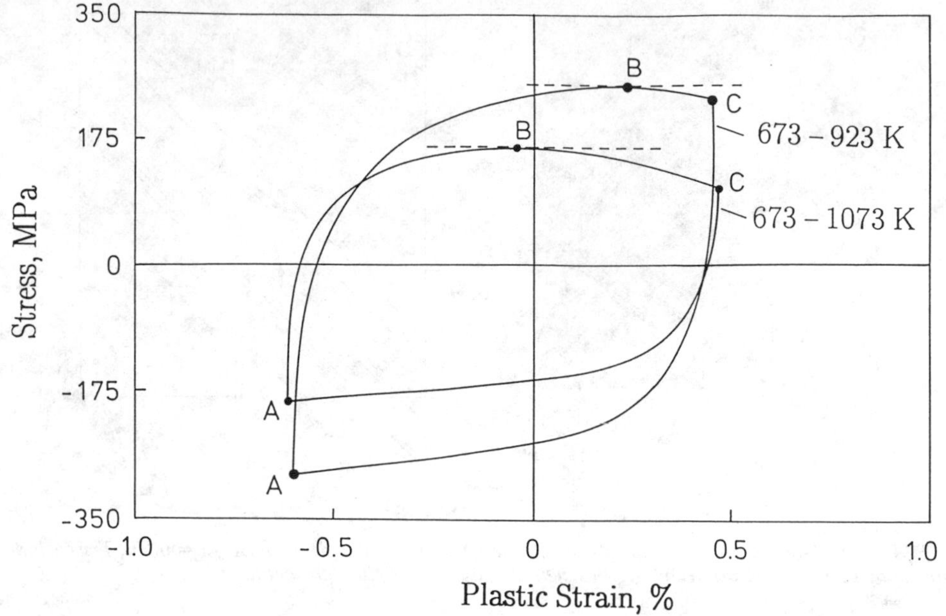

FIG. 8—*Saturation TMF hysteresis loops of two tests with temperature intervals of 673 to 1073K and 673 to 923K. Three characteristic points are marked and labelled with A, B and C.*

Thermomechanical Fatigue

Mechanical Behavior—Figure 8 shows two stabilized hysteresis loops of two different in-phase tests with temperature ranges of 673K–1073K and 673K–923K, respectively, registered during the saturation state. In the high-temperature half cycle a stress maximum (point B in Fig. 8) is reached at a temperature lower than the peak temperature and at a plastic strain value lower than the amplitude value of 0.5%. Obviously, a softening of the material occurs as a consequence of the increasing importance of thermally activated dislocation motion (dynamic recovery), which leads to a decrease of the stress even under the prevailing condition of $\dot{\epsilon}_{pl}$ = constant. The maximum temperature and maximum plastic strain are reached at point C. At point A the minima of stress, plastic strain and temperature coincide due to the decreasing temperature during the compression half cycle. The points A, B and C define three characteristic values of the hysteresis loop under thermomechanical conditions and will be used later in further representations.

Figure 8 documents clearly the effect of the peak temperature on the shape of the hysteresis loop. Even at point A where identical temperatures (and plastic strain values) exist for the two cases considered, the stress values differ strongly. In both tests, a dynamic recovery at high temperatures takes place. The higher the maximum temperature T_{max}, the more severe is the influence of dynamic recovery, so that the steel is accordingly softer also in the compression part of the cycle. A comparison of the hysteresis loops of various tests shows that the points B and C of the loops coincide (that means the maximum of stress occurs at the maxima of temperature and plastic strain), if the peak temperature lies below a value of about 873K. Under this condition, the dynamic recovery processes are less important regarding their influence on the loop shape.

Some examples of cyclic hardening/softening curves as measured in TMF tests are represented in Fig. 9. In addition to the stress amplitude $\Delta\sigma/2$, the corresponding mean stress is

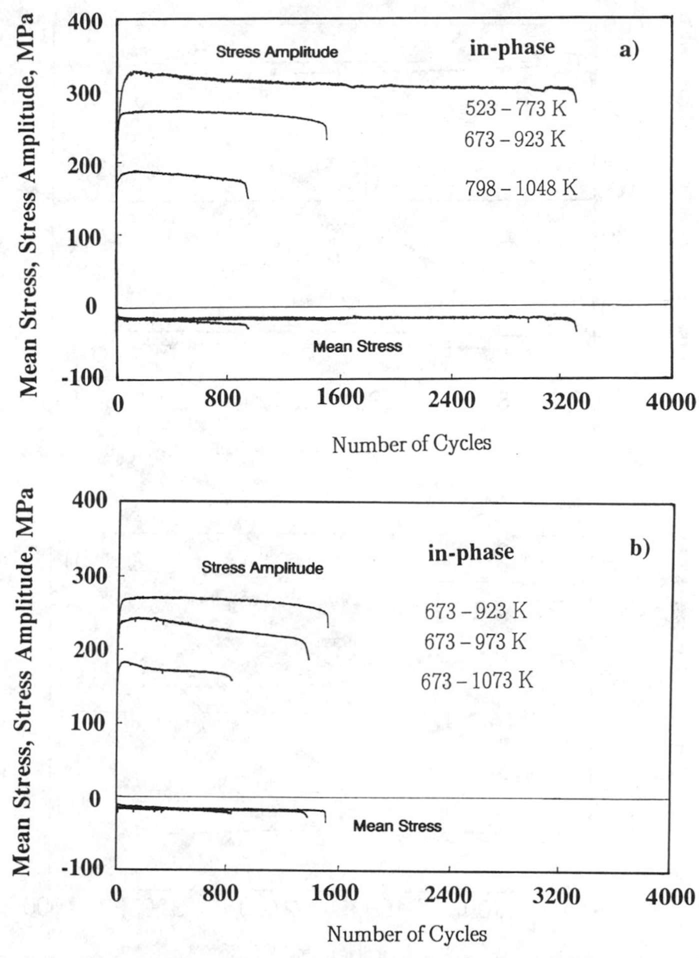

FIG. 9—*Cyclic hardening/softening curves and mean stress of various TMF tests* (a) *with a temperature interval of 250°K and different mean temperatures and* (b) *with a minimum temperature of 673°K and different widths of the temperature interval.*

plotted versus the numbers of cycles. Figure 9*a* shows the results of three different experiments. In all cases, the temperature interval ΔT was 250K, but the mean temperatures were varied (648, 798, and 923°K). In contrast, the tests represented in Fig. 9*b* refer to different temperature intervals but identical minimum temperature values (673K).

Most curves exhibit a pronounced saturation state or a slight softening after an initial primary hardening. Only the test with the temperature range of 673 to 1073K shows a weak maximum of the stress amplitude prior to saturation. It is evident (and easy to understand) that the stress amplitude decreases with increasing peak temperature (Fig. 9*b*) and increasing mean temperature (Fig. 9*a*). The in-phase TMF loading leads to asymmetric hysteresis loops (Fig. 8), manifested with a compressive mean stress. In spite of a variation of mean and peak temperature, the value of the mean stress is nearly unchanged (about -17 MPa). The cyclic lifetime is not determined by the stress amplitude, since in all cases shown in Figs. 9*a* and 9*b* a reduction of N_f results from an increase of the peak temperature (decreasing $\Delta\sigma/2$). Later it will be shown that the mean temperature does not govern the lifetime.

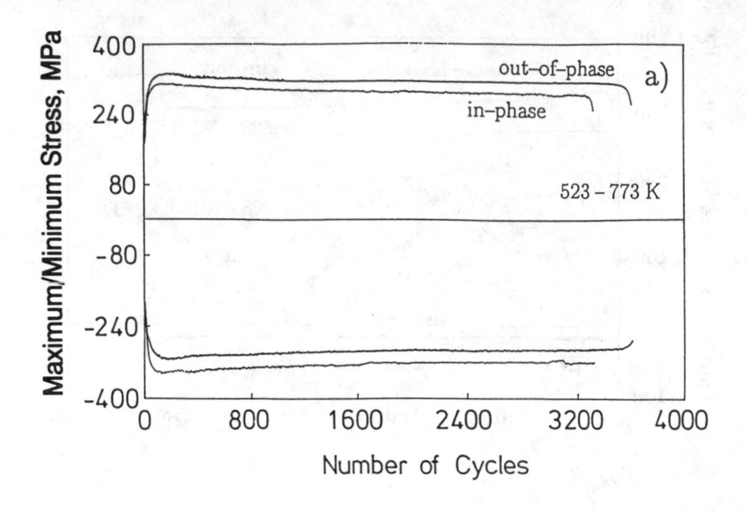

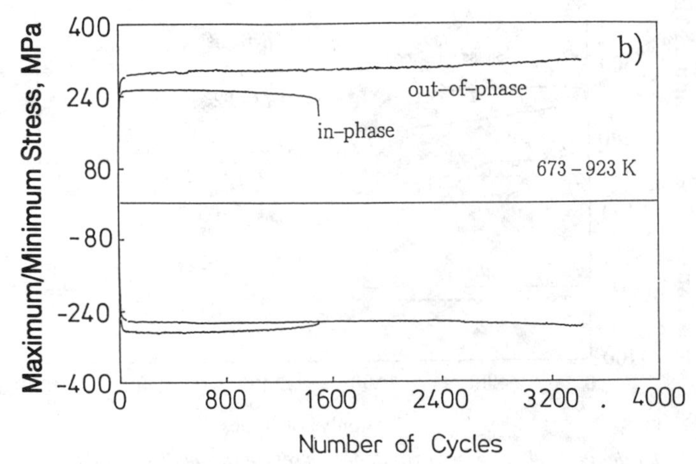

FIG. 10—*Comparison of the cyclic softening/hardening curves of in-phase and out-of-phase TMF tests with temperature intervals of* (a) *523 to 773K and* (b) *673 to 923K.*

Figure 10 allows a comparison of the cyclic hardening/softening curves obtained in in-phase and out-of-phase tests. The maximum and minimum stress values per cycle are plotted separately with positive and negative sign, respectively, as a function of the number of cycles. In Fig. 10*a* the temperature interval covered during each cycle (523 to 773K) corresponds well with the temperature range of the dynamic strain ageing (see Fig. 4). This explains why the stress amplitudes are relatively high and not strongly dependent on the testing mode. In contrast to the negative mean stress developing during in-phase TMF, the out-of-phase loading leads to a positive mean stress in the same magnitude (the two related curves of each test can be found easily by referring to the number of cycles to fracture). Since in this special case at constant $\Delta\epsilon_{pl}/2$ the stress values in tension/compression and compression/tension in in-phase/out-of-phase cycling, respectively, are nearly identical and the maximum temperature is relatively low, a similar cyclic lifetime is not unexpected.

The situation changes drastically if the mean temperature of the temperature interval is shifted to higher values (Fig. 10*b*). In spite of the fact that the stress range $\Delta\sigma$ is slightly influ-

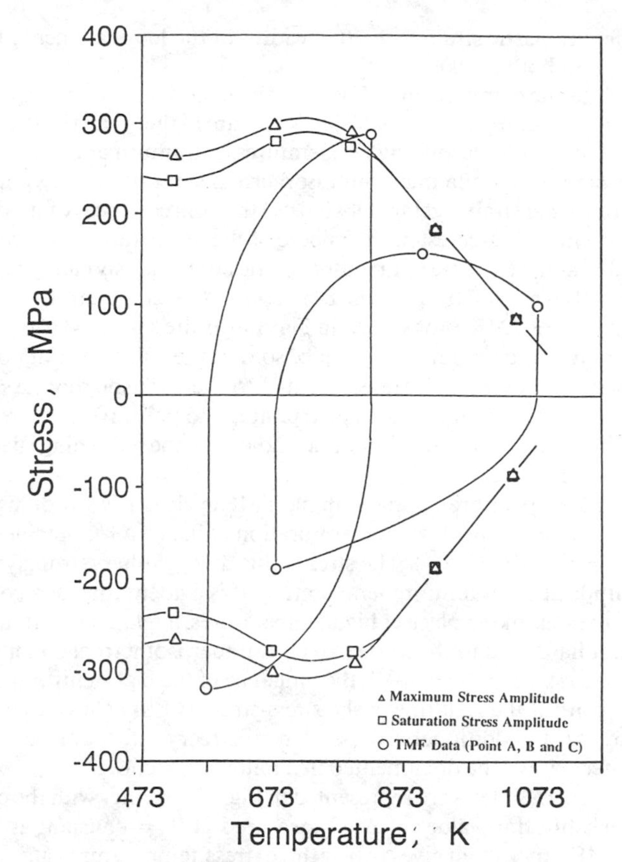

FIG. 11—*Two TMF hysteresis loops in a representation of stress versus temperature compared with the values of maximum and saturation stress amplitude obtained under isothermal conditions.*

enced only by the testing mode, the specimen fails as a consequence of in-phase cycling at a number of cycles to fracture which is less than half of the value observed in the corresponding out-of-phase experiment. Obviously, the peak temperature is high enough to allow creep deformation to take place during a part of each cycle (visible in the shape of the hystersis loop in Fig. 8). In the case of in-phase TMF, creep is restricted to tensile loading and causes damage to the steel by the formation of wedge cracks and cavities (as described in a following section). A strong reduction of N_f results. During the out-of-phase test, the high temperatures exist in combination with compressive loading and creep damage is reduced or even totally inhibited. Interestingly, the cycle numbers to fracture of out-of-phase tests are affected only slightly by the location of the temperature interval.

A comparison of the isothermal fatigue stress amplitude values and two hysteresis loops of in-phase TMF tests in saturation is given in Fig. 11, whereby, the plastic strain rate in both the isothermal and TMF tests is $\dot{\epsilon}_{pl} = 6.7 \times 10^{-5}\ s^{-1}$. The data measured under isothermal conditions are plotted against temperature in the form of the maximum and saturation stress amplitudes separately for tension and compression. An isothermal (saturation) hysteresis loop represented in this figure would lead to a straight vertical line extending between the two curves of $\Delta\sigma/2$. Due to the test technique used, with $\dot{\epsilon}_{pl}$ = constant and dT/dt = constant, a linear relation between ϵ_{pl} and T exists for the TMF tests. As a consequence, the shape of the hysteresis loops plotted in Fig. 11 corresponds to the representation of stress versus plastic strain, though with two different plastic strain scales (the widths of the loops are hence not identical, although $\Delta\epsilon_{pl}/2$ is 0.5% in both cases).

An important result of the comparison in Fig. 11 is the fact that the extreme values of stress of the TMF test with a temperature range of 573 to 823K are higher than the stress amplitudes in the isothermal case at the corresponding temperatures (minimum and maximum temperature), and are even higher than the maximum isothermal saturation stress within the temperature interval of dynamic strain ageing. Obviously, the consequences of hardening due to dynamic strain ageing which is greatest in the middle of the temperature interval of the TMF test remain observcable at the extreme temperatures. The continuously changing temperature supports the influence of dynamic strain ageing processes. If the temperature interval is shifted to higher values, e.g., 673 − 923K (not shown in Fig. 11 for the sake of clarity), the dynamic recovery at the high temperatures gains importance so that the values of σ at points A and B of the hysteresis loop (Fig. 8) are approximately equal to the corresponding $\Delta\sigma/2$ of the isothermal tests (at T_{min} and at the temperature corresponding to point B). The stress at T_{max} lies considerably above the isothermal saturation value, indicating the hardening effect of the plastic deformation at low temperatures.

When the maximum temperature T_{max} used in the TMF cycle is 1048°K or higher, only the stress value at T_{max} is similar to the isothermal saturation value of $\Delta\sigma/2$ obtained at the same plastic strain rate ($\dot{\epsilon}_{pl} = 6.7 \times 10^{-5}\ s^{-1}$). The stress value at T_{min} differs strongly from the isothermal stress amplitude at the minimum temperature. It is evident that, as a consequence of dynamic recovery processes taking place at high temperatures, a relatively soft microstructure is formed which is not hardened to the same extent as under isothermal conditions at lower temperatures. In these cases ($T_{max} \geq 1048$K), the influence of the high-temperature part of the TMF cycle seems to control the resulting cyclic stress-strain (CSS) behavior even during the low-temperature part. As an additional process, dynamic recrystallization can contribute to the softening of the steel, as will be documented in a following section.

The results of in-phase TMF tests are represented in Fig. 12 together with the curve describing the course of the saturation values of $\Delta\sigma/2$ measured at T = constant as a function of temperature. Each TMF experiment gives two or three stress-temperature data points, accord-

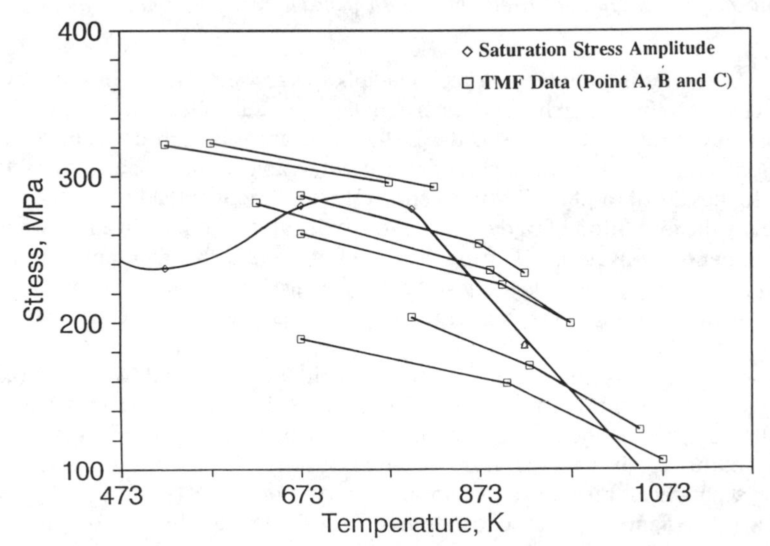

FIG. 12—*Comparison of stress amplitude values of TMF and isothermal tests plotted against temperature.*

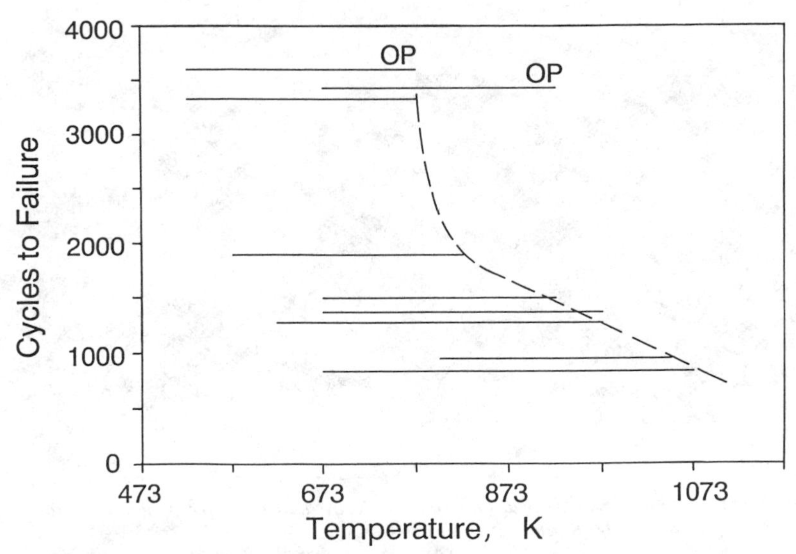

FIG. 13—*Number of cycles to failure* N_f *in dependence on the temperature interval. Lines not labelled correspond to in-phase tests, OP refers to out-of-phase cycling.*

ing to the points A, B and C in the hysteresis curve, which are plotted in absolute values of stress and connected with straight lines (if σ in tension is highest at T_{max}, point B does not exist). The above described fact, namely that the peak temperature controls the stresses during the whole TMF cycle and the relation between stress amplitudes in isothermal and TMF cycling (both with the same strain rate of $\dot{\epsilon}_{pl} = 6.7 \times 10^{-5}\ s^{-1}$) are displayed clearly.

In spite of the decreasing stress amplitude of the TMF tests with increasing maximum temperature, the number of cycles to fracture N_f is lowered. In Fig. 13 the measured values of N_f are shown for various in-phase and two out-of-phase tests. Each experiment is characterized by a straight horizontal line at N_f extending from T_{min} to T_{max}. The observations can be interpreted as follows. If the maximum temperature is high ($T_{max} > 873$K), a linear relationship between N_f and T_{max} exists in a good approximation. Obviously, at the strain range and strain rate used in this study the cyclic life is determined by the upper temperature limit used in the test, indicating that creep damage processes play a dominant role. The corresponding out-of-phase test (with a high value of T_{max}) confirms this interpretation. High temperatures in connection with compressive stresses cause less damage. The value of N_f is increased strongly, if the maximum temperature is too low to permit creep deformation. Then the type of test (in-phase or out-of-phase) has only a minor influence on cyclic life (see the two lines for the temperature interval 523K–773K in Fig. 13). Informations about the effect of the strain rate and strain range which were not varied in this study are given in [*30*].

Dislocation Arrangement—In Fig. 14 typical TEM micrographs showing the dislocation substructures developed during TMF loading are represented. The specimens investigated had been deformed cyclically until fracture using a temperature range ΔT of 250K and different mean temperatures (*a*: 573K–823K, *b*: 673K–923K, *c*: 798K–1048K).

Figure 14*a* documents that the test at relatively low temperatures (573K–823K) leads to arrangements of dislocation walls of high dislocation density and to dislocation layers along the trace of $\{111\}_{tr}$ of a possible glide plane. The trace is indicated by a dashed line in Fig. 14*a* (and also in Fig. 14*b*). This structure is partly reminiscent of observations on fatigued wavy-slip fcc metals in which dislocation walls perpendicular to the active Burgers vector are a common feature [*32*]. In addition, regions exist which show on the one hand disorderly planar

FIG. 14—*TEM micrographs of dislocation structures formed during TMF with a temperature interval of* (a) *573 to 823K,* (b) *673 to 923K, and* (c) *798 to 1048K.*

dislocation arrays and, on the other hand, labyrinth structures. Obviously, in the latter case multiple (wavy) slip occurs [*32,33*]. Since the dislocation substructure of the isothermal fatigue test at the mean temperature (Fig. 5*b*) is determined by dynamic strain ageing and by largely planar slip, it is concluded that in the TMF test a more wavy slip mode occurs as a consequence of cyclic deformation at temperatures above and below the regime of dynamic strain ageing during each TMF cycle. It is reported in the literature [*34,35*] that under isothermal monotonic conditions with $T = T_{max}$ (823°K), AISI 304 austenitic stainless steel experiences dynamic recovery (dislocation climb). But the deformation rate used in the TMF tests seems to be too fast and the time fraction spent at sufficiently high temperature too small in this case that a softening could play an important role. In the TEM study only negligible misorientations between neighboring areas were found, indicating that the dislocation walls cannot be referred to as subgrain walls.

If the mean temperature is increased to 798K (Fig. 14*b*), elongated cells develop. The cell walls are formed of dislocations arranged in networks and, in some areas in layers oriented along the traces of {111} planes. Differences in the orientations between neighboring cells exist. A comparison of this substructure with the dislocation arrangement formed as a consequence of isothermal fatigue at $T = T_{max}$ documents that the dislocation density within the cells is higher in the case of TMF. Nevertheless, the global dislocation substructure resulting from TMF seems to be determined mainly by the high temperature part of the cycle. No resemblance to the planar slip configuration typical for fatigue in the temperature range of dynamic strain ageing is observable, although these temperatures are traversed during TMF cycling. The important role of dynamic recovery has already been documented by the shape of the hysteresis loop (Fig. 8), showing that the maximum stress is reached about 50°K before the maximum temperature (and plastic strain). During the high temperature part of the cycle, the dislocations are partly annihilated and rearranged, forming a high-temperature arrangement which persists also at lower temperature.

The microstructural investigations of a specimen tested at high temperatures (798 to 1048K) showed that a clear subgrain structure is formed which is evidently finer in subgrain size than in the isothermal test at $T = T_{max} = 1048$K. This can simply be attributed to the higher stress amplitude at the high temperature connected with TMF. Inside the subgrains, a very low dislocation density was observed (Fig. 14*c*).

FIG. 15—*Evidence of grain boundary migration observed after TMF with a temperature interval of 673 to 1073K,* (a) *TEM micrograph,* (b) *optical micrograph (Beraha I color etching), stress axis lies horizontal.*

If the maximum temperature is very high ($T_{max} \geq 1073K$), dynamic recrystallization comes into play as an additional recovery mechanism. As an example, the TMF experiment with a temperature interval of 673 to 1073K is considered in more detail.

The size of the subgrains is very similar to that observed after isothermal fatigue at 1048K. By means of optical microscopy as well as of TEM, short segments of grain boundaries could be found which have migrated. Figure 15*a* shows a TEM micrograph in which a straight line of carbide precipitates can be seen. It is suggested that these carbides have been formed at a grain boundary existing formerly at this position. The grain boundary has migrated over a length of about 3 μm into the neighboring grain, leaving behind it a line of carbides inside a grain. Electron diffraction analysis clearly proves that at the precipitates no grain boundary exists. In Fig. 15*b* an optical micrograph (prepared with Beraha I color etching) shows the movement of a part of a grain boundary with a length of about 30 μm over a distance of about 5 μm.

Grain boundary migration or dynamic recrystallization effects have not been found in isothermal fatigue tests which were carried out up to a temperature of 1048K. Hence, it can be concluded that the occurrence of dynamic recrystallization during TMF results from the different deformation mechanisms which are active, when the temperature is varied within a large interval.

Grain boundary migration during high-temperature fatigue loading of fcc materials has been reported by many authors [*36–38*], but the phenomena observed in this study resemble more the effects which are typical of dynamic recrystallization, as also reported in papers on high temperature fatigue of a Pb alloy [*39*] or on creep of AISI 304 at relatively low temperatures [*40*]. For the special heat of AISI 304L used in this study, it could be shown that dynamic recrystallization influences the creep behavior strongly at test temperatures of 998 and 1073K [*41*].

Cracks and Cavities—As stated above, the isothermal cycling at low and intermediate temperature is connected with the initiation and propagation of predominantly transcrystalline cracks. With increasing temperature the number of intergranular cracks increases, and this crack type dominates at high temperatures (923 and 1028K). This observation is in good accordance with other investigations taking the influence of the strain rate into account [*3,30,31*].

The in-phase TMF tests with the temperature interval of 250K also show a systematic change of the failure mode with increasing maximum temperature. In-phase cycling with a mean temperature of 698K (or with lower mean temperatures) leads to a relatively high fraction of transgranular cracks on the surface and also the fatal crack is transgranular. At higher maximum temperatures (e.g. T_{max} = 923K), grain boundary cracking at the surface and within the specimen becomes more important. Cavities and wedge cracks which are typical for creep damage are formed and are responsible for a drastic reduction of N_f [*30*]

Figure 16 shows cavities and wedge cracks observed after TMF with ΔT = 798 to 1048K in a SEM micrograph using the channelling contrast. The advantage of this technique should be noted. The surrounding subgrain structure is visible by the different brightness of the subgrains. The wedge crack shown in Fig. 16*b* is obviously formed at a grain boundary triple point which is a location of stress concentration. The subgrain size near the triple point is smaller than within the grains. It can be assumed that at these locations a stress concentration is a consequence of a strain concentration which itself results from grain boundary sliding. Grain boundary sliding has been proposed in many studies to be a dominant creep-fatigue interaction mechanism acting, for instance, during elevated-temperature isothermal LCF [*42,43*] and to be damaging in austenitic stainless steels, if the strain rate is slow [*3,44*]. Some authors [*9,13*] emphasize that during TMF the direction of grain boundary sliding is not reversed, causing wedge cracking and lifetime reduction.

If the maximum temperature of TMF exceeds the temperature at which creep by dislocation climb is activated, creep damage is severe in in-phase cycling, because tensile loading takes place at high temperature. In contrast, out-of-phase cycling does not show the typical damage effects, because under compression at high temperature the formation of cavities and wedge cracks is inhibited. Transgranular fracture dominates, in accordance with the results reported by other authors [*7,8,11,14,30*].

Figure 16*a* shows another well-accepted creep-fatigue interaction mechanism [*45,46*], namely cavitation along the grain boundaries which occurs mainly in the regions which are located near the crack tip.

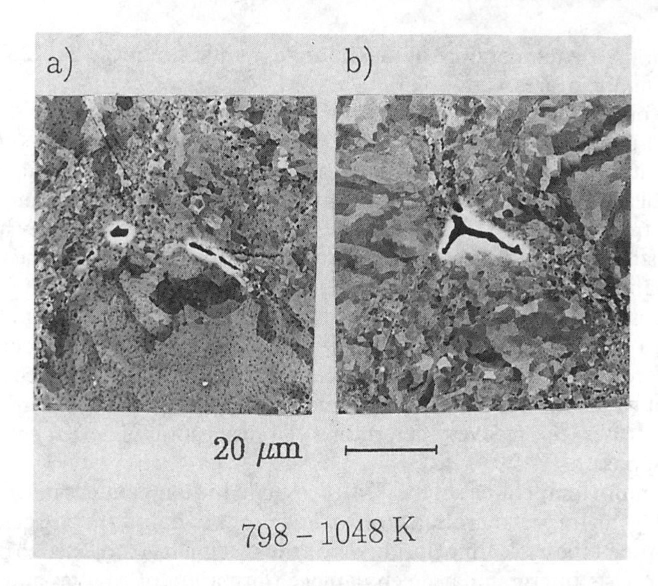

FIG. 16—(a) *Cavities at grain boundaries and* (b) *wedge crack formed at a grain boundary triple point during TMF cycling with 798 to 1048K. Notice the small subgrain size in front of the crack tip (SEM, channelling contrast).*

Carbide Precipitation—Although the thermal pretreatment should lead to a thermally stable carbide structure at the grain boundaries, TEM and SEM observations on the deformed specimens which had been subjected to temperatures of 923K or higher show a further precipitation of carbides within the grains and a coarsening of the grain boundary carbides. The nucleation of matrix carbides takes place in the early stage of deformation at sites with high dislocation density. The carbides were identified by means of electron diffraction and energy-dispersive X-ray (EDX) analysis as type $M_{23}C_6$ with a high content of chromium. These carbides are considered to have a negligible influence on the cyclic deformation behavior, since their size is relatively large, probably as a consequence of the relatively high maximum temperature used in the present TMF study. This finding shows that, in spite of the thermal pretreatment, further carbide precipitation occurs in presence of cyclic deformation at high temperatures.

In isothermal tests at lower temperatures (623K), very small coherent precipitates are formed which show an interaction with the dislocations and must be considered as strong obstacles to cyclic deformation (as described above in detail). In TMF tests with a maximum temperature not higher than about 823K, it is suspected that similar small carbide precipitation may occur. However, this could not be verified by TEM so far.

Conclusions

Isothermal and thermomechanical fatigue experiments were carried out in vacuum using true plastic strain control, a plastic strain amplitude of $\Delta\epsilon_{pl}/2 = 0.5\%$ and a constant plastic strain rate of $\dot{\epsilon}_{pl} = 6.7 \cdot 10^{-5}\ s^{-1}$. The results of the thermo-mechanical tests were compared with the obtained isothermal high-temperature fatigue data. The main results can be summarized as follows:

1. Under isothermal fatigue conditions, AISI 304L austenitic stainless steel shows dynamic strain ageing in a temperature range between 523K and 823K. This leads to a maximum of the stress amplitude at about 723K and to a change of the dislocation slip mode from wavy (at

523K) to planar (at the temperature of maximum dynamic strain ageing, 723K) and back to wavy at higher temperatures.

2. If TMF deformation is carried out at relatively low temperatures in the regime of dynamic strain ageing, in-phase and out-of-phase cycling lead to very similar numbers of cycles to fracture. The dislocation arrangement shows the typical appearance for low and intermediate temperature multiple slip, namely wall, labyrinth and cell structures. The fracture mode is transgranular. The stress amplitudes in tension and compression are even somewhat higher than the saturation stress amplitude observed under isothermal conditions at the temperature of maximum dynamic strain ageing (about 723K).

3. With increasing maximum temperature of the TMF cycle the stress amplitude decreases. If the maximum temperature is sufficiently high, a maximum of the stress occurs in the hysteresis loop before the maximum temperature and maximum strain are reached. It characterizes the point at which the dynamic recovery outweighs the deformation hardening (at about 873K). In these cases, the recovery determines the dislocation arrangement and the cyclic stress-strain response.

4. If the maximum temperature of the TMF cycle is in the temperature range in which creep deformation plays an important role, a more or less clear subgrain structure develops, depending on the fraction of the cycle time during which the specimen is subjected to dynamic recovery. In in-phase tests intergranular creep damage (formation of cavities and wedge cracks) occurs during tensile deformation at high temperatures and reduces the cyclic life drastically in comparison to out-of-phase cycling. In the latter case, high temperatures are connected with compressive loading which inhibits cavitation.

5. During TMF tests with very high maximum temperatures (1073K), dynamic recrystallization associated with grain boundary migration occurs contributing to the softening.

Further TMF tests are in progress with the aim to obtain additional informations on the TMF behavior of AISI 304L. The influence of the plastic strain amplitude, of the plastic strain rate and of larger temperature intervals will be studied in detail.

Acknowledgment

The financial support of the "Volkswagenstiftung" (AZ I/61021) and of the "Deutsche Forschungsgemeinschaft" (DFG Mu 502/5-1) and the valuable technical assistance of Mr. K. Eckert are gratefully acknowledged.

References

[*1*] Brinkman, C. R., "High-Temperature Time-Dependent Fatigue Behaviour of Several Engineering Structural Alloys," *International Metals Review*, Vol. 3, 1985, pp. 235–258.

[*2*] Tsuzaki, K., Hori, T, Maki, T., and Tamura, I., "Dynamic Strain Ageing during Fatigue Deformation in Type 304 Austenitic Stainless Steel," *Materials Science and Engineering*, Vol. 61, 1983, pp. 247–260.

[*3*] Kanazawa, K., Yamaguchi, K., and Nishijima, S., "Mapping of Low-Cycle Fatigue Mechanisms at Elevated Temperatures for an Austenitic Stainless Steel," in *Low Cycle Fatigue, ASTM STP 942*, H. D. Solomon, G. R. Halford, L. R. Kaisand, and B. N. Leis, Eds., American Society for Testing and Materials, Philadelphia, 1988, pp. 519–530.

[*4*] Kim, D. J. and Nam, S. W., "Strain-Rate Effect on High Temperature Low-Cycle Fatigue Deformation of AISI 304L Stainless Steel," *Journal of Materials Science*, Vol. 23, 1988, pp. 1024–1029.

[*5*] Abdel-Raouf, H., Plumtree, A., and Topper, T. H., "Temperature and Strain Rate Dependence of Cyclic Deformation Response and Damage Accumulation in OFHC Copper and 304 Stainless Steel," *Metallurgical Transactions*, Vol. 5, 1974, pp. 267–277.

[*6*] Ermi, A. M. and Moteff, J., "Correlation of Substructure with Time-Dependent Fatigue Properties of AISI 304 Stainless Steel," *Metallurgical Transactions A*, Vol. 13A, 1982, pp. 1577–1588.

[7] Kuwabara, K. and Nitta, A., "Thermal Fatigue Behavior and Strength Estimation of Power Plant Component Materials under Creep-Fatigue Interaction," CRIEPI Report E 278003, 1978.
[8] Kuwabara, K. and Nitta, A., "Thermal-Mechanical Low-Cycle Fatigue under Creep-Fatigue Interaction on Type 304 Stainless Steel," *Fatigue of Engineering Materials and Structures*, Vol. 2, 1979, pp. 293–304.
[9] Fujino, M. and Taira, S., "Effect of Thermal Cycle on Low-Cycle Thermal Fatigue Life of Steels and Grain Boundary Sliding Characteristics," in *Proceedings, Third International Conference on Mechanical Behavior of Materials,* Vol. 2, Pergamon Press, Oxford, 1979, pp. 49–58.
[10] Halford, G. R., McGaw, M. A., Bill, R. C., and Fanti, P. D., "Bithermal Fatigue: A Link between Isothermal and Thermomechanical Fatigue," in *Low Cycle Fatigue: Future Directions, ASTM STP 942,* American Society for Testing and Materials, Philadelphia, 1988, pp. 625–637.
[11] Nitta, A., Kuwabara, K., and Kitamura, T., "Prediction of Thermal Fatigue Life in High Temperature Component Materials for Power Plants," CRIEPI Report E 282015, 1983.
[12] Westwood, H. J., "Tensile Hold Time Effects on Isothermal and Thermal Low-Cycle Fatigue of 304 Stainless Steel," in *Proceedings, Third International Conference on Mechanical Behavior of Materials,* Vol. 2, Pergamon Press, Oxford, 1979, pp. 59–67.
[13] Sheffler, K. D., "Vacuum Thermal-Mechanical Fatigue Behavior of Two Iron Base Alloys," in *Thermal Fatigue of Materials and Components,* ASTM STP 612, D. A. Spera and D. F. Mowbray, Eds., American Society for Testing and Materials, Philadelphia, 1976, pp. 214–226.
[14] Neu, R. W. and Sehitoglu, H., "Thermomechanical Fatigue, Oxidation and Creep: Part I. Damage Mechanisms," *Metallurgical Transactions A,* Vol. 20A, 1989, pp. 1755–1767.
[15] Petry, F., "Ermüdungsverhalten des austenitischen Stahles X3 CrNi 18 9 bei isothermer und thermomechanischer Beanspruchung," Doctorate Thesis, Universität Erlangen-Nürnberg, 1989.
[16] Zauter, R., "Thermomechaniches Ermüdungsverhatten des austenitischen Edelstahls X3uCrNiu/8u9," Doctorate Thesis, Universität Erlangen-Nürnberg, 1992.
[17] Petry, F., Christ, H.-J., and Mughrabi, H., "Cyclic Deformation of AISI 304L as a Function of Temperature," *Proceedings, Symposium Gefüge und mechanische Eigenschaften von Werkstoffen,* Nov. 1990, Bad Nauheim, DGM Informationsgesellschaft, Oberursel, 1991, p.79.
[18] Bayerlein, M., Christ, H.-J., and Mughrabi, H., "Plasticity-Induced Martensitic Transformation During Cyclic Deformation of AISI 304L Stainless Steel," *Materials Science and Engineering*, Vol. A114, 1989, pp. L11–L16.
[19] Manganon Jr., P. L. and Thomas, G., "The Martensite Phase in 304 Stainless Steel," *Metallurgical Transactions,* Vol. 1, 1970, pp. 1577–1586.
[20] Katz, Y., Bussiba, A., and Mathias, H., "The Role of Induced Phase Transformations on Fatigue Processes in AISI 304L," in *Fracture and the Role of Microstructure, Proceedings, Fourth European Conference of Fracture*, Leoben, Sept 1982, Vol. 2, K. L. Maurer and F. E. Matzer, Eds., Chameleon Press Ltd., London, 1982, pp. 503–511.
[21] Hennessy, D., Steckel, G., and Altstetter, C., "Phase Transformation of Stainless Steel During Fatigue," *Metallurgical Transactions A,* Vol. 7A, 1976, pp. 415–424.
[22] Tsuzaki, K., Maki, T., and Tamura, I., "Kinetics of α'-Martensite Formation During Fatigue Deformation of Metastable Austenitic Stainless Steel," *Journal de Physique, Colloque C4, supplement*, Vol. 12, 1982, pp. C4-423–C4-428.
[23] Kashyap, B. P., McTaggart, K., and Tangri, K., "Study on the Substructure Evolution and Flow Behaviour in Type 316L Stainless Steel Over the Temperature Range 21–900°C," *Philosophical Magazine*, Vol. A57, 1988, pp. 97–114.
[24] Jenkins, C. F. and Smith, G. V., "Serrated Plastic Flow in Austenitic Stainless Steel," *Transactions, The Metallurgical Society of AIME,* Vol. 245, 1969, pp. 2149–2156.
[25] Rose, K. S. B. and Glover, S. G., "A Study of Strain Ageing in Austenite," *Acta Metallugica*, Vol. 14, 1966, pp. 1505–1516.
[26] Ohnishi, K. and Ishizaka, J., "The Effect of Temperature and Strain Rate on the Tensile Deformation Behaviour of Austenitic Stainless Steel," *Journal of the Iron and Steel Institute Japan*, Vol. 63, 1977, pp. 2362–2371.
[27] Stanley, J. T. and Cost, J. R., "Short-Range Ordering Kinetics in 316 Austenitic Stainless Steel," *Metallurgical Transactions A,* Vol. 13A, 1982, pp. 1915–1919.
[28] Douglass, D. L., Thomas, G., and Roser, W. R., "Ordering, Stacking Faults and Stress Corrosion Cracking in Austenitic Alloys," *Corrosion*, Vol. 20, 1961, pp. 15–28.
[29] Gerold, V. and Karnthaler, H. P., "On the Origin of Planar Slip in F.c.c. Alloys," *Acta Metallurgica*, Vol. 37, 1989, pp. 2177–2183.
[30] Majumdar, S., "Thermomechanical Fatigue of Type 304 Stainless Steel," in *Thermal Stress, Material Deformation, and Thermomechanical Fatigue,* H. Sehitoglu and S. Y. Zamrik, Eds., ASME, PVP-Vol. 123, New York, 1987, pp. 31–36.

[*31*] Yamaguchi, K. and Kanazawa, K., "Influence of Grain Size on the Low-Cycle Fatigue Lives of Austenitic Stainless Steels at High Temperatures," *Metallurgical Transactions A,* Vol. 11A, 1980, pp. 1691–1699.
[*32*] Mughrabi, H. and Wang, R., "Cyclic Deformation of Face-Centred Cubic Polycrystals: A Comparison with Observations on Single Crystals," in *Deformation of Polycrystals: Mechanisms and Microstructure, Proceedings, Second Risø International Symposium on Metallurgy and Materials Science,* Roskilde, Sept. 1981, N. Hansen, A. Horsewell, T. Leffers, and H. Lilholt, Eds., Risø National Laboratory, Roskilde, Denmark, 1982, pp. 87–98.
[*33*] Jin, N. Y. and Winter, A. T., "Dislocation Structures in Cyclically Deformed [001] Copper Single Crystals," *Acta Metallurgica,* Vol. 32, 1984, pp. 1173–1176.
[*34*] Frost, H. J. and Ashby, M. F., "Deformation Mechanism Maps," Pergamon Press, Oxford, 1982, p. 66.
[*35*] Zauter, R., Petry, F., Christ, H.-J., and Mughrabi, H., "High Temperature Creep Behaviour and Microstructure Development of AISI 304L Stainless Steel," *Materials Science and Engineering,* Vol. A124, 1990, pp. 125–132.
[*36*] Williams, H. D. and Corti, C. W., "Grain-Boundary Migration and Cavitation during Fatigue," *Metal Science Journal,* Vol. 2, 1968, pp. 28–31.
[*37*] Snowden, K. U., Stathers, P. A., and Hughes, D. S., "Grain-Boundary Cavitation and Migration During High Temperature Fatigue," *Res Mechanica,* Vol. 1, 1980, pp. 129–147.
[*38*] Raman, V., Watanabe, T., and Langdon, T. G., "A Determination of the Structural Dependence of Cyclic Migration in Polycrystalline Aluminum Using Electron Channeling Pattern Analysis," *Acta Metallurgica,* Vol. 37, No. 2, 1989, pp. 705–714.
[*39*] Raman, V. and Reiley, T. C., "Dynamic Recrystallization During High Temperature Fatigue," *Scripta Metallurgica,* Vol. 20, 1986, pp. 1343–1346.
[*40*] Biss, V. A. and Sikka, V. K., "Metallographic Study of Type AISI 304 Stainless Steel Long-Term Creep-Rupture Specimen," *Metallurgical Transactions A,* Vol. 12A, 1981, pp. 1360–1362.
[*41*] Großhäuser, M., "Kriechen des austenitischen Stahles X3 CrNi 18 9 bei hohen Temperaturen," Diploma Thesis, Universität Erlangen-Nürnberg, 1991.
[*42*] Baik, S. and Raj, R., "Wedge Type Creep Damage in Low Cycle Fatigue," *Metallurgical Transactions A,* Vol. 13A, 1982, pp. 1207–1214.
[*43*] Min, B. K. and Raj, R., "A Mechanism of Intergranular Fracture During High-Temperature Fatigue," in *Fatigue Mechanisms, Proceedings, ASTM-NBS-NSF Symposium,* Kansas City, Mo., May 1978, *ASTM STP 675,* J. T. Fong, Ed., American Society for Testing and Materials, Philadelphia, 1979, pp. 569–591.
[*44*] Driver, J. H., "The Effect of Boundary Precipitates on the High Temperature Fatigue Strength of Austenitic Stainless Steels," *Metal Science Journal,* Vol. 5, 1971, pp. 47–50.
[*45*] Majumdar, S. and Maiya, P. S., "A Mechanistic Model for Time-Dependent Fatigue," *Journal of Engineering Materials and Technology,* Vol. 102, 1980, pp. 159–167.
[*46*] Kirkwood, B. and Weertman, J. R., "Cavity Nucleation During Fatigue Crack Growth Caused by Linkage of Grain Boundary Cavities," in *Micro and Macro Mechanics of Crack Growth, Proceedings,* The Metallurgical Society of AIME, K. Sadananda, B. B. Rath, and D. J. Michel, Eds., Warrendale, 1982, pp. 199–212.

Peter L. Hacke,[1] Arnold F. Sprecher,[1] and Hans Conrad[1]

Modeling of the Thermomechanical Fatigue of 63Sn-37Pb Alloy

REFERENCE: Hacke, P. L., Sprecher, A. F., and Conrad, H., **"Modeling of the Thermomechanical Fatigue of 63Sn-37Pb Alloy,"** *Thermomechanical Fatigue Behavior of Materials, ASTM STP 1186,* H. Sehitoglu, Ed., American Society for Testing and Materials, Philadelphia, 1993, pp. 91–105.

ABSTRACT: The stress-strain hysteresis behavior of 63Sn-37Pb solder joints in thermomechanical cycling was determined experimentally with a bimetallic loading frame for temperature excursions between −30 and 130°C applied in triangular waveform for periods of 12, 24, 48, and 300 min. A computer simulation based on a constitutive equation previously formulated for steady-state creep conditions yielded reasonable agreement between the predicted and measured hysteresis behavior. An approximately linear relationship was found between the drop in maximum load in a hysteresis cycle and the measured fatigue cracked area.

KEYWORDS: solder, fatigue, thermomechanical fatigue, fatigue modeling, fatigue crack growth, steady-state creep, constitutive equation

Fatigue failure is an important problem in solder joints used in microelectronic packaging because components connected with solder often have differing coefficients of thermal expansion. Temperature fluctuations, which result from heat dissipated by electronic devices or environmental temperature variations, cause a complex and cyclic state of stress and strain in the solder joints, which in turn leads to fatigue damage and failure. It is therefore desired to understand, minimize, and estimate the time to fatigue failure in solder joints subjected to thermal cycling. For a given package assembly configuration, it would be useful to have a model with which to determine, with minimum costly experimentation, the lifetime of a solder joint exposed to thermal fluctuations. To this end, much effort has been made to develop models which describe the stress-strain behavior of solder joints subjected to fatigue cycling [*1–5*]. Crack growth rate and fatigue life approaches can be included in such models, as for example in the damage integral approach [*2*], or the integrated matrix creep approach [*4,6*].

Since solder joints in electronic packaging experience temperatures greater than one-half their homologous temperature (that is, in the range of what is considered high temperature creep), it is a reasonable assumption that the same deformation mechanisms observed in creep tests are operative in an electronic package subjected to temperature changes. A semi-empirical equation in the form

$$\dot{\gamma} = \frac{ADGb}{kT}\left(\frac{b}{d}\right)^{p}\left(\frac{\tau}{G}\right)^{n} \tag{1}$$

has been shown to successfully describe the steady-state shear rate ($\dot{\gamma} = d\gamma/dt$) for a number of alloys [*7–9*]. A, n, and p are dimensionless constants. $n[= \partial \ln \dot{\gamma}/\partial \ln \tau]$ is the stress expo-

[1] Graduate student, senior research associate, and professor, respectively, Materials Science and Engineering Dept., North Carolina State University, Raleigh, NC 27695-7907.

nent and $p[= -\partial \ln \dot{\gamma}/\partial \ln d)]$ is the grain size exponent. D is a diffusion coefficient $[= D_0 \exp(-Q/kT)$, where D_0 is the pre-exponential factor, Q is the activation energy, k is Boltzmann's constant, T is temperature (Kelvin)]. G is the shear modulus, b the Burger's vector, and d the mean grain size.

Grivas et al. [*8*] investigated the deformation behavior of 62Sn-38Pb eutectic solder in the superplastic region (II) and the conventional region (III). Region II deformation is attributed to a grain boundary sliding process, whereas Region III deformation is considered to be controlled by dislocation climb and glide. Region III deformation is sometimes referred to as matrix creep. Details concerning these regions can be found in [*7,8*]. The alloy in [*8*] was rolled and annealed to achieve the fine and equiaxed grains that permit grain boundary sliding. Steady-state creep rates were determined experimentally in shear on bulk specimens for several grain sizes and temperatures. Grives et al. [*8*] suggested that the Pb-Sn eutectic alloy exhibited both conventional and superplastic deformation with the assumption that both deformation modes occur simultaneously via independent mechanisms but that superplastic deformation is rate controlling at lower stresses and conventional deformation is rate controlling at higher stresses. Combining the rates of the two deformation mechanisms, that is, taking

$$\dot{\gamma} = \dot{\gamma}_{II} + \dot{\gamma}_{III} \tag{2}$$

a semi-empirical equation was developed summing the shear strain rates in Regions II and III, so that

$$\dot{\gamma} = \frac{D_0 G b}{kT}\left[A_{II}\left(\frac{\tau}{G}\right)^{1.96}\left(\frac{b}{d}\right)^{1.8}\exp(-48{,}157/RT) + A_{III}\left(\frac{\tau}{G}\right)^{7.1}\exp(-81{,}239/RT)\right] \tag{3}$$

This equation was applied to eutectic solders [*8*], where d is the mean phase size, ranging in their tests between 5.5 and 9.9 μm determined by the mean intercept method. The gas constant R has units $J \cdot \text{mol}^{-1} \cdot K^{-1}$.

Hall [*5*] performed creep tests on solder joints as they occur in an actual surface mount packaging configuration. He found that the stress exponent n is in the range of about 5 to 20, not 2 as is observed in superplasticity. In contrast, Shine and Fox [*6*], having also measured the stress exponent of actual solder joints on chip carriers, did find superplastic behavior at strain rates below 10^{-5} s^{-1}. Such differences between the results of different investigators could come from different solder microstructures obtained from differing reflow techniques and cooling rates.

The strain in a solder joint assembly has been given by [*3,5*]

$$h(\gamma_{el} + \gamma_p) + \frac{\tau A}{K} = L\,\Delta\alpha(T - T_0) \tag{4}$$

which states that (on the right side of the equation) any temperature excursion T, from some reference temperature T_0, where the load and strain in the system is zero, will create a strain in the system with scaling factors $\Delta\alpha$, the difference in coefficients of thermal expansion of the materials being joined. L is the distance between the solder joint and the neutral point, h is the thickness of the solder joint, and A is its cross-sectional area. The strain created by the thermal mismatch is accommodated by some combination of elastic and plastic shear strains of the solder, γ_{el} and γ_p, respectively, and the assumed elastic deformation of the assembly governed by the shear stress over the solder joint τ, and the assembly stiffness K. The anelastic strain component is not distinguished because, in the relatively slow loading rate employed here, anelastic strain is assumed to be transferred to plastic strain.

If the solder joint and package assembly is instantaneously subjected to a small increment in temperature ΔT above T_0, the assembly will change dimensions according to their respective coefficients of thermal expansion, thus creating a stress and strain state in the system. At that instant, the strain must be absorbed elastically in the solder γ_{el}, or the surrounding assembly as described by the term $\frac{\tau A}{K}$. The resultant instantaneous stress change is given by

$$\Delta\tau = \frac{L\,\Delta\alpha\,\Delta T}{A}\left[\frac{1}{\left(\frac{1}{K} + \frac{h}{AG(T)}\right)}\right] \tag{5}$$

The first term on the right side expresses the temperature-induced displacement per unit contact area of the solder joint. The second term in the brackets represents the combined stiffness of the solder joint and surrounding assembly considered as two springs in series [*10*]. The change in elastic strain in the solder joint $\Delta\gamma_{el}$ is given by

$$\Delta\gamma_{el} = \frac{\Delta\tau}{G(T)} \tag{6}$$

where the shear modulus $G(T)$ has some dependence on temperature. For this study, the value for $G(T)$ is taken from [*7*].

Following the instantaneous increment in temperature ΔT, with the assumption that all the resulting deformation will be absorbed elastically in the solder and assembly, the solder will deform plastically over an increment of time Δt as a function of the applied stress, the temperature, and kinetics of the creep-relaxation of the solder. The magnitude of plastic strain can be determined via the product of Δt and the shear rate given by a constitutive equation such as Eq 3. The direction of the plastic shear strain in the solder joint will be such that the magnitude of stress in the system is reduced. The change in stress due to plastic shear strain over time Δt is

$$\Delta\tau_p = -\,\mathrm{sgn}\,(\tau)\,|\Delta\gamma_p|\left[\frac{1}{\left(\frac{A}{hK} + \frac{1}{G(T)}\right)}\right] \tag{7}$$

where sgn (τ) returns the sign (direction) of the shear stress.

In such a manner, the mechanical state of the system after an increment in temperature ΔT and time Δt can be characterized. These calculations are iterated using the rate of temperature change and parameters of the solder joint and bimetallic assembly described below to simulate the stress-strain behavior of a thermomechanical cycle.

In this work, the deformation of near-eutectic Pb-Sn alloy solder joints subjected to thermomechanical cycling were examined both experimentally and via modeling with a constitutive equation. Empirical or semi-empirical equations based on observed mechanical response and theoretical considerations were used to describe the deformation behavior of solder. To characterize the mechanical response of solder joints in thermomechanical fatigue, a specially designed loading frame was built. With use of this loading frame in a temperature cycling chamber the stress-strain behavior of solder joints was measured. Hysteresis loops were

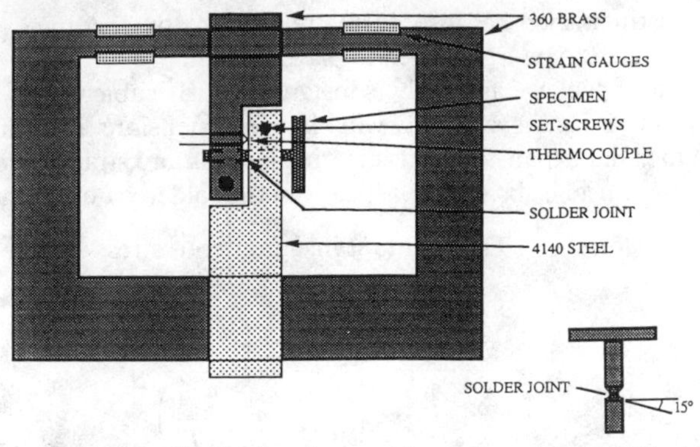

FIG. 1—*Schematic of the bi-metallic loading frame with specimen inserted. The inset is a schematic of the test specimen consisting of the copper cylinders and the reflowed solder joint.*

generated showing the mechanical response of a 63Sn-37Pb alloy in thermomechanical fatigue at several frequencies. The applicability of a constitutive equation [*8*] based on steady-state creep of the bulk eutectic Pb-Sn alloy in shear to the experimentally obtained hysteresis loops was investigated. Since eventual failure of the solder joint will take place by initiation and growth of cracks, these features of fatigue were also investigated during the course of the thermomechanical cycling.

Experimental Procedure

Apparatus

A system was built to characterize the thermomechanical fatigue of solder joints in shear. The system consists of a loading frame (Fig. 1) and a temperature cycling chamber (Fig. 2). The bimetallic loading frame operates on the principle of mismatch between thermal coefficients of expansion α. Displacement is imparted to the solder specimen by virtue of the mismatch between the 4140 steel, which is one leg of the frame, and 360 brass, which constitutes the remainder of the frame. The geometry of the frame is such that the specimen is loaded in shear, although there are some normal components of stress due to bending moments and

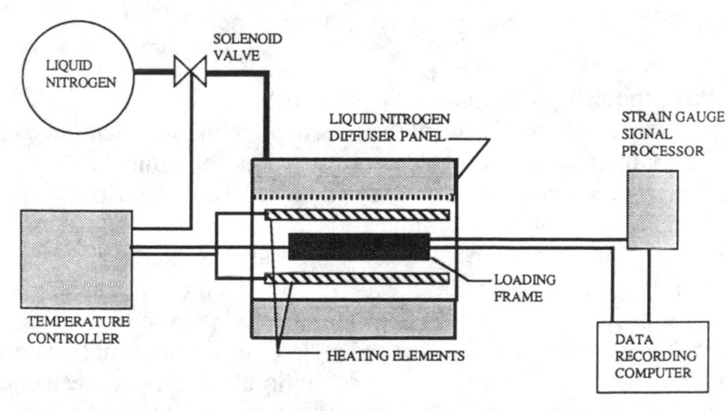

FIG. 2—*Schematic of the thermal cycling apparatus.*

lateral thermal mismatch between the copper, steel, and brass. The normal strain imparted to the specimen resulting from lateral deflection was estimated to be on the order of 0.5% of the applied shear strain. Bending moments are inherent to the geometry of solder joints like those used in the present investigation. It was assumed that the effects of moments on the plastic deformation of the specimen are comparable to those found in numerous other investigations using solder specimens of similar thickness, and that they do not detract from the analysis.

The system was built so that accurate measures of load and displacement in the system could be obtained. A resistance strain gage bridge is attached to the top member of the frame so that, with calibration, the load and displacement to which the specimen is subjected can be determined. The dimensions of the beam were chosen so that the output signal of the strain gage bridge was maximized. However, it was decided to limit the maximum deflection in the frame to about 50% of the displacement over the solder joint, that is, it was desired that the frame not be too compliant.

Calibration of the loading frame was carried out to experimentally determine its mechanical properties, that is, the frame stiffness (K) and the effective thermally induced displacement over the specimen grips. The calibration procedure consisted of determining the following relationships:

(1) load (kg) and strain gage bridge output (V),
(2) temperature (°C) and load on a dummy hard steel pin specimen,
(3) load and displacement (m) over the specimen holder (with no specimen mounted), and
(4) temperature and displacement over the specimen holder (with no specimen mounted).

Equations 1 and 3, the strain gage output and displacement resulting from an applied load, were determined by applying known loads on the loading frame. The strain gage bridge output was monitored and the displacement over the specimen holder area was determined using a strain gage extensometer. The strain gage bridge response to the applied load was determined to be 0.0498 V/kg, and the displacement response to load was determined to be 9.26×10^{-7} m/kg ($K = 1.08 \times 10^6$ kg/m).

Equation 2 was determined by placing a hard steel pin (assumed not to yield) in the specimen grips and running the loading frame through cycles between the temperature extremes of -30 and 130°C in triangular waveform. The steel pin impedes displacement between the specimen grips. Thus, this test gives the maximum load that can be imparted to the specimen by the bimetallic frame in this temperature regime. The slope was found to be 0.144 kg/°C.

Equation 4, displacement over the specimen grips with no specimen mounted as a function of temperature, is determined from the other relationships. Given Eq 3 (displacement versus applied load) and Eq 2 (applied load versus temperature), displacement as a function of temperature is determined to be 1.33×10^{-7} m/°C under the condition that the frame is allowed to expand or contract unimpeded. In a temperature excursion of 160°C, as is the case here, the temperature-induced displacement is 2.13×10^{-5} m. Since the solder specimen thickness is 2×10^{-4} m, this displacement corresponds to a temperature-induced uninhibited total strain range of $\Delta\gamma_T = 0.107$.

During a fatigue test, load is measured with the strain gage bridge and by using Eq 1. Displacement over the specimen is calculated via

$$\delta_{spec} = \delta_T - \delta_P \tag{8}$$

where δ_T is the temperature-induced displacement determined from Eq 4 knowing the temperature. δ_P is the load-induced deflection in the frame caused by the specimen's resistance to deformation determined from Eqs 3 and 1 and knowing the strain gage bridge output. In this way, the stress and strain that the specimen undergoes can be determined throughout the fatigue test.

The strain gages selected are matched to the material on which they are fixed (brass) for self-temperature compensation; that is, resistance of the strain gage changes only minimally with temperature when bonded to the unstressed brass. The temperature range in this thermomechanical fatigue testing is considerable, and there are temperature gradients in the lead wires producing artifact signals. This artifact signal was subtracted from the measured signal after a thermomechanical fatigue experiment was performed to yield the real signal. The artifact signal was determined by measuring the strain gage output of the frame with no specimen mounted. This was done for each of the temperature cycle profiles.

Specimen Preparation

The test specimens used in these tests, shown in Fig. 1, were based on the Iosipescu pure shear specimen used by Solomon [*11*] and adapted by Bae *et al.* [*12*]. The test specimens consist of two cylindrical copper grips, one with a base for use in the solder reflow. The solder pad is a 2.0-mm diameter round and is bordered by a sloping portion 15° to the pad. Near-eutectic 63Sn-37Pb alloy solder joints were prepared in a manner similar to that described in Ref *12* but of one half the diameter. Solder pieces were produced by rolling bulk solder and punching out cylinders 2.0-mm diameter by 0.2-mm thick. The solder disc was fluxed and placed between cylindrical copper grips using a jig to maintain alignment and to establish reproducible solder joints of thickness 0.203 ± 0.013 mm. This joint thickness is near that used by Solomon [*11*] and Shine and Fox [*6*]. It is thin compared to the area thereby minimizing tensile stresses associated with bending moments. The specimen-jig assembly was then placed on a hot plate and heated to 35°C above the liquidus of the solder alloy. After the sample reached the desired temperature the assembly was placed on a steel plate to cool to room temperature. The time to cool from just above the liquidus (184°C) to halfway to room temperature was 57 s. Shown in Figs. 3*a* and *b* is the resultant microstructure at two levels of magnification. The mean grain size measured by the mean phase intercept method [*8*] is $d = 0.9$ μm with standard deviation 0.9 μm. The standard deviation is large because the distribution of sizes is greatly skewed. Occasionally, large lead-rich dendrites are seen in the micrographs (Fig. 3*a*), however, most often the microstructure consists of smaller dispersed lead rich regions within the tin matrix. The eutectic structure appears to be strongly degenerate, though some colony structure could be seen.

Thermomechanical Fatigue Tests

The specimen was inserted and clamped into place as shown in Fig. 1. The fatigue tests were performed by ramping in triangular waveform between the temperature extremes of −30 and 130°C. This range of temperature is similar, albeit slightly greater than that employed by other researchers. For example, Hall [*13*] employed the range of −25 to 125°C, whereas Liljestrand and Andersson [*14*] used extremes of −10 and 100°C. A triangular waveform is a simplification of the actual temperature-time profile experienced in a real solder joint. It was chosen as a starting point for experimental and modeling simplicity. Dwells at the low temperature extreme would result in a slow transfer of elastic energy from the loading frame to plasticity in the solder joint, an increase in cyclic plastic strain range, and a lower number of cycles to failure. A dwell at the high temperature extreme would probably do little, other than coarsen the microstructure, since the joint is essentially stress-relaxed at the end of the ramp up to 130°C.

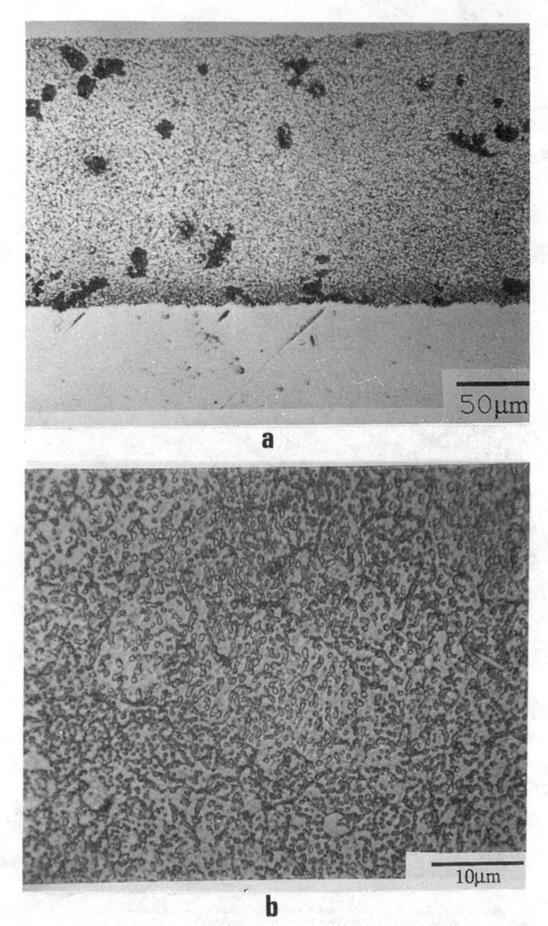

FIG. 3—*Initial microstructure of the as reflowed 63Sn-37Pb alloy at two magnifications:* (a) *the etched with, by volume, 1 concentrated nitric acid: 1 glacial acetic acid: 8 glycerol, and* (b) *etched with 25 mL H_2O, 5 mL concentrated HCL, 5 g NH_4NO_3. Dark phase is lead-rich; light phase is tin-rich.*

Variation in temperature-ramp rates was chosen to examine the applicability of Eq 3 to simulate the mechanical response of the 63Sn-37Pb alloy over the indicated temperature range. Cycling periods of 12, 24, 48, and 300 min were chosen for examining the initial stress-strain behavior of the solder joints with known microstructure before the growth of cracks (Figs. 4 to 7). In these experiments, specimens were mounted into the loading frame at room temperature and the tests begun by first ramping down to the low temperature extreme of −30°C. Results for the third cycle of the test are presented, with the exception of those for the 12-min period. For the 12-min period, the average of the second through the tenth cycles are presented to reduce scatter in the data.

Tests employing the 12-min period were extended to characterize changes in stress-strain behavior as a function of crack growth. A series of tests was performed with specimens cycled to various fractions of fatigue life. Following removal from the test fixture, the specimens were manually fractured in liquid nitrogen and examined by scanning electron microscopy (SEM) to determine the nature and size of the fatigue cracks.

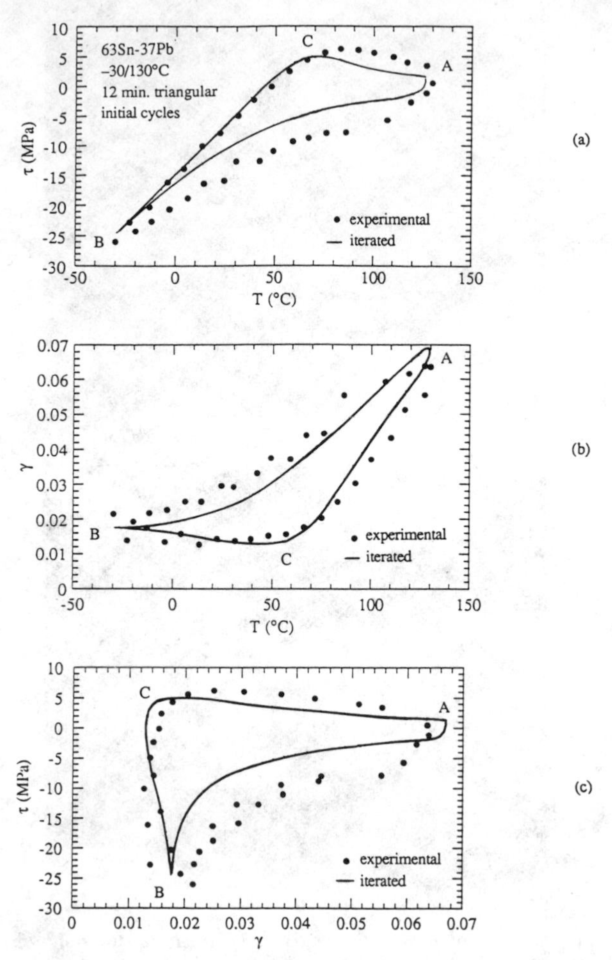

FIG. 4—*Initial hysteresis loops for the 12-min period cycle between −30 and 130°C:* (a) τ *versus* T*;* (b) γ *versus* T*, and* (c) τ *versus* γ*.* τ *is shear stress,* γ *is shear strain, and* T *temperature.*

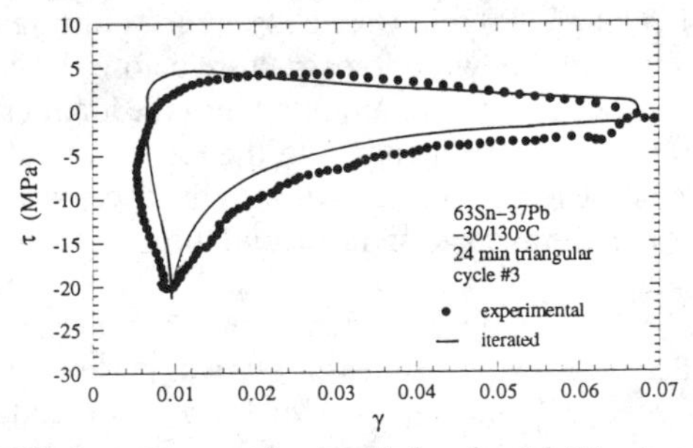

FIG. 5—τ *versus* γ *for the third cycle with 24-min period.*

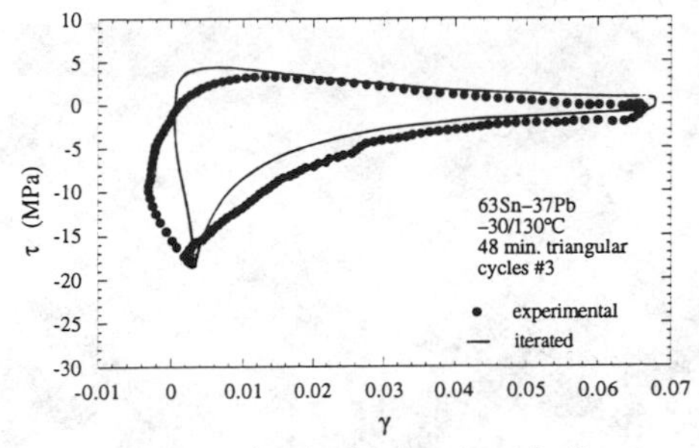

FIG. 6—τ *versus* γ *for the third cycle with 48-min period.*

Results and Discussion

Initial Cycle Mechanical Response

Experimentally obtained and superimposed computer-generated curves describing the asymmetric hysteresis behavior in thermomechanical cycling for the 12-min period are shown in Fig. 4. Included are (a) apparent shear stress τ versus temperature T, (b) shear strain γ versus T, and (c) τ versus γ. Examining Fig. 4*a*, at Point A, the high temperature extreme of the cycle (130°C in the present tests), the flow stress is significantly lowered and the solder is largely stress-relaxed, since the alloy is at 0.88 T_m. As the temperature is ramped down to -30°C (Point B), the displacement due to the mismatch in coefficients of thermal expansion of the assembly is gradually transferred from plastic deformation in the solder joint to elastic deformation in the assembly as the solder flow stress increases. Upon heating to 130°C between Points B and C, the conditions are such that the flow stress of the solder is above the applied stress and the slope reflects the elastic deformation of the machine and solder. Plastic deformation in the solder is minimal until Point C, at which the temperature is high enough for significant plastic deformation to take place as the flow stress of the solder is exceeded. On further increase in temperature, plastic flow takes place and the load-bearing ability of the solder decreases. Additional descriptions of thermomechanical hysteresis behavior of solder alloys can be found in [*1,2,5,15,16*].

With increased cycle period, more plastic flow occurs and the magnitude of the maximum load, occurring at -30°C in each case, decreases and the plastic strain range increases.

For purposes of modeling, the plastic strain rate equation based on Eqs 1 and 3 was imple-

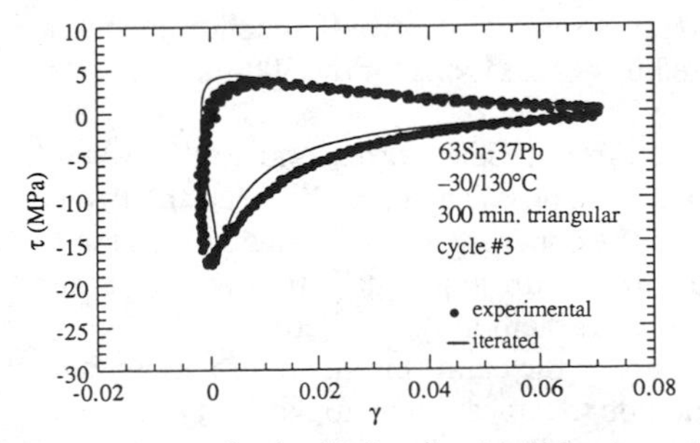

FIG. 7—τ *versus* γ *for the third cycle with 300-min period.*

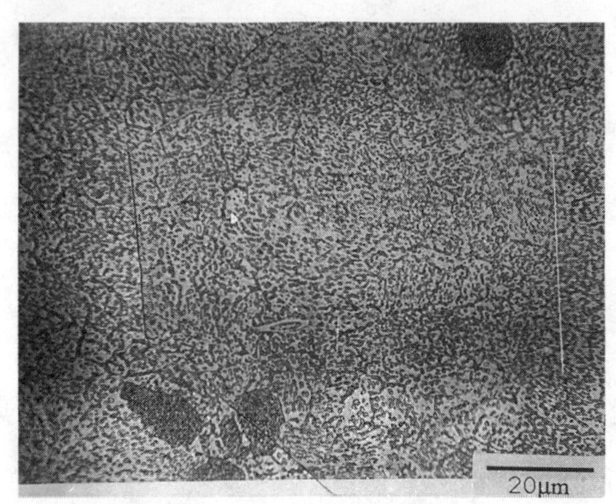

FIG. 8—*Microstructure after three 24-min cycles. Etched with 25-mL H_2O, 5-mL concentrated HCL and 5-g NH_4NO_3. Dark phase is lead-rich; light phase is tin-rich.*

mented with the pre-exponential constants chosen to fit the experimental results. The equation employed to simulate our test results was

$$\dot{\gamma} = \frac{2.12 \times 10^{-16}}{T} \frac{\tau^{1.98}}{d^{1.8}} \exp(-48\,157/RT) + \frac{6.55 \times 10^{-33}}{T} \tau^{7.1} \exp(-81{,}239/RT) \quad (9)$$

where the shear stress τ has units $kg \cdot m^{-2}$.

Since Eq 3 was formulated with the grain size parameter d being the mean phase intercept, this measurement was also used here for d. It is acknowledged that this may not be the most mechanistically correct grain-size parameter; however, it is assumed that some effective relationship exists between the mean phase size and the size of the grains involved in the grain boundary sliding, which is considered in the pre-exponential constant. The mean value d was measured after the third 24-min period cycle and was found to be 1.0 μm (standard deviation = 0.7 μm), Fig. 8. This value was then used as input in the 12-, 24-, and 48-min period modeling. In the 300-min period tests, where significant grain coarsening was observed (Fig. 9), the mean value of d was measured to be 2.4 μm (standard deviation = 2.2 μm) and this value was used in the modeling in Fig. 7.

Grain coarsening reduces the contribution of grain boundary sliding to the plastic deformation, that is, as the grain size increases, the fraction $\dot{\gamma}_{II}/\dot{\gamma}_{III}$ decreases. From Eq 9, it can be seen that a larger value of d reduces $\dot{\gamma}_{II}$, so significantly less plastic flow occurs per cycle for the 300-min period than if d remained ~1.0 μm as in the shorter cycles. Taking into account grain coarsening, the calculated hysteresis loops for the 300-min period now have a form nearly the same as those for the other cycle periods.

There is reasonable, though not ideal, correspondence between the iterated and the experimentally obtained hysteresis loops. Deviations between the two could result from a number of sources. With respect to the experimentally obtained hysteresis loop, irregularities that may skew the results include temperature gradients in the cycling frame and the solder joint (which is expected to be more of a problem at the higher frequency tests) and imperfect gripping of the copper/solder specimen by the clamps in the frame. Issues surrounding the results of the computer simulation include the question of applicability of the particular constitutive equation chosen (Eq 9, based on Eq 3) and the inherent assumptions made in its derivation. Differences in microstructure and the slight difference in composition between the specimens

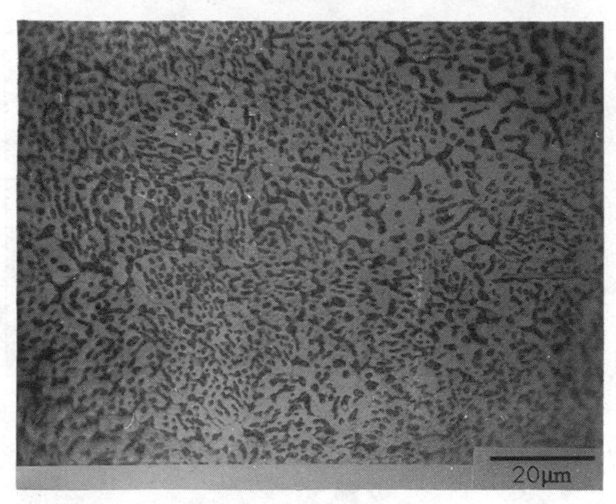

FIG. 9—*Microstructure after three 300-min cycles. Etch is the solution same as that for Fig. 8.*

used in Ref *8* and the present study are likely to account for some incompatibility. Further, in Ref *8* Eq 3 referred to steady-state creep, whereas in the present work it is applied to thermomechanical fatigue.

The present computer model was also applied to a thermomechanical hysteresis loop undergone by a solder joint in a ceramic chip carrier/printed wired board assembly obtained by Hall [*13*]; see Fig. 10. The parameters derived from that publication and employed as input for the present simulation are the assembly stiffness, $K = 2.8 \times 10^4$ kg/m, and the effective unimpeded displacement per °C between the materials being joined, $\Delta\alpha \cdot L = 1.3 \times 10^{-7}$ m/°C. The only unpublished parameter that is assumed for this simulation is the mean phase intercept size, taken here as $d_{eff} = 4.8$ μm. This is taken as an effective value because the microstructure and the active deformation mechanism of the solder in Hall's experiment may differ from that for which the constitutive equation used in this simulation was formulated. Nevertheless, the simulation shows reasonably good agreement with the experimentally obtained results if the above-mentioned assumption is made. Modeling of Hall's data in Fig. 10 was also done by Stone *et al.* [*1*] and Wilcox *et al.* [*15*] with nearly perfect correspondence using the state variable approach.

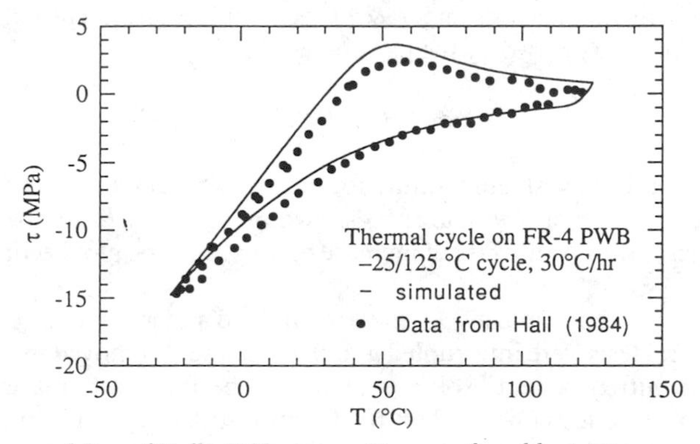

FIG. 10—*Our modeling of Hall's* [13] *τ versus* T *curve of a solder joint in a package assembly.*

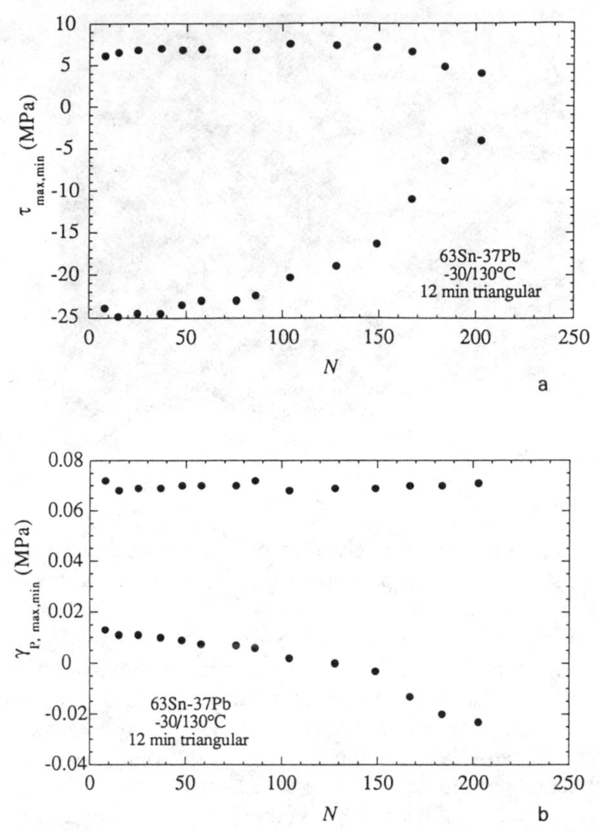

FIG. 11—*Maximum and minimum values of apparent shear stress* (a) *and plastic shear strain* (b) *versus number of 12-min period cycles. These values are taken from the hysteresis loops obtained during the thermomechanical fatigue cycling.*

Crack Growth

It was found that the difference between the apparent shear stress values at the maximum and minimum temperatures became less with increase in number of cycles N. Simultaneously, the difference in shear strain became larger; see for example Fig. 11. The change in load with number of cycles can be expressed by the parameter

$$\phi = 1 - (P/P_0) \tag{10}$$

where P_0 is the load at the lowest temperature for the first several stable cycles and P is that for the Nth cycle. The variation of ϕ with log N is shown in Fig. 12 for a number of tests. Within the scatter, ϕ increases with N, the rate of increase on this semi-log plot becoming greater as N becomes larger.

To investigate the relationship between the drop in load and any fatigue cracking that may have occurred, several tests were interrupted and the specimens removed from the test fixture, immersed in liquid nitrogen, and broken manually. The fractured area was characterized using SEM. Typical structures observed on the fracture surface of a specimen broken in this manner are presented in Fig. 13 for a specimen cycled to $N = 182$ (12-min period) and $\phi =$

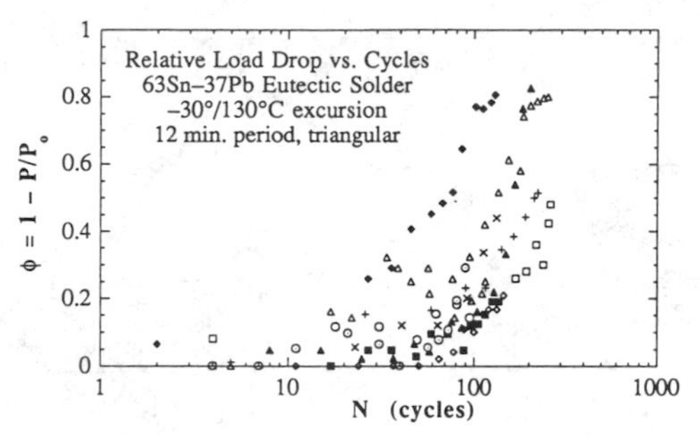

FIG. 12—*Load drop parameter* ϕ *versus number of thermal cycles (−30/130°C, 12 min) for a number of specimens. Each set of data points is for a given specimen.*

17%. The low magnification overview in Fig. 13*a* exhibits three regions: (i) a central region, (ii) a transition region, and (iii) a region near the circumference of the specimen. Higher magnification views of the three regions are given in Figs. 13*b*, *c*, and *d*, respectively. These higher magnification photomicrographs indicated the following regarding the three regions.

Region (i) in Fig. 13*b* is identified as solder that has not been cracked during the fatigue test.

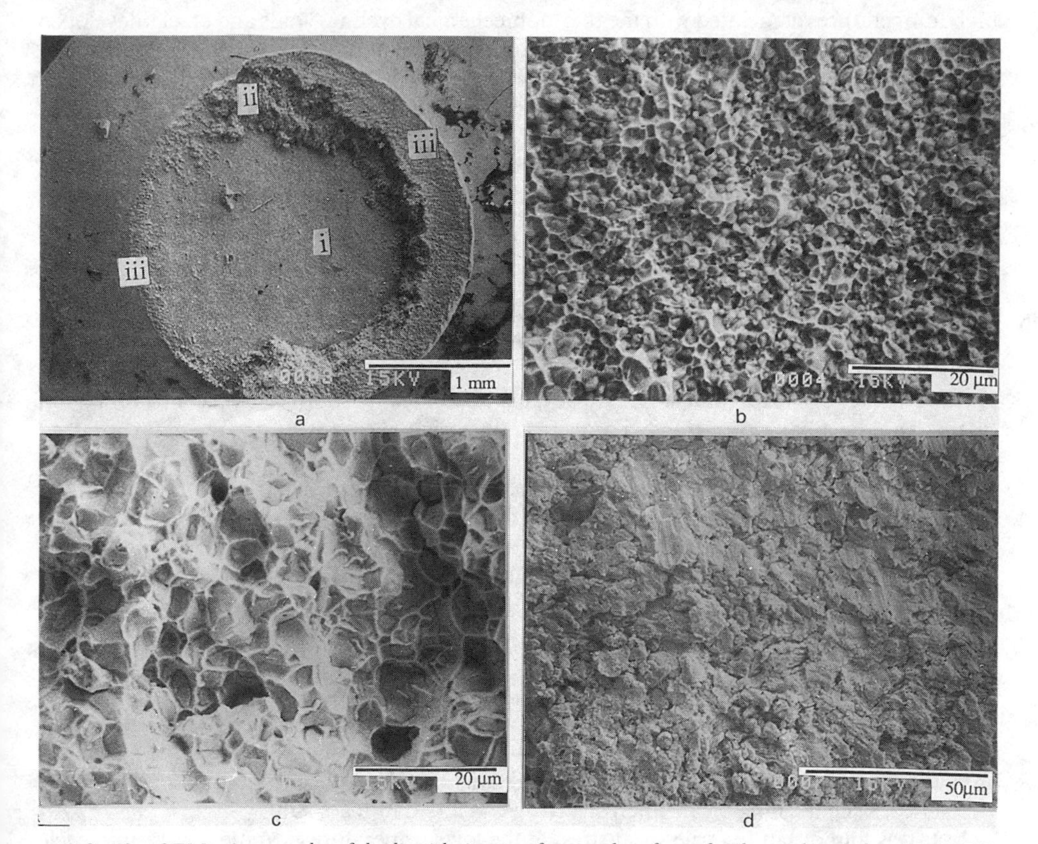

FIG. 13—*SEM micrographs of the liquid nitrogen-fractured surface of a thermal-cycled (−30/130°C, 12 min) specimen to* N = *182 giving a 17% drop in load:* (a) *overall view,* (b) *higher magnification of Region i,* (c) *Region ii, and* (d) *Region iii.*

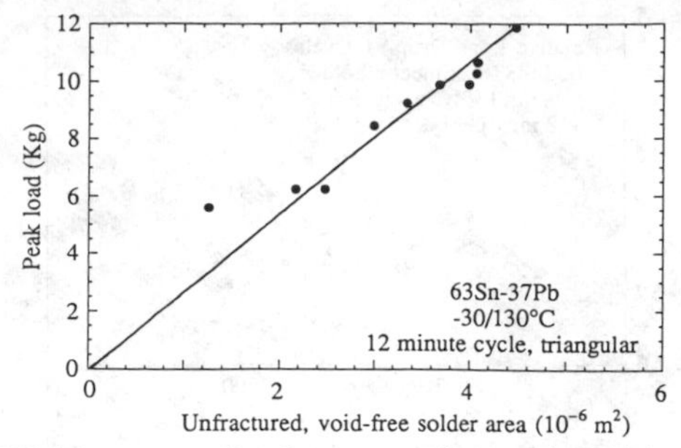

FIG. 14—*Load at −30°C (maximum load) versus unfractured, void-free area* A *given by* $A = A_o - A_{iii} - A_{voids}$.

This structure was observed previously in [*17*] and energy-dispersive spectroscopy of this surface indicated the presence of a copper-tin intermetallic compound.

Region (ii) in Fig. 13*c*, which connects Regions (i) and (iii), was only observed in specimens that had been thermally cycled. Its structure has the appearance of intergranular fracture of a two-phase material. It is tentatively concluded that because of grain boundary sliding at the high temperatures associated with the thermomechanical cycling, voids and cracks developed along the boundaries, resulting in their separation upon breaking the specimen at liquid nitrogen temperature.

Region (iii) in Fig. 13*d* occurred along the perimeter of the specimens. This region was only observed in specimens that had been thermally cycled.

In Solomon's [*11*] and Guo et al.'s [*17*] analysis of low cycle fatigue of 60Sn-40Pb solder joints, it was determined that the decrease in ΔP is due entirely to a decrease in the load bearing area, that is

$$P/P_0 = A_c/A_o = 1 - \phi \tag{11}$$

For the present tests, the maximum load at the lowest temperature of the cycle was plotted versus the area of sound material in Fig. 14. It is here assumed that region (iii) is fatigue-cracked material. The area of sound material is calculated by subtracting the area of Region (iii) A_{iii} and the area occupied by voids A_v from the original cross-section area of the joint A_o. Except for the one data point at the smallest unfractured area, the present results are in good accord with Eq 11. This indicates that the area over which the shear load was transmitted decreased due to fatigue cracking, so that the true magnitude of the shear stress τ at the low temperature extreme of the thermal cycles remained approximately constant during a thermal cycling test. The slope of the curve in Fig. 14 yields $\tau = 26.2$ MPa for the maximum stress at the low temperature extreme.

Conclusions

1. An equation describing the steady state creep rate of bulk eutectic Pb-Sn alloy in shear successfully modeled the experimental stress-strain behavior of solder joints in shear during the initial cycles of thermomechanical fatigue in the range of −30 to 130°C when the effect of coarsening of the microstructure was taken into account.
2. It was found that the true shear stress at the low temperature extreme of a 12-min cycle remained essentially constant as growth of the fatigue crack occurred.

Acknowledgments

This work was sponsored by the Electronics Division of E. I. Du Pont de Nemours and Co., with Dr. J. Dorfman as technical monitor. The authors would also like to thank Drs. K. L. Murty and P. M. Hall for valuable discussions.

References

[*1*] Stone, D., Hannula, S.-P., and Li, C.-Y., "The Effects of Service and Material Variables on the Fatigue Behavior of Solder Joints during the Thermal Cycle," *Proceedings of the 35th Electronic Components Conference,* Institute of Electrical and Electronics Engineers, New York, 1985, pp. 46–51.

[*2*] Subrahmanyan, R., Wilcox, J. R., and Li, C.-Y., "A Damage Integral Approach to Thermal Fatigue of Solder Joints," *IEEE Transactions on Components, Hybrids, and Manufacturing Technology,* Vol. CHMT-12, No. 4, 1989, pp. 480–491.

[*3*] Clech, J.-P. and Augis, J. A., "Engineering Analysis of Thermal Cycling Accelerated Tests for Surface-Mount Attachment Reliability Evaluation," *Proceedings of the 1987 International Electronic Packaging Conference,* Boston, Nov. 9–11, 1987, pp. 385–410.

[*4*] Knecht, S. and Fox, L., "Integrated Matrix Creep: Application to Accelerated Testing and Lifetime Prediction," *Solder Joint Reliability Theory and Applications,* J. H. Lau, Ed., Van Nostrand Reinhold, New York, 1991, pp. 508–544.

[*5*] Hall, P. M., "Creep and Stress Relaxation in Solder Joints of Surface-Mounted Chip Carriers," *IEEE Transactions on Components, Hybrids, and Manufacturing Technology,* Vol. CHMT-12, No. 4, Dec. 1987, pp. 356–365.

[*6*] Shine, M. C. and Fox, L. R., "Fatigue of Solder Joints in Surface Mount Devices," *Low Cycle Fatigue, ASTM STP 942,* H. D. Solomon, G. R. Halford, L. R. Kaisand, and B. N. Leis, Eds., American Society for Testing and Materials, Philadelphia, 1988, pp. 588–610.

[*7*] Mohamed, F. A. and Langdon, T. G. "Creep Behavior in the Superplastic Pb-62% Sn Eutectic," *Philosophical Magazine,* Vol. 32, 1975, pp. 697–709.

[*8*] Grivas, D., Murty, K. L., and Morris, J. W., Jr. "Deformation of Pb-Sn Eutectic Alloys at Relatively High Strain Rates," *Acta Metallurgica,* Vol. 27, 1979, pp. 731–737.

[*9*] Bird, J. E., Mukherjee, A. K., and Dorn, J. E., "Correlations Between High-Temperature Creep Behavior and Structure," *Quantitative Relation Between Properties and Microstructure,* D. G. Brandon and A. Rosen, Eds., Israel University Press, Jerusalem, 1969, pp. 255–342.

[*10*] Meyers, M. A. and Chawla, K. K., *Mechanical Metallurgy Principles and Applications,* Prentice-Hall, New York, 1987, pp. 575–576.

[*11*] Solomon, H., "Low Cycle Fatigue of 60/40 Solder-Plastic Strain Limited vs. Displacement Limited Testing," *Electronic Packaging-Materials and Processes,* J. A. Sartell, Ed., American Society for Metals, Metals Park, OH, 1986, pp. 29–47.

[*12*] Bae, A., Sprecher, A. F., Conrad, H., and Jung, D. Y., "Fatigue of 63Sn-37Pb Solder used in Electronic Packaging," *14th International Symposium for Testing and Failure Analysis,* American Society for Metals, Metals Park, OH, 1988, pp. 53–61.

[*13*] Hall, P. M., "Forces, Moments, and Displacements during Thermal Chamber Cycling of Leadless Ceramic Chip Carriers Soldered to Printed Boards," *IEEE Transactions on Components, Hybrids, and Manufacturing Technology,* Vol. CHMT-7, No. 4, 1984, pp. 314–327.

[*14*] Liljestrand, L. G. and Andersson, L. O., "Accelerated Thermal Fatigue Cycling of Surface Mounted PWB Assemblies in Telecom Equipment," *Proceedings of the 1987 International Electronic Packaging Conference,* Boston, Nov. 9–11, 1987, pp. 411–424.

[*15*] Wilcox, J. R., Subrahmanyan, R., and Li, C.-Y., "Thermal Stresses and Inelastic Deformation of Solder Joints," *Microelectronic Packaging Technology: Materials and Processes, Proceedings of the 2nd International Electronic Materials and Processing Congress,* American Society for Metals, Metals Park, OH, 1989, pp. 203–211.

[*16*] Li, C.-Y., Subrahmanyan, R., Wilcox, J. R., and Stone, D., "A Damage Integral Methodology for Thermal and Mechanical Fatigue of Solder Joints," *Solder Joint Reliability Theory and Applications,* J. H. Lau, Ed., Van Nostrand Reinhold, New York, 1991, pp. 361–383.

[*17*] Guo, Z., Sprecher, A. F., and Conrad, H., "Crack Initiation and Growth During Low Cycle Fatigue of Pb-Sn Solder Joints," *Proceedings of the 41th Electronic Components Conference,* Institute of Electrical and Electronics Engineers, 1990, pp. 658–666.

Michael G. Castelli,[1] Robert V. Miner,[2] and David N. Robinson[3]

Thermomechanical Deformation Behavior of a Dynamic Strain Aging Alloy, Hastelloy X®

REFERENCE: Castelli, M. G., Miner, R. V., and Robinson, D. N., **"Thermomechanical Deformation Behavior of a Dynamic Strain Aging Alloy, Hastelloy X®,"** *Thermomechanical Fatigue Behavior of Materials, ASTM STP 1186,* H. Sehitoglu, Ed., American Society for Testing and Materials, Philadelphia, 1993, pp. 106–125.

ABSTRACT: An experimental study was performed to identify the effects of dynamic strain aging (solute drag) and metallurgical instabilities under thermomechanical loading conditions. The study involved a series of closely controlled thermomechanical deformation (TMD) tests on the solid-solution-strengthened nickel-base superalloy, Hastelloy X.®[4] This alloy exhibits a strong isothermal strain aging peak at approximately 600°C promoted by the combined effects of solute drag and precipitation hardening. Macroscopic thermomechanical hardening trends are correlated with microstructural characteristics through the use of transmission electron microscopy. These observations are also compared and contrasted with isothermal conditions. Thermomechanical behavior unique to the isothermal database is identified and discussed. The microstructural characteristics were shown to be dominated by effects associated with the highest temperature of the thermomechanical cycle. Results clearly reveal that the deformation behavior of Hastelloy X is thermomechanically path dependent. In addition, guidance is given pertaining to deformation modeling in the context of a macroscopic unified thermoviscoplastic constitutive theory. An internal state variable is formulated to qualitatively reflect the isotropic hardening trends identified in the TMD experiments.

KEY WORDS: thermomechanical deformation (TMD), dynamic strain aging, thermoviscoplastic modeling, solute drag, Hastelloy X®, precipitation hardening, microstructural instabilities

Introduction

Structural components used in high temperature applications will often experience inelastic deformation resulting from mechanical loading, thermal transient cycles, and thermal gradients. Such forms of combined thermal and mechanical loading can potentially promote material behavior which is different from that exhibited under idealized isothermal conditions. The potential for these differences is greatly increased if the material is inclined to experience various temperature-stress-time dependent microstructural changes, such as those promoted by thermal, static, and dynamic aging mechanisms.

One such mechanism of key interest in this investigation is termed dynamic strain aging [*1*]. Dynamic strain aging (solute drag) effects are exhibited by many alloys of common use, including austenitic stainless steels and many Co- and Ni-base alloys. At temperatures where this phenomenon is present, the material will experience a relative increase in flow stress and work hardening rate, and various forms of inhomogeneous deformation. For the Ni-base

[1] Research Engineer, Sverdrup Technology Inc., LeRC Group, Brook Park, OH 44142.
[2] Chief, Advanced Metallics Branch, NASA Lewis Research Center, Cleveland, OH 44135.
[3] Professor, Department of Civil Engineering, University of Akron, Akron, OH 44325.
[4] Hastelloy X is a trademark of Haynes International, Inc., Kokomo, IN.

superalloy, Hastelloy X®, dynamic strain aging has been noted to contribute to complex isothermal hardening trends [*2*], raising obvious questions about its influence under more prototypical thermomechanical conditions.

In addition, many materials prone to dynamic strain aging effects are also subject to various forms of metallurgical instabilities. Typically, such instabilities result in the precipitation of a new microstructural phase [*3–7*], the introduction of which can promote deformation behavior different from that observed prior to the precipitation. Microstructural changes of this nature often lead to an increase in dislocation density and dislocation pinning, thereby enhancing the strain hardening behavior of the material. Consequently, at any given isothermal loading condition, the combined effects of precipitation hardening and dynamic strain aging can produce a highly complex macroscopic hardening behavior. This behavior is, in turn, potentially further complicated through the introduction of thermomechanical conditions.

This work examines the hardening response of a dynamic strain aging alloy under thermomechanical loading. A series of closely controlled thermomechanical deformation (TMD) tests were conducted on Hastelloy X. Hastelloy X was chosen as a model material because of its common use in high temperature power generation applications. Also, the strong dynamic strain aging effects in Hastelloy X are experienced at temperatures of most common use. Isothermal and thermomechanical material responses are examined and compared on the basis of deformation trends and microstructural observations. The microstructural mechanisms responsible for the cyclic hardening will be briefly discussed; however, a more detailed examination of the relative contributions between solute drag and precipitation hardening in Hastelloy X is addressed elsewhere [*8*]. Observations made from both the macroscopic and microscopic arenas are linked to help explain unique phenomenological trends experienced under thermomechanical conditions.

Upon gaining an understanding of the microstructural physics and their influence on TMD behavior, an extension is proposed to an established macroscopic unified thermoviscoplastic constitutive theory [*9,10*] to qualitatively reflect the observed trends. This extension incorporates an internal state variable to track the phenomenological effects of precipitation hardening and dynamic strain aging. The proposed framework is established with emphasis placed on modeling the isotropic hardening behavior.

Material

Hastelloy X is a nickel-base solid-solution-strengthened alloy with good corrosion and oxidation resistance at temperatures up to 1100°C. Its composition is nominally 22Cr, 18.5Fe, 9Mo, 0.6W, 1.5Co, and 0.1C, all in weight percent, with the balance being Ni. Typically supplied in a solution treated state, the as received microstructure consists of an FCC solid-solution matrix with a sparse distribution of Mo-rich M_6C carbides. The material heat specific to this investigation was produced in accordance with Aerospace Material Specification 5754H. Mechanical test specimens were machined from solution treated 19 mm diameter bar stock.

Hardening Mechanisms in Hastelloy X

Dynamic Strain Aging

The phenomenon of dynamic strain aging [*1*] occurs in solid solutions where solute atoms are particularly free to diffuse (either interstitial or substitutional solutes may contribute at appropriate temperatures) through the parent lattice. It is energetically preferable for these solute atoms to occupy sites in the vicinity of dislocations where they form solute (Cottrell) atmospheres [11] around the dislocations. This temporary solute atmosphere impedes subsequent dislocation movement and thus causes strengthening, as revealed by a relative increase in flow stress. With increasing strain rate, the dislocations begin to move too fast to allow the atmosphere to diffuse with them and as a result, the solute atmosphere becomes progressively more dilute. The overall effect reduces the solute drag force with increasing dislocation velocity, giving rise to an inverse strain rate/flow stress sensitivity. Dynamic strain aging is also associated with several forms of inhomogeneous deformations, termed Portevin-le Chatelier effects [12]. Mechanisms promoting such behavior include the sudden creation and/or mobilization of formerly pinned dislocations. These fast moving dislocations (free of solute atmospheres) give rise to sudden stress drops, more commonly referred to as jerky flow or serrated yielding.

Cyclic loading at temperatures where dynamic strain aging mechanisms are active results in marked material hardening. This behavior is evidenced by an increased hardening rate and hardening range. For any given dynamic strain aging alloy, an intermediate temperature range exists where the hardening rate and cyclic saturation strength is maximized. This maxima is interpreted as a manifestation of dynamic strain aging, and is typically referred to as the dynamic strain aging peak. At temperatures below the dynamic strain aging range, solute diffusion will not take place. At temperatures above, normal thermal recovery processes (for example, dislocation climb) will dominate.

Precipitation Hardening

In general, many alloys experience various forms of microstructural instabilities leading to the precipitation of a wide variety of new phases. However, to maintain applicability to the work at hand, this brief generalized discussion will be limited to the precipitation of the $M_{23}C_6$ carbide phase in Hastelloy X. In the solution treated condition, Hastelloy X is found to contain only the solid solution matrix and a sparse distribution of Mo-rich M_6C carbides. The M_6C phase is unstable in a Cr-rich environment and consequently, thermal aging at intermediate temperatures results in the precipitation of a Cr-rich $M_{23}C_6$ carbide phase [13,14]. In addition to the effects associated with a static thermal age, experiments [7] have suggested that a significant amount of deformation-assisted $M_{23}C_6$ precipitation occurs as a result of the dislocations' powerful effect on lowering the nucleation barrier [15] and becoming preferred carbide nucleation sites.

In its initial state, the $M_{23}C_6$ phase is relatively small (10 to 100 mm [8]) and highly effective at dislocation blocking and pinning; hence notable material hardening is experienced. Under isothermal loading conditions, the peak effectiveness of the precipitation hardening in Hastelloy X occurs in the temperature range of about 500 to 650°C, thus partially overlapping the 200 to 700°C range of dynamic strain aging. This precipitation strain aging peak is also typically associated with a ductility minimum. Aging at temperatures above the precipitation strain aging peak, the $M_{23}C_6$ carbide undergoes considerable growth. In this enlarged, overaged state, the phase appears to be considerably less effective as a hardening mechanism.

Test Methods

Cyclic Isothermal Deformation

Isothermal data taken from a previous experimental study conducted by Ellis et al. [2] are presented to enable comparisons with the TMD data generated in this study. All isothermal deformation tests were performed in air under strain control with constant cyclic axial-strain amplitudes of ± 0.003. A constant strain rate of 0.0001 s^{-1} was used with a triangular command waveform. Strains were controlled over a 25.4 mm parallel gage section with a high

TABLE 1—*Thermomechanical test matrix.*

	Temperature Range, °C					
Phase	200 to 400	400 to 600	600 to 800	800 to 1000	300 to 800	300 to 1000
In-phase	1	1	1	1	1	—
Out-of-phase	1	1	1	1	—	1

temperature water cooled axial extensometer. Specimen heating was accomplished through the use of direct induction coils. Isothermal conditions from room to 1000°C were investigated. Temperature gradients were maintained in accordance with ASTM Practice for Constant-Amplitude Low-Cycle Fatigue Testing (E 606). Unless otherwise noted, the isothermal tests were terminated at 10^4 cycles or evidence of the first macro-crack, whichever occurred first.

Cyclic Thermomechanical Deformation

The thermomechanical test matrix is shown in Table 1. A series of in-phase and out-of-phase tests were conducted with temperature ranges of 200°C. Two relatively large temperature ranges were also examined to establish the effects of cycling completely through the strain aging peak regime. Phasing is defined by the relationship between the mechanical strain waveform and the temperature waveform. Thus, a test is defined to be "in-phase" when these two waveforms are coincidental, and "out-of-phase" when the two waveforms are phased with a time shift equal to ½ the cycle period. The mechanical strain (ε^m) is defined as the difference between the total (e^t) and thermal (ε^{th}) strains

$$\varepsilon^m = \varepsilon^t - \varepsilon^{th}$$

All TMD tests were conducted in air with cyclic mechanical strain amplitudes of ± 0.003 and a constant temperature rate of 100°C/min, independent of temperature range. Given the fixed mechanical strain range and temperature rate, the mechanical strain rate was calculated based on temperature range. Note that the strain rate for the thermomechanical tests (with $\Delta T = 200$°C the mechanical strain rate is 5.0×10^{-5}/s) was slightly less than that used for the isothermal data taken from [*2*] (1.0×10^{-4}/s). The effects of strain rate from 1.0×10^{-3}/s to 3.0×10^{-5}/s were examined for this material (unpublished data by Allen et al. at NASA Lewis) and found to have a negligible effect on the magnitude of isotropic hardening. Therefore, the factor-of-two difference between the thermomechanical and isothermal data was assumed to have a negligible effect on the isotropic hardening behavior.

The test specimens were heated with direct induction heating. Specimen cooling was accomplished through the use of water cooled grips; no forced air was used. A triangular waveform was used for both the temperature and strain. The strains were controlled over a 12.7 mm parallel gage section with a high temperature water-cooled axial extensometer. A thin-walled tube specimen design was developed and used for this study. Details and the relative advantages of this specimen design under thermomechanical conditions can be found in Ref *16*.

Microstructural Examination

Specimens for transmission electron microscopy (TEM) were obtained from the wall in the gage section of the tubular test specimens by electrodischarge machining. The specimens were prepared by fine grinding and electropolishing to perforation. Electropolishing was accomplished in a solution of 10% Perchloric acid, 45% Butanol and 45% Methanol at 0°C and 30 V.

Results and Discussion

Cyclic Isothermal Deformation

Shown in Fig. 1 are plots of cyclic peak stress versus cycle number, revealing the isothermal cyclic hardening behavior of Hastelloy X. The isothermal data did not reveal a noteworthy mean stress during fully-reversed strain cycling; thus the peak stress values shown are representative of the excursions from zero in both tension and compression. Figure 1 displays hardening curves over three generalized temperature domains approximately bounded by room temperature-200, 200 to 650, and 650 to 1000°C, subsequently referred to as "low," "middle," and "high," respectively.

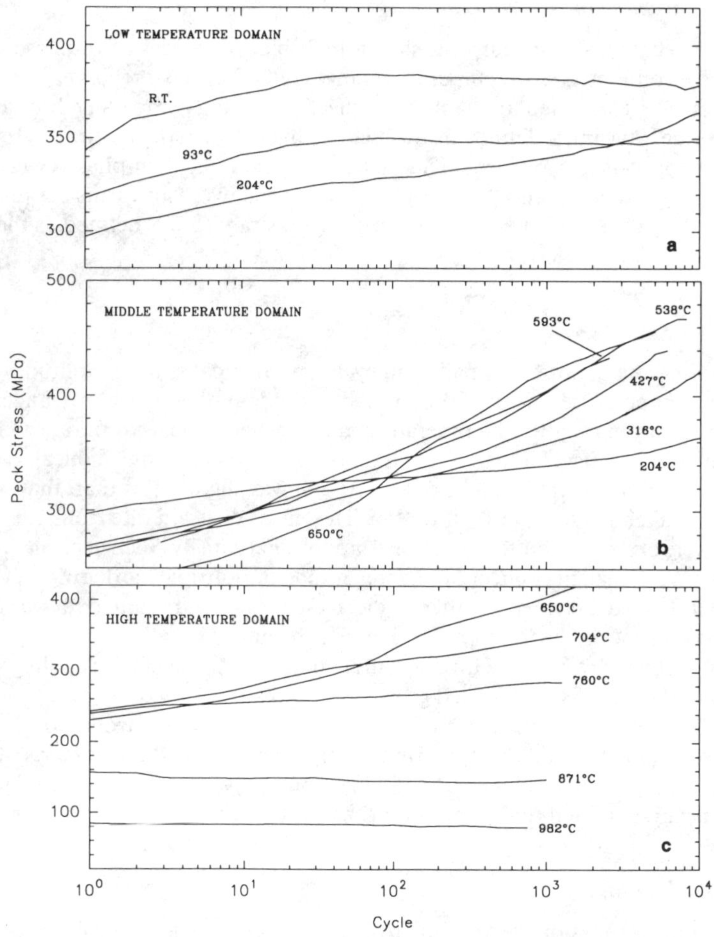

FIG. 1—*Isothermal hardening behavior of Hastelloy X with* $\varepsilon = \pm 0.003$ *and* $\dot{\varepsilon} = 0.0001/s$ [2] *over three generalized temperature domains:* (a) *low,* (b) *middle, and* (c) *high.*

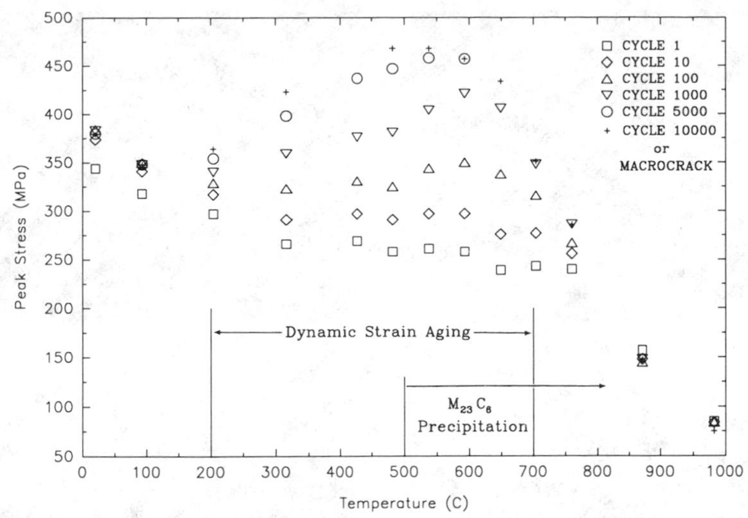

FIG. 2—*Isothermal data from Fig. 1 plotted to illustrate the strain aging peak.*

In the low temperature domain, the isothermal cyclic response of Hastelloy X is essentially neutral with respect to hardening. Peak stress increases of only 10 to 15% (with respect to first cycle values) were realized before a state of cyclic saturation was reached.

This trend quickly changed as temperatures increased into the middle temperature domain, Fig. 1*b*, encompassing the strain aging peak. Tests performed at 316°C and 427°C exhibit an increase in peak stress of 60 and 66%, respectively. Hardening continues to increase through the middle temperature domain until maximums are realized at 500 to 650°C. Here, increases greater than 85% can be experienced. Serrated yielding was noted in a surprisingly small percentage of the specimens tested in this temperature domain. However, other evidence indicating dynamic strain aging, such as an inverse strain rate/flow stress sensitivity, were consistently and clearly identified. Further, the dynamic strain aging range is felt to begin at temperatures as low as 200°C, where marked hardening and increased dislocation densities were observed [*8*].

The transition from the middle to the high temperature domain occurs when the strain aging peak temperature is exceeded. In the high temperature domain, Fig. 1*c*, cyclic hardening drops significantly with a relatively small increase in temperature. The percent increase in peak stress drops from over 85 at 650°C to 44 at 704°C. The decrease continues until cyclic softening occurs during tests performed at 871°C and higher temperatures.

The phenomenological hardening effects of dynamic strain aging on Hastelloy X are best summarized in Fig. 2. Peak stresses are plotted at intermittent cycles over the full working temperature range. The test conducted at 427°C was terminated prematurely at approximately 5000 cycles because of a heater problem. This figure clearly reveals that the strain aging is maximized in the general temperature range of 400 to 600°C.

Isothermal Microstructures

TEM was used to identify the physical characteristics associated with the various hardening behaviors exhibited in the isothermal tests. Examination revealed general microstructural features uniquely associated with the three temperature domains (low-middle-high) discussed

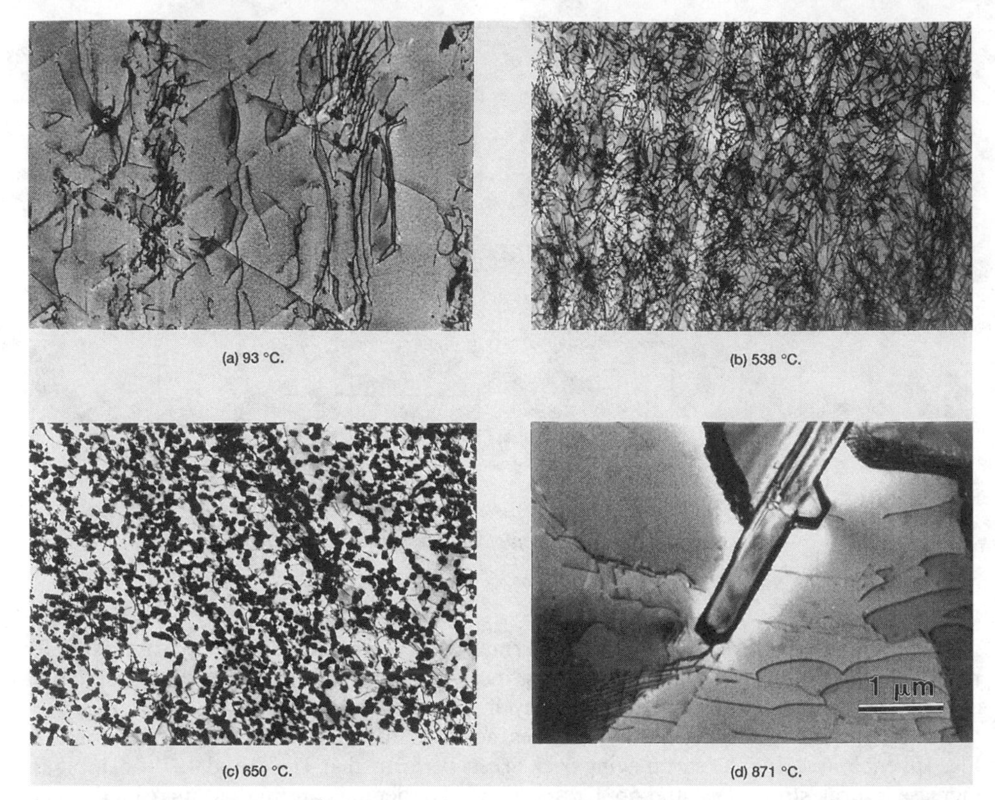

FIG. 3—*Microstructures developed under isothermal loading conditions at* (a) *93°C,* (b) *538°C,* (c) *650°C, and* (d) *871°C. See Fig. 1 for respective number of loading cycles.*

above. Dislocation substructures representative of these features were taken from specimens tested at 93, 538, 650, and 871°C and are shown in Fig. 3.

Micrographs taken from specimens tested in the low temperature domain, such as that shown in Fig. 3*a*, reveal a relatively low dislocation density. The imposed temperature and mechanical loading conditions do not appear to have affected the stability of the M_6C carbide phase (present in the as-received condition), as no new phases can be identified. Slip is inhomogeneous and the dislocations are found primarily concentrated in a few planar slip bands. These general features are consistent with the relatively low cyclic hardening observed in the low temperature domain.

In the specimens examined from the middle temperature domain, there are two distinct physical changes observed, (*1*) a dramatically increased dislocation density and (*2*) the precipitation of the $M_{23}C_6$ phase. The first visually detectable change in the microstructure is a relative increase in dislocation density, such as that revealed in Fig. 3*b*. Dislocation densities (at 10^4 cycles or failure, whichever occurred first) steadily increase from the values found at 200°C to those observed at temperatures of 500 to 600°C, representing the maximum values (quantitative results are addressed elsewhere [*8*]). This increased dislocation production can result from either hardening mechanism. Both (*1*) dynamic strain aging (solute drag) and (*2*) $M_{23}C_6$ precipitation, will hinder the movement of existing dislocations and enhance the production of additional dislocations to accommodate the applied strain.

Given that the dynamic strain aging effects have been identified over the temperature range of 200 to 700°C, it is important to note at what point precipitation of the $M_{23}C_6$ phase begins;

this information is critical to the understanding of the thermomechanical deformation. TEM observations and mathematical modeling of the $M_{23}C_6$ precipitation kinetics [*8*] both agree and show with certainty that this secondary carbide phase did not precipitate during the isothermal tests conducted below 500°C. Therefore, the marked hardening experienced below this temperature, such as that observed at 316, 427, and 482°C can be attributed to the effects of dynamic strain aging (see Fig. 2). At temperatures above 500°C, $M_{23}C_6$ precipitation will occur (given sufficient time) and the hardening will be influenced by the combined effects of dynamic strain aging and precipitation hardening.

An examination of the 650°C microstructure, shown in Fig. 3*c*, reveals a slightly reduced dislocation density and a slightly larger $M_{23}C_6$ phase than those found in the 538 and 593°C specimens. At this size (≈60 nm in diameter), the $M_{23}C_6$ phase is visible in a conventional bright-field image, pictured in Fig. 3*c* as the numerous dark circular objects. This phase is present but not visible in Fig. 3*b* because of the extremely small carbide size (≈5 nm in diameter). The increased size of the $M_{23}C_6$ carbide found at 650°C suggests an early stage of microstructural overaging.

Microstructures of specimens tested in the high temperature domain ($T > 650$°C) begin to reveal features of thermally activated processes. The high temperature conditions promote dissolution of the original M_6C carbides, which become less numerous. The $M_{23}C_6$ carbides have undergone considerable physical growth and appear concentrated at cell boundaries and twin boundaries. At the higher temperatures of this domain, Fig. 3*d*, large cells and subcells appear with low intracellular dislocation densities. The majority of dislocations are located at cell and twin boundaries, displaying well ordered arrangements.

Cyclic Thermomechanical Deformation

Thermomechanical and isothermal data comparisons are made by presenting the thermomechanical data along with "bounding" isothermal data, where "bounding" refers to the upper and lower test temperatures. For ease of presentation, all peak stresses are plotted according to their absolute values.

Displayed in Fig. 4 are results from TMD tests performed with a temperature range of 200

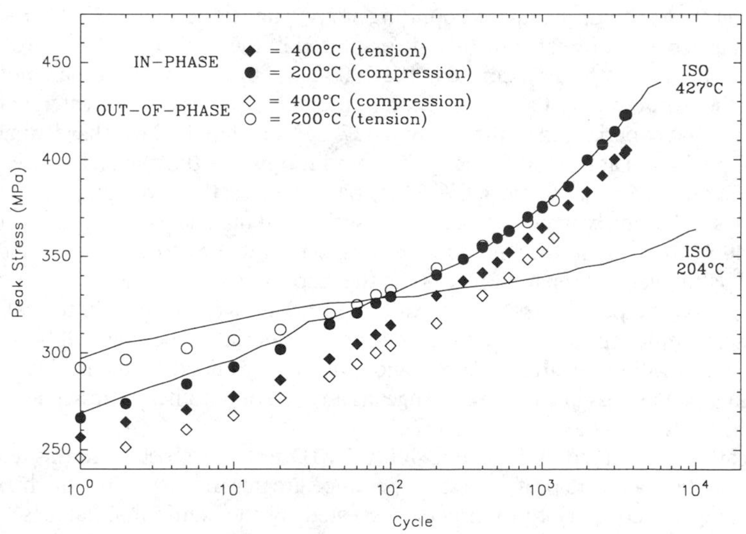

FIG. 4—*Peak stress versus cycle plot for 200 to 400°C in- and out-of-phase TMD and bounding isothermal tests.*

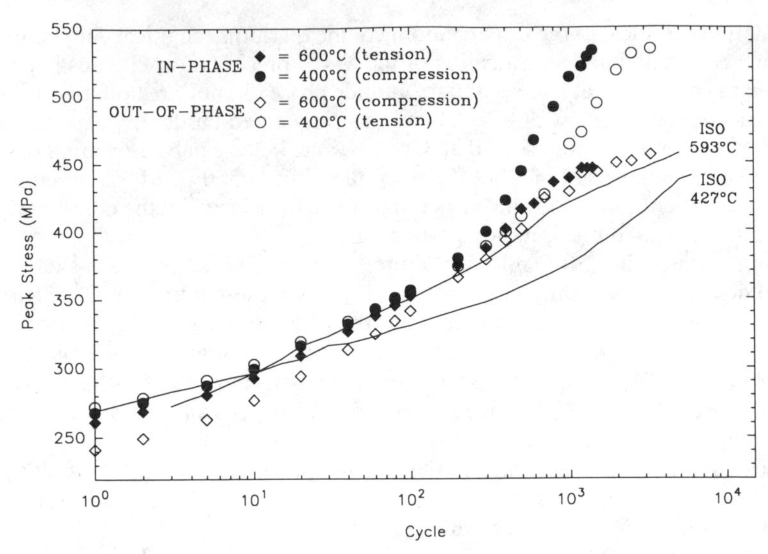

FIG. 5—*Peak stress versus cycle plot for 400 to 600°C in- and out-of-phase TMD and bounding isothermal tests.*

to 400°C. The six curves represent data taken from four tests. The in-phase and out-of-phase TMD tests are each represented by two curves; one represents the peak tensile stresses and the other represents the peak compressive stresses. The two remaining curves are the bounding isothermal peak stresses.

As a first approximation, it seems reasonable to anticipate that the TMD hardening trends would be bound by the bounding isothermal data. In fact, one might expect the hardening to be similar to that displayed isothermally at the average temperature of the thermomechanical cycle. This "anticipated" scenario is a fairly accurate description of what occurred at the 400°C TMD peak stresses. The 400°C TMD peak stresses (both in- and out-of-phase conditions) increased slightly less than the upper bounding curve (the isothermal 427°C data). Also, the hardening rate is best represented by the rate exhibited in the isothermal 427°C test.

In contrast, the peak stresses incurred at the 200°C end of the TMD cycle are not well represented by the isothermal test at 204°C. Rather, the hardening is much greater, and actually very similar to that experienced isothermally at 427°C. Note that the TMD hardening is greater at the 200°C peak than at the 400°C peak. This trend is opposite that which is suggested by the isothermal database. Similar to the 400°C TMD peak stresses, the hardening rate at the 200°C TMD peaks is well approximated by the 427°C isothermal rate. The hysteresis experienced in the 200 to 400°C tests exhibited a mild mean stress which tended to decrease with cycling, as represented by the vertical separation between the respective peak stresses.

Figure 5 shows the peak stress values exhibited under TMD in the 400 to 600°C experiments. This particular condition promoted dramatic material hardening well beyond that of any other condition investigated. The behavior was not bounded by the isothermal data in terms of either peak stress or total stress range, as a result of both peak stresses surpassing the greater of the isothermal conditions (593°C).

Certain patterns observed in the 200 to 400°C TMD tests are repeated in this temperature range. The low-temperature peak stresses experience greater increases than the high-temperature values, again, contrary to the trends suggested by the isothermal database. Also, the TMD hardening rates are better represented by the hot-temperature bounding isothermal test,

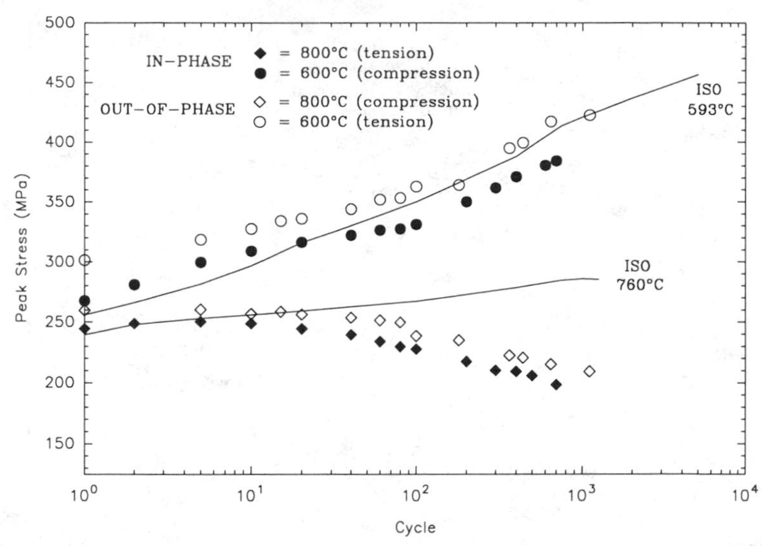

FIG. 6—*Peak stress versus cycle plot for 600 to 800°C in- and out-of-phase TMD and bounding isothermal tests.*

although the 400°C TMD peak stresses eventually surpasses both the hardening rate and range of the 593°C isothermal test.

The 600°C TMD peak stresses are well represented by those in the bounding isothermal test at 593°C. This applies to both the hardening rate and magnitude. However, the 400°C TMD peak stresses clearly show significant deviation from the 427°C isothermal test. Further, the hardening rates and magnitudes experienced by the 400°C TMD peaks far exceed those displayed isothermally at 593°C. This result has added significance because the 593°C data is representative of the isothermal strain aging peak; thus the hardening behaviors observed in the 400 to 600°C TMD tests exceed the entire isothermal database to the degree shown in Fig. 5. This unbounded behavior is clear evidence of thermomechanical path dependence, as the material's behavior at 400°C is profoundly influenced by the previous thermomechanical history.

Shown in Fig. 6 are the data obtained from in-phase, out-of-phase, and bounding isothermal tests in the temperature range of 600 to 800°C. Under this condition the isothermal data is successful at bounding the TMD material response. The 600°C peak stresses of the TMD agree well with the rate and magnitude observed in the isothermal 593°C experiment. Likewise, the hardening (or in this case softening) behavior experienced at the hot end of the TMD cycle is comparable to that observed during isothermal deformation. Although the 600°C TMD peak stress values experience the excessive increase associated with the isothermal strain aging peak, the total hardening range is subdued because the majority of the loading involves higher temperatures. At these slightly higher temperatures the hardening mechanisms quickly become less effective and thermal recovery processes begin to dominate. Note the large mean stress that develops as a result of conditions which promote hardening at one end of the cycle and softening at the other.

The final TMD condition examined with a temperature range of 200°C is that of 800 to 1000°C and is shown in Fig. 7. The thermomechanical hardening trends are typified by the isothermal data, with respect to both stress magnitude and hardening rate. At these high homologous temperatures the material will tend to flow freely and offer little resistance to

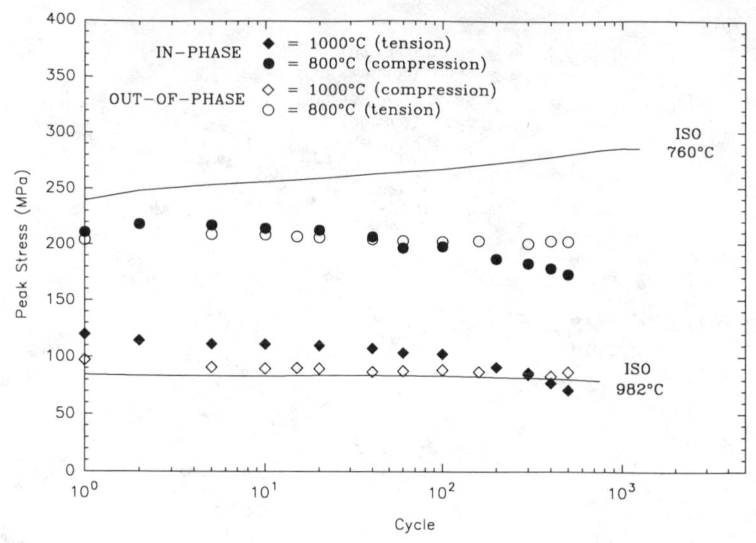

FIG. 7—*Peak stress versus cycle plot for 800 to 1000°C in- and out-of-phase TMD and bounding isothermal tests.*

mechanical deformation, hence the response is essentially saturated in the first few cycles. Here thermal recovery effects dominate the behavior.

The two remaining TMD conditions are unique in the sense that they examine the effects of cycling completely through the dynamic strain aging regime. These two conditions are 300 to 800°C and 300 to 1000°C TMD, shown in Figs. 8 and 9, respectively.

The 300 to 800°C TMD resulted in hardening trends which again show evidence of ther-

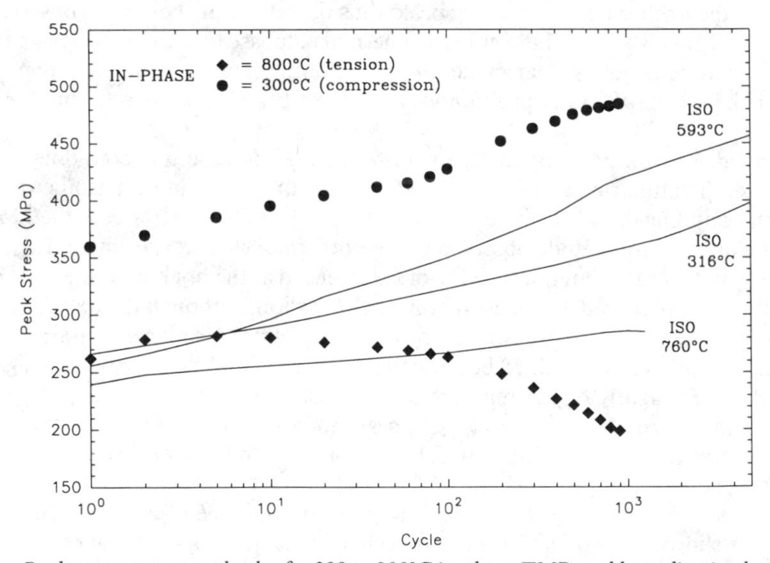

FIG. 8—*Peak stress versus cycle plot for 300 to 800°C in-phase TMD and bounding isothermal tests.*

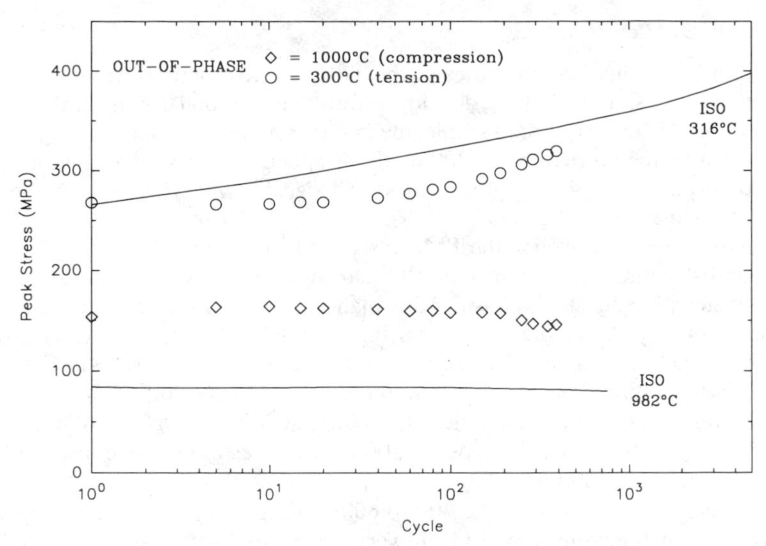

FIG. 9—*Peak stress versus cycle plot for 300 to 1000°C out-of-phase TMD and bounding isothermal tests.*

momechanical path dependence. The hardening behavior at the 300°C peak stresses was similar only in rate to that experienced isothermally at 316°C. In contrast, the magnitude of hardening was found to be much greater than the isothermal 316°C data. As shown in Fig. 8, the peak stresses were considerably higher than those found at the isothermal dynamic strain aging peak temperature of 593°C. These high TMD peak stresses at 300°C resulted from a continually increasing mean stress. The steady softening experienced at the 800°C end of the cycle appears to have enhanced a stress ratchetting effect in the direction of the 300°C end, enabling the relatively high stresses to be achieved in the fixed mechanical strain range. However, it is important to note that unlike the 400 to 600°C TMD, the total stress range experienced is not atypical. In fact, the stress range experienced by the 300 to 800°C TMD is comparable to that observed under isothermal conditions at 316°C.

The TMD cycle from 300 to 1000°C is shown in Fig. 9. Although the full dynamic strain aging range was spanned, it appears that the upper temperature of 1000°C had a dominant effect on the overall deformation behavior. This condition revealed a relatively small amount of cyclic hardening at 300°C and mild softening at 1000°C. In contrast to the 300 to 800°C TMD, a strong mean stress did not develop because of the lack of hardening by the 300°C peak stress values.

In general, thermomechanical hardening rates appear to be dominated by effects associated with the upper-temperature of the cycle. In TMD loading cases where the isothermal data suggests that the upper-temperature condition hardens at a higher rate (200 to 400°C and 400 to 600°C), both TMD peak stresses (tensile and compressive) increased at the higher rate. In TMD cases where the isothermal data suggests that the upper-temperature condition is inclined to harden less (all remaining TMD conditions), the peak stresses revealed rates of increase similar to their isothermal counterparts, and the hardening rate of the total stress range was better reflected by the upper-temperature isothermal test. In addition, no significant differences or obvious trends were revealed between in-phase and out-of-phase hardening behaviors. A possible exception may exist at the 400 to 600°C test condition (Fig. 5) where a slight deviation at the 400°C peak was observed.

Thermomechanical Microstructures

A description of the physical attributes of the TMD microstructures will be brief, as they were found to be very similar to those developed during isothermal cycling at the maximum temperature of the TMD cycle. For example, the microstructure developed during a TMD test conducted with a temperature range of 200 to 400°C appeared very similar to that developed during an isothermal test at 400°C. With this knowledge, we will briefly review the thermomechanical behavior.

The microstructures developed in the TMD tests conducted from 200 to 400°C consisted of relatively high dislocation densities as a result of strong dynamic strain aging effects (but no $M_{23}C_6$). This dislocation density is considerably higher than that found isothermally at 204°C where the hardening effects of dynamic strain aging are much reduced. The TMD test develops a "400°C microstructure" which is then subjected to loading at temperatures as low as 200°C. Given this high dislocation density, the hardening exhibited by the 200°C TMD peak stresses is a reasonable result. Also, since the material response at 200°C is slightly stiffer than that at 400°C ($E_{200°C} > E_{400°C}$), the magnitudes of the 200°C peak stresses remain consistently higher than those experienced at 400°C (see Fig. 4).

The microstructures developed in the 400 to 600°C TMD tests are similar to those developed during isothermal loading at 593°C and very similar to that which is shown in Fig. 3*b*. This microstructure contains small precipitated $M_{23}C_6$ carbides which serve to effectively pin dislocations, and at this size clearly add to the hardening mechanisms of dynamic strain aging (solute drag). The extreme "unbounded" hardening can be at least qualitatively rationalized on the basis of this microstructure and loading temperature.

Recall that isothermal deformation below 500°C did not promote precipitation of the $M_{23}C_6$, thus, the significant hardening (shown in Fig. 2) at 427 and 482°C is an indication of the effectuality of solute drag without the aid of the $M_{23}C_6$ phase. The 400 to 600°C TMD condition is unique in the sense that this small carbide is introduced into the microstructure (requiring temperatures above 500°C) and then subjected to loads at temperatures in the vicinity of 400°C, where the solute drag effects remain very strong and the material has greater stiffness. The result (shown in Fig. 5) is extreme hardening, as revealed by the 400°C TMD peak stress values. This condition of "small carbides" in a microstructure loaded at 400°C cannot be represented in a strict isothermal test and as a result, the isothermal database is not capable of accurately representing the thermomechanical response.

In contrast, the 600°C peak stresses from the 400 to 600°C TMD tests exhibit hardening trends similar to that displayed isothermally at 593°C. This is a sensible result, given that the microstructures developed in the TMD tests are indistinguishable from that which was developed in the 593°C isothermal test.

Microstructures taken from TMD tests with upper temperatures of 800 or 1000°C exhibit the typical overaging effects cited earlier in the high temperature isothermal tests. This was the case even for the 300 to 800°C TMD microstructure, where the vast majority of the cyclic loading occurs at temperatures below which overaging effects occur. In the TMD conditions involving 1000°C peaks, the microstructures were very similar to that which is shown in Fig. 3*d*. This state clearly supports the lack of cyclic hardening experienced by the 300 to 1000°C TMD specimen.

Thermoviscoplastic Modeling

Having established a basic understanding of the microstructural changes and loading conditions leading to the "unbounded" thermomechanical hardening, it is desirable to represent this behavior in the context of a macroscopic deformation theory. The approach taken is to extend an established unified thermoviscoplastic constitutive theory proposed by Robinson [9,10]. Specifically, an evolutionary law for an internal state variable β is formulated to account for the isotropic hardening effects induced by precipitation hardening and dynamic strain aging under thermomechanical conditions. This proposed state variable β is incorporated into the cyclic (isotropic) hardening variable (that is, the drag stress, K) in the Robinson Model. As a result, the modeling discussion will be limited to a concise presentation of the drag stress K, and the proposed extension β; however, for convenience, a brief description of the mathematical framework is given in Appendix A.

The present modeling objective is intended as guidance and one of presenting a mathematical formulation capable of qualitatively representing the observed trends; the objective is not one of establishing a fully characterized form for the purposes of quantitative predictions. The approach presented represents one possible methodology; other approaches to modeling deformation behavior in the presence of metallurgical changes and instabilities can be found in Refs *17* and *18*.

Isotropic Hardening

The drag stress K is a scalar internal state variable associated with the isotropic hardening of the material, and physically represents a measure of the microstructural dislocation density. The evolutionary law, specifying the growth of K, is obtained by differentiating a chosen dissipation potential function Ω with respect to K (see Appendix A), resulting in

$$\dot{K} = -H(K, T)\frac{\partial \Omega}{\partial K} = \Gamma(K, T)\dot{W} \tag{1}$$

where

$$\dot{W} = \Sigma_{ij}\dot{\varepsilon}_{ij}$$

This form indicates that the evolution of K is proportional to the effective inelastic work W where the proportionality is affected by the current temperature T. More recently, this evolution law was extended [*19*] by the inclusion of a nonassociative term Y, yielding

$$\dot{K} = \Gamma(K, T)\dot{W} + Y(T)\dot{T} \tag{2}$$

The term Y was introduced to allow a spontaneous change in the drag stress resulting from a change in temperature (that is, a spontaneous change in the yield surface associated with a change in temperature). The need for this spontaneous change was clearly revealed by a unique set of "temperature-step" experiments conducted on Hastelloy X [*19*].

The functions Γ and Y are as follows:

$$\Gamma(K, T) = \Gamma_0(T)_p(Z), \quad \Gamma(T)_p(0) = 0$$

where

$$Z(K, T) = \frac{K_s(T) - K}{K_s(T) - K_0(T)} \quad \text{and} \quad K_0(T) = K_0(T_*) - [1 - e^{-Q_0(T_*^{-1} - T^{-1})}] \tag{3}$$

$$Y(T) = \frac{-Q_0}{T}\, e^{-Q_0[T_*^{-1} - T^{-1}]} \tag{4}$$

The function $K_0(T)$ is the initial value of the state variable K (before mechanical cycling) and $K_s(T)$ is the saturated value of K after extended cycling. The function $\Gamma_0(T)$ is a measure of the rate of hardening. The value T_* is a reference temperature, and Q_0 is an "activation energy." The function $p(Z)$ can be taken as

$$p(Z) = AZ^n \quad \text{or} \quad p(Z) = A[1 - e^{-z}]$$

By selecting appropriate forms for $K_s(T)$ and $\Gamma_0(T)$, such as those shown in Fig. 10, the isothermal hardening (strain aging) behavior of Hastelloy X can be represented. Specifically, the maximum hardening (K or $\Delta\sigma$) and hardening rate $\left[\frac{dK}{dW} \text{ or } \frac{d(\Delta\sigma)}{dN}\right]$ are experienced at some intermediate "peak" hardening temperature T_p.

Proposed β Extension

Two general trends observed in the TMD data are in need of representation. First, the thermomechanical hardening must be capable of exceeding the baseline isothermal hardening under certain loading paths (for example, TMD at 400 to 600°C). Second, thermomechanical paths involving temperatures above the strain aging peak must allow for a form of thermal recovery of the strain aging effects. To this end, an internal state variable β is introduced to track the combined hardening effects of solute drag and $M_{23}C_6$ precipitation in Hastelloy X under thermomechanical conditions. An evolutionary law featuring "competing mechanisms" is proposed as follows

$$\dot{\beta} = \Psi(T)\dot{W} - \Phi(T) \tag{5}$$

where

$$\Psi(T) = Ae^{[-(T-T_p)^2/p]}$$

and

$$\Phi(T) = Be^{[Q_r(T_r^{-1} - T^{-1})]}$$

where T_p represents the strain aging peak temperature, Q_r and T_r are the activation energy and reference temperature for recovery effects, respectively, and A, B and ρ are material constants. The functions $\Psi(T)$ and $\Phi(T)$ are plotted in Fig. 11.

The hardening term $\Psi(T)$ is non-zero only near the strain aging peak temperature T_p. Thus, the first term in Eq 5 is also non-zero only when the temperature is in the vicinity of T_p. Further, this term contributes to β only if effective inelastic work is being incurred ($\dot{W} \neq 0$). This structure is consistent with the "dynamic" nature of the strain aging phenomenon and also represents the macroscopic hardening effects promoted by the presence of the $M_{23}C_6$ carbide phase, where time spent in the vicinity of the peak temperature accompanied by inelastic work enhance the precipitation of the carbide in the non-overaged state.

The function $\Phi(T)$ is an Arrhenius form that allows for thermal recovery of β when the temperature exceeds the peak temperature T_p and mechanisms such as dislocation climb become significant. Time at temperature above T_p leads to $\beta < 0$ and allows for partial or complete erasure of hardening that may have occurred by exposure to cycling at temperatures near T_p. Thus, the framework reflects the observed thermal recovery and microstructural overaging effects exhibited in Hastelloy X where the $M_{23}C_6$ carbides are seen to coalesce and enlarge, and well ordered dislocations are found at cell boundaries.

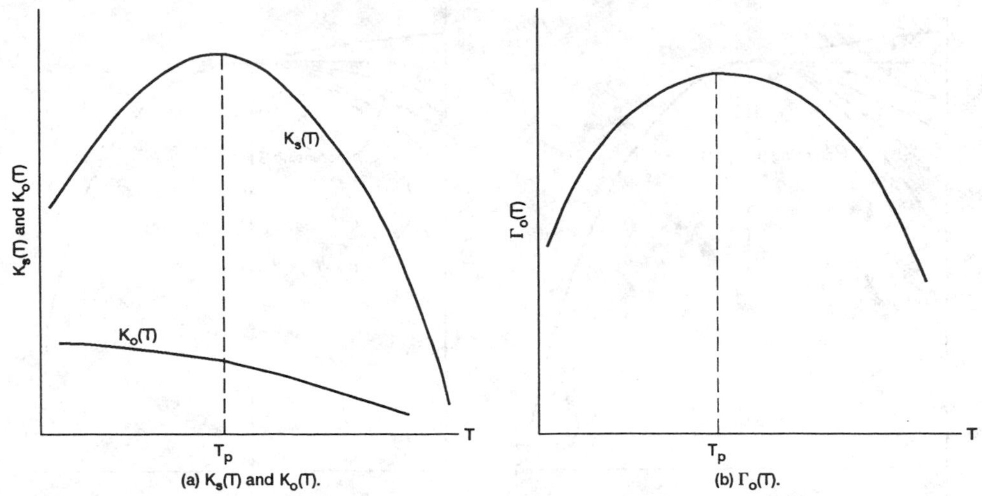

FIG. 10—*Appropriate forms of a)* $K_s(T)$ *and* $K_0(T)$*, and b)* $\Gamma_0(T)$ *which represent the isothermal isotropic hardening behavior associated with dynamic strain aging and precipitation hardening effects in Hastelloy X.*

The dependence of the drag stress on β enters in the evolution law (Eq 2) through the function Γ. Here, Γ is taken as $\Gamma(K,T,\beta)$ where dependence on β enters through the functions $K_s(\langle\beta\rangle,T)$ and $\Gamma_0(\langle\beta\rangle,T)$

$$\langle\beta\rangle = \begin{bmatrix} \beta; & \beta > 0 \\ 0; & \beta \leq 0 \end{bmatrix}$$

This dependence is proposed as illustrated in Fig. 12. With $\beta = 0$, K_s and Γ_0 will maintain their respective forms illustrated in Fig. 10. As β increases according to its evolution equation, K_s

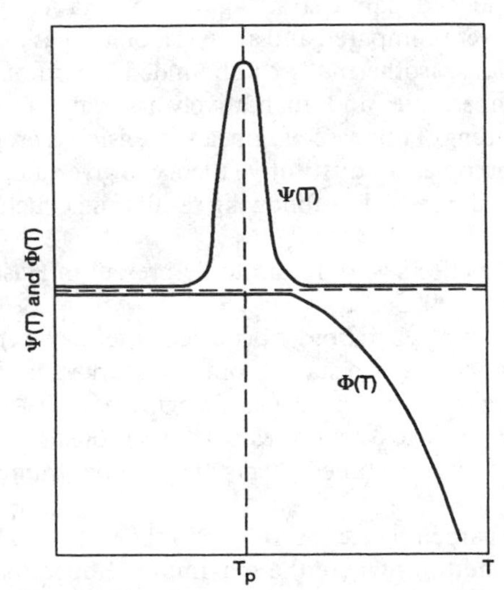

FIG. 11—*Appropriate forms of* $\Psi(T)$ *and* $\Phi(T)$ *capable of representing the unbounded thermomechanical isotropic hardening features associated with dynamic strain aging and precipitation hardening in Hastelloy X.*

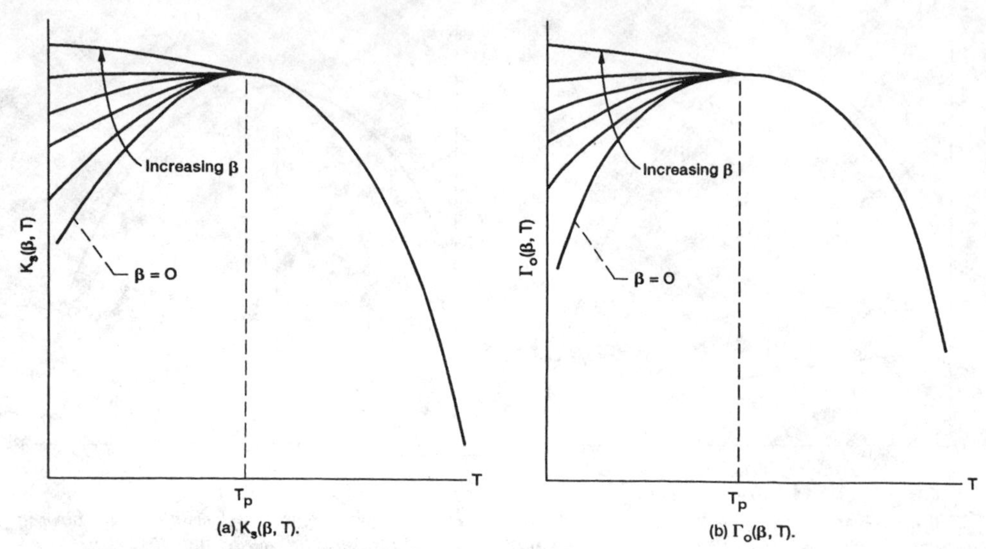

FIG. 12—*Evolutionary trends of* (a) $K_s(\beta,T)$ *and* (b) $\Gamma_0(\beta,T)$ *associated with increasing* β.

and Γ_0 evolve as shown in Fig. 12 and are capable of reflecting the isothermally "unbounded" thermomechanical hardening trends observed in this set of experiments. Further, this mathematical structure allows the thermomechanical hardening to be dominated by effects associated with the "hot" end of the cycle.

Summary and Conclusions

Closely controlled thermomechanical deformation tests were conducted to investigate the hardening behavior of a model dynamic strain aging alloy, Hastelloy X. Isothermal and thermomechanical behaviors were compared and analyzed on the basis of phenomenological and microstructural observations. Isothermally "unbounded" thermomechanical behavior was observed over certain temperatures and qualitatively associated to microstructural changes. With a working understanding of this association, an extension is proposed to an existing macroscopic unified thermoviscoplastic constitutive theory to give guidance and insight on modeling the thermomechanical trends. The following results and conclusions were established:

1. Dynamic strain aging effects were found to be present in Hastelloy X at temperatures ranging from ≈200 to 700°C.
2. Precipitation of Cr-rich $M_{23}C_6$ carbides occurred at temperatures above ≈500°C where their presence, in a non-overaged state, promoted marked hardening.
3. Isothermal hardening trends revealed a strain aging peak at ≈500 to 650°C with maximum increases above first cycle peak stresses in excess of 85%. These maxima were determined to be promoted by combined effects of dynamic strain aging and $M_{23}C_6$ precipitation hardening.
4. Thermomechanical hardening trends unique to the isothermal database were found to exist under cyclic conditions where the maximum temperature did not exceed that at which microstructural overaging began ($T_{max} \leq \approx 650$°C). Of the TMD conditions investigated, this included the 200 to 400 and 400 to 600°C cycles.

5. Microstructures of material subjected to TMD cycles were similar to those developed during isothermal cycling at the maximum temperature of the TMD cycle.
6. A mathematical structure was formulated as an extension to a macroscopic thermoviscoplastic deformation theory to reflect the thermomechanical path dependent isotropic hardening in Hastelloy X.

Acknowledgments

The authors wish to thank Rod Ellis, Steve Arnold, and Paul Bartolotta for their assistance and discussions concerning this research. Also, they wish to thank Chris Burke and Ron Shinn for their technical assistance in the high-temperature fatigue and structures laboratory at NASA Lewis Research Center.

References

[1] Baird, J. D., "Dynamic Strain Aging," *The Inhomogeneity of Plastic Deformation,* ASM, Metals Park, OH, 1973, pp. 191–221.
[2] Ellis, J. R., Bartolotta, P. A., Allen, G. P., and Robinson, D. N., "Thermomechanical Characterization of Hastelloy X Under Uniaxial Cyclic Loading," NASA CP-2444, 1986, pp. 293-305.
[3] Kotval, P. S. and Hatwell, H., "Discontinuous Precipitation of $M_{23}C_6$ Carbide in a Nickel-Base Superalloy," *Transactions of the Metallurgical Society of AIME,* Vol. 245, Aug. 1969, pp. 1821–1823.
[4] Lewis, M. H. and Hattersley, *Acta Metallurgica,* Vol. 13, 1965, pp. 1159–1168.
[5] Tu, K. N. and Turnbull, D., *Acta Metallurgica,* Vol. 15, 1967, pp. 369–376.
[6] Singhal, L. K. and Martin, J. W., *Transactions of the Metallurgical Society of AIME,* Vol. 242, May 1968, pp. 814–819.
[7] Arkoosh, M. A. and Fiore, N. F., "Elevated Temperature Ductility Minimum in Hastelloy Alloy X," *Metallurgical Transactions,* Vol. 3, Aug. 1972, pp. 2235–2240.
[8] Miner, R. V. and Castelli, M. G., "Hardening Mechanisms in a Dynamic Strain Aging Alloy, Hastelloy X, During Isothermal and Thermomechanical Cyclic Deformation," *Metallurgical Transactions,* Vol. 23A, Feb. 1992, pp. 551–562.
[9] Robinson, D. N., "A Unified Creep-Plasticity Model for Structural Metals at High Temperatures," ORNL TM-5969, 1978.
[10] Robinson, D. N. and Swindeman, R. W., "Unified Creep-Plasticity Constitutive Equations for Structural Alloys at Elevated Temperatures," ORNL TM-8444, 1982.
[11] Cottrell, A. H., *Philosophical Magazine,* Vol. 44, No. 355, 1953, pp. 829–832.
[12] Portevin, A. and le Chatelier, F., *Academie des Sciences—Comptes Rendus,* Vol. 176, Feb. 1923, pp. 507–510.
[13] Tawancy, H. M., "Long-term Aging Characteristics of Hastelloy X," *Journal of Materials Science,* Vol. 18, Oct. 1983, pp. 2976–2986.
[14] Lai, G. Y., "An Investigation of the Thermal Stability of a Commercial Ni-Cr-Fe-Mo Alloy (Hastelloy Alloy X)," *Metallurgical Transactions A,* Vol. 9, June 1978, pp. 827–833.
[15] Kotval, P. S., *Transactions of the Metallurgical Society of AIME,* Vol. 242, Aug. 1968, pp. 1651–1656.
[16] Castelli, M. G. and Ellis, J. R., "Improved Techniques for Thermomechanical Testing in Support of Deformation Modeling," in this publication, pp. 195–211.
[17] Miller, A. K., "Unified Constitutive Equations for Creep and Plasticity," *Elsevier Applied Science,* 1987, pp. 139–212.
[18] Cailletaud, G. and Chaboche, J. L., "Macroscopic Description of the Microstructural Changes Induced by Varying Temperature: Example of IN 100 Cyclic Behaviour," Third International Conference on Mechanical Behaviour of Materials, K. J. Miller and R. F. Smith, eds., Vol. 2, Pergamon Press, New York, 1980, Paper ONRPA TP No. 1979-112.
[19] Robinson, D. N. and Bartolotta, P. A., "Viscoplastic Constitutive Relationships with Dependence on Thermomechanical History," NASA CR 174836, 1985.
[20] Robinson, D. N., "Constitutive Relationships for Anisotropic High Temperature Alloys," *Nuclear Engineering and Design,* Vol. 83, 1983, pp. 389–396.
[21] Robinson, D. N., Duffy, S. F., and Ellis, J. R., "A Viscoplastic Constitutive Theory for Metal Matrix

Composites at High Temperatures," *Thermal Stress, Material Deformation, and Thermo-Mechanical Fatigue,* PVP ASME, Vol. 123, 1987, pp. 49–56.
[22] Bingham, E. C., *Fluidity and Plasticity,* McGraw-Hill, NY, 1922.
[23] Ponter, A. R. S., "General Theorems for the Dynamic Loading of Structures for a State Variable Description of Material Behavior," (Institute of Physics Conference Series No. 47), Bristol, England, 1980, pp. 130–141.
[24] Walker, K. P., "Research and Development Program for Nonlinear Structural Modeling with Advanced Time-Temperature Dependent Constitutive Relationships," NASA CR-165533, 1981.
[25] Miller, A. K., "An Inelastic Constitutive Model for Monotonic, Cyclic and Creep Deformation, Part I & II," *ASME Journal of Engineering Materials and Technology,* Vol. 98, Apr. 1976, pp. 97–113.

APPENDIX A

Mathematical Framework of the Robinson Model

The general mathematical framework used in the Robinson model evolves from a class of constitutive equations originally derived from the gradient of a complementary dissipation potential function, Ω. This function is defined as

$$\Omega = \Omega(\sigma_{ij}, \alpha_{ij}, K, T) \tag{6}$$

$$\Omega = \Omega(F, G, T) = \theta_f(T)\int f(F)\, dF + \theta_g(T)\int g(G)\, dG \tag{7}$$

where

$$F = \frac{J_2}{K^2} - 1 \quad \text{and} \quad G = \frac{\hat{J}_2}{K^2}$$

$$J_2 = \tfrac{1}{2}\Sigma_{ij}\Sigma_{ij} \quad \Sigma_{ij} = S_{ij} - a_{ij}$$

$$\hat{J}_2 = \tfrac{1}{2}a_{ij}a_{ij} \quad a_{ij} = \alpha_{ij} - \tfrac{1}{3}\alpha_{kk}\delta_{ij}$$

$$S_{ij} = \sigma_{ij} - \tfrac{1}{3}\sigma_{kk}\delta_{ij}$$

Here σ_{ij} is the applied stress, α_{ij} is the internal (or back) stress, K is the drag stress, and T is the temperature. For initially isotropic materials, Ω can be taken [*9,20,21*] to depend upon the principal invariants of deviatoric stress quantities through scalar functions F and G. The applied stress dependence enters through F which plays the role of a Bingham-Prager yield function; the drag stress K plays the role of a Bingham threshold shear stress [*22*], below which the inelastic stain rate vanishes. Models developed from this potential/normality structure have been shown [*23*] to be consistent with a simple thermodynamic formalism. Also, this general framework has been used [*9,10,20,21*] as a starting point for various viscoplastic models of isotropic and anisotropic metals.

Shown in the following equations are the flow law $\dot{\varepsilon}_{ij}$ for the inelastic strain, and the evolutionary laws for the internal stress (a_{ij}) and drag stress (K), obtained by differentiating the dissipation potential function (Ω) with respect to the applied stress, internal stress, and drag stress, respectively.

$$\dot{\varepsilon}_{ij} = \frac{\partial\Omega}{\partial\sigma_{ij}} = \frac{\theta_f(T)}{K^2} f(F)\Sigma_{ij} \tag{8}$$

$$\dot{a}_{ij} = -h(\alpha_{kl}, T)\frac{\partial\Omega}{\partial\alpha_{ij}} = h(G)\dot{\varepsilon}_{ij} - \theta_g(T)r(G)a_{ij} \tag{9}$$

where

$$r(G) = h(G)g(G)$$

$$\dot{K} = -H(K, T)\frac{\partial \Omega}{\partial K} = \Gamma(K, T)\dot{W} \tag{10}$$

where

$$\dot{W} = \Sigma_{ij}\dot{\varepsilon}_{ij}$$

The functions θ_f and θ_g are typically taken with Arrhenius forms. Various specializations of this framework can be found in the references cited.

Sreeramesh Kalluri[1] *and Gary R. Halford*[2]

Damage Mechanisms in Bithermal and Thermomechanical Fatigue of Haynes 188

REFERENCE: Kalluri, S. and Halford, G. R., **"Damage Mechanisms in Bithermal and Thermomechanical Fatigue of Haynes 188,"** *Thermomechanical Fatigue Behavior of Materials, ASTM STP 1186,* H. Sehitoglu, Ed., American Society for Testing and Materials, Philadelphia, 1993, pp. 126–143.

ABSTRACT: Post failure fractographic and metallographic studies were conducted on Haynes 188 specimens fatigued under bithermal and thermomechanical loading conditions between 316 and 760°C. Bithermal fatigue specimens examined included those tested under high strain rate in-phase and out-of-phase, tensile creep in-phase, and compressive creep out-of-phase loading conditions. Specimens tested under in-phase and out-of-phase thermomechanical fatigue were also examined. The nature of the failure mode (transgranular versus intergranular), the topography of the fracture surface, and the roles of oxidation and metallurgical changes were investigated for each type of bithermal and thermomechanical fatigue test.

KEYWORDS: damage mechanisms, bithermal fatigue, thermomechanical fatigue, fractography, metallography, oxidation, cobalt-base superalloy

Nomenclature

t_f	Time to failure
CCOP	Compressive creep out-of-phase bithermal fatigue test
HRIP	High strain rate inphase bithermal fatigue test
HROP	High strain rate out-of-phase bithermal fatigue test
N_f	Number of cycles to failure
TCIP	Tensile creep in-phase bithermal fatigue test
TMIP	Thermomechanical in-phase fatigue test
TMOP	Thermomechanical out-of-phase fatigue test
σ_{comp}	Stress at peak compressive mechanical strain
σ_m	Mean stress, $(\sigma_{tens} + \sigma_{comp})/2$
σ_{tens}	Stress at peak tensile mechanical strain
$\Delta\epsilon_{el}$	Elastic mechanical strain range
$\Delta\epsilon_{in}$	Inelastic mechanical strain range
$\Delta\epsilon_t$	Total mechanical strain range, $\Delta\epsilon_{el} + \Delta\epsilon_{in}$
$\Delta\sigma$	Stress range, $\sigma_{tens} - \sigma_{comp}$

Introduction

Thermomechanical fatigue (TMF), where temperature as well as mechanical loading are applied simultaneously to engineering materials, can involve damage mechanisms that differ

[1] Senior research engineer, Sverdrup Technology, Inc., NASA Lewis Research Center Group, 21000 Brookpark Rd., Mail Stop 49-7, Cleveland, OH 44135.

[2] Senior scientific technologist, NASA Lewis Research Center, 21000 Brookpark Rd., Mail Stop 49-7, Cleveland, OH 44135.

significantly from those in isothermal fatigue (IF). To understand these mechanisms, it is imperative that extremes in the phasing of temperature and mechanical loading be explored. Investigators commonly perform in-phase (maximum temperature synchronized with the imposed peak tensile strain) and out-of-phase (maximum temperature synchronized with the imposed peak compressive strain) TMF experiments on material specimens [*1*]. Even though the task of conducting TMF tests has been greatly simplified due to advances in computer-controlled fatigue testing, it is still very time-consuming, labor-intensive, and consequently expensive, especially at very small mechanical strains. These circumstances have hindered generation and usage of conventional TMF test data as a basis for TMF life prediction modeling.

Bithermal fatigue (BF) testing was originally employed to selectively impose time-dependent inelastic strain (creep) either in tension or compression during a creep-fatigue cycle [*2,3*]. Halford [*1*] summarized bithermal fatigue testing techniques, the simplicity of BF tests over conventional TMF tests, and the prior research by several investigators who had employed BF tests to characterize the fatigue behavior of numerous alloys. A few years ago Halford and his coworkers [*4*] proposed BF as a potential link between IF and TMF. The rationale was that the simplified BF tests containing two different isothermal segments (tensile and compressive halves of the hysteresis loop) could emulate the damage mechanisms of TMF tests with the same temperature limits.

To verify this philosophy, load-controlled, strain-limited BF tests and strain-controlled TMF tests were conducted on a wrought cobalt-base superalloy, Haynes 188, between the temperatures of 316 and 760°C. Both in-phase and out-of-phase BF and TMF tests were conducted between these temperature limits. The fatigue life data and the cyclic stress-strain response of Haynes 188 in the BF and TMF tests were previously reported [*5*]. In this study, fractography and metallography were used on a few fatigued specimens to failure in an effort to establish the similarities as well as differences in the damage mechanisms of BF and TMF for Haynes 188 superalloy. The roles of surface oxidation, grain boundary activity (for example, grain boundary sliding), microstructural changes, and modes of failure in BF and TMF of Haynes 188 are discussed.

Material Details

The cobalt-base superalloy, Haynes 188, is a solid solution strengthened alloy that exhibits high strength, excellent oxidation resistance, and good ductility. Annealed Haynes 188 has a face-centered cubic matrix, a random distribution of M_6C carbides, and a lanthanum-rich phase that is associated with some of the M_6C carbides [*6,7*]. If Haynes 188 is thermally aged between 704 and 982°C then $M_{23}C_6$ type of carbides are precipitated first at the grain boundaries and later at the twin boundaries as the exposure time is increased [*6–8*]. After prolonged exposure times there is a tendency for the formation of Laves phase in Haynes 188 [*6,7*]. These metallurgical instabilities increase the hardness and decrease the ductility of Haynes 188 in the post-aged condition compared to the annealed condition. However, ductility of the superalloy can be restored and retained by the material for extended periods [*6*] by reannealing at 1178°C.

Wrought Haynes 188 was purchased in the form of 16-mm thick rolled plates from a commercial vendor. The as-received material had a hardness of Rockwell 'C' 22, which implied a solution-annealed condition [*6*]. The chemical composition of the superalloy in weight percent was determined to be the following: $<$0.0005-S; $<$0.2 ppm-P; 0.06-N; 0.1-C; 0.4-Si; $<$0.3 ppm-La; 1.6 ppm-Fe; 3.6-Mn; 17.8-W; 21.0-Ni; 22.3-Cr and the balance was cobalt. Microstructure of the as-received material (Fig. 1) exhibited grains with a large number of twins and the nominal diameters of the grains varied from 15 to 120 μm. Typical elastic mod-

FIG. 1—*Microstructure of as-received Haynes 188.*

uli and tensile properties of Haynes 188 are available in a handbook [*9*] and are also listed in Refs *5* and *10*.

Fatigue Experiments

Bithermal and Thermomechanical Tests

Typical BF tests [*1*,*4*,*5*] consist of two load-controlled, strain-limited, isothermal segments of loading, one each for tensile and compressive halves of the hysteresis loop (Fig. 2). Temperature of the specimen is changed between the higher and lower values under zero load in a BF test to facilitate free thermal expansion of the specimen. During an in-phase BF test, the tensile half of the hysteresis loop occurs at the higher isothermal temperature, whereas during an out-of-phase BF test it occurs at the lower isothermal temperature. The mechanical and thermal strains are applied to the specimen sequentially in load-controlled BF tests as opposed to the conventional strain-controlled TMF tests in which the mechanical and thermal strains are applied simultaneously to the specimen (Fig. 2).

In a BF test at the higher temperature, time-independent inelastic strain (high loading rate, Fig. 2) or time-dependent inelastic strain (high loading rate followed by a load-hold, Fig. 2) or a combination of the two can be imposed on the specimen. Thus, by properly choosing the temperature limits, loading rates, and load levels in a BF test, it is possible to induce the time-independent and time-dependent damage mechanisms separately under non-isothermal conditions [*1*]. Conventional TMF tests do not permit such a separation of damage mechanisms because of the relatively long cycle times that are required to accommodate the continuous temperature change in a cycle. The BF tests can capture the potential thermal free expansion mismatch straining effect that can occur between the surface oxide layers and the substrate material during a non-isothermal cycle [*1*]. However, there is a possibility of thermal recovery of the material during the temperature changes under zero load in BF tests that is not necessarily present in TMF tests. If simultaneously applied thermal and mechanical strains precipitate some synergistic damage mechanisms in a material, then BF tests would not be able to

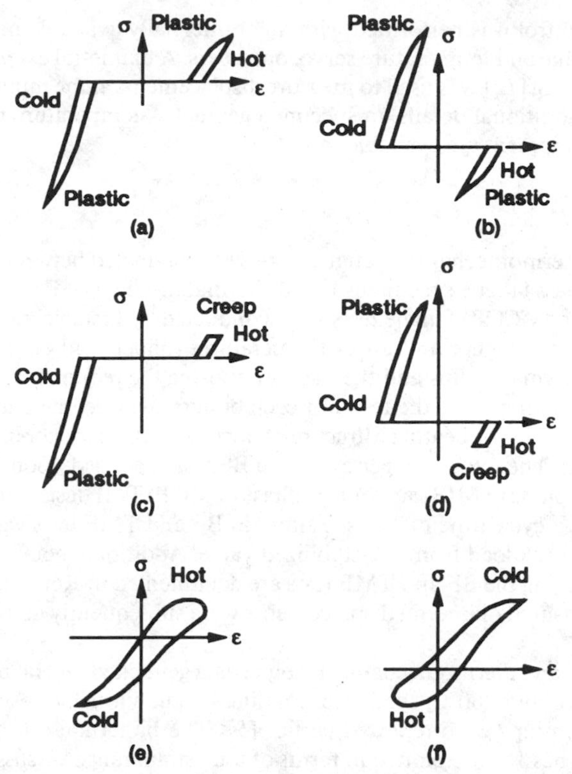

FIG. 2—*Schematic hysteresis loops of bithermal and thermomechanical fatigue tests:* (a) *Bithermal high rate in-phase test (HRIP);* (b) *Bithermal high rate out-of-phase test (HROP);* (c) *Bithermal tensile creep in-phase test (TCIP);* (d) *Bithermal compressive creep out-of-phase test (CCOP);* (e) *Thermomechanical in-phase test (TMIP); and* (f) *Thermomechanical out-of-phase test (TMOP).*

capture such damage mechanisms [*11*]. The main advantage of BF tests over TMF tests are the simplicity with which BF tests can be conducted and the ease of interpreting the results of BF tests. However, the total time required to execute a BF cycle is of the same order as that of a TMF test. In summary, BF tests can capture some of the damage mechanisms that prevail in conventional TMF tests. For additional details on bithermal and thermomechanical fatigue tests, the reader is referred to Refs *1*, *4*, and *5*.

Specimens and Test Equipment

Hourglass fatigue specimens [*10*] with a minimum test section diameter of 6.4 mm were machined with the longitudinal axis of the specimen parallel to the final rolling direction of the Haynes 188 plates. Test specimens were heated to the required test temperature in air by a direct resistance method. Control of temperature at the minimum cross section of the hourglass specimen was accomplished by chromel-alumel thermocouples spot welded at a calibrated location of 5-mm away from the minimum cross section of the specimen. This was done to prevent premature cracking from thermocouple spot welds near the minimum cross section of the fatigue specimen. All the tests were conducted on a uniaxial servohydraulic fatigue rig. Test control was performed with a commercially available minicomputer interfaced to the hydraulic and temperature servocontrollers. A diametral extensometer described in detail by Hirschberg [*12*] was used to measure displacements at the minimum cross-section of the specimen. Additional details on specimen geometry, temperature measurement techniques, and test equipment are available in Ref *10*.

Fatigue Data

Bithermal and thermomechanical fatigue tests were conducted between 316 and 760°C on Haynes 188 hourglass fatigue specimens [*5*]. Bithermal in-phase (HRIP and TCIP) and out-of-phase (HROP and CCOP) fatigue tests were conducted by first cycling the specimen for a few cycles between the higher and lower temperatures (under load control at zero load) to obtain stabilized thermal strains and then superimposing the required mechanical strains to obtain the total strain limits for the test. For each bithermal cycle, the temperature variation time was 180 s, with 75 s for heating (direct resistance) and 105 s for cooling (conduction and natural convection). The total time per cycle in a BF test depended upon the specific loading condition. Both in-phase (TMIP) and out-of-phase (TMOP) TMF tests were conducted under strain-control with a cycle time of 180 s. Failure in BF and TMF tests was defined as a 50% drop in the tensile peak load from the stabilized value. Additional details on the procedures adopted for performing the BF and TMF tests are documented in Refs *5* and *10*. For BF and TMF tests, the measured diametral displacements were subsequently converted to longitudinal strains.

All the bithermal and thermomechanical fatigue data generated on Haynes 188 between the temperatures of 316 and 760°C, the corresponding fatigue life relationships, and the cyclic stress-strain relationships were reported earlier [*5*]. The bithermal and thermomechanical fatigue data of Haynes 188 are plotted in terms of total strain range versus cyclic life in Fig. 3. The fatigue life curves are shown only for the bithermal HRIP and HROP tests and serve as the reference curves for comparing the fatigue lives of other types of BF and TMF tests. The constants for the bithermal HRIP and HROP fatigue life curves were obtained from Ref *5*. The bithermal HRIP life curve is lower than the corresponding bithermal HROP life curve indicating that even when no time-dependent inelastic strains are present, in-phase bithermal

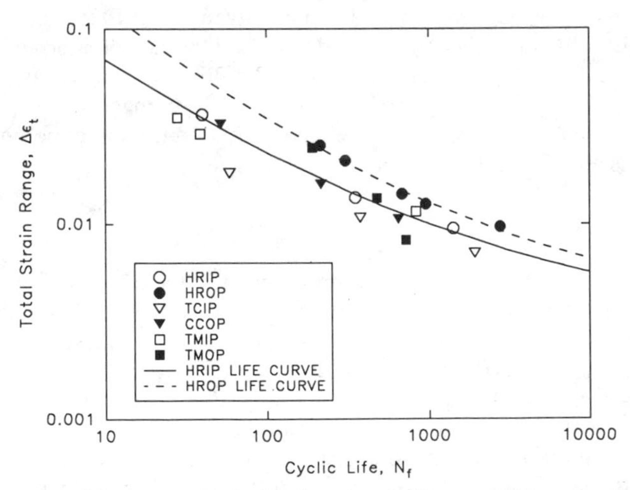

FIG. 3—*Bithermal and thermomechanical fatigue data of Haynes 188 between 316 and 760°C (from Ref 5).*

TABLE 1—*Selected Haynes 188[a] fatigue test specimens for fractographic and metallographic investigations.*

Test Type	Temperature, °C[b]	$\Delta\epsilon_{el}$, %	$\Delta\epsilon_{in}$, %	$\Delta\epsilon_t$, %	σ_{tens}, MPa	σ_{comp}, MPa	$\Delta\sigma$, MPa	σ_m, MPa	N_f	t_f^c, h
HRIP	760 <-> 316	0.60	0.35	0.95	541	−598	1139	−28.5	1421	78.9
HROP	316 <-> 760	0.63	0.34	0.97	620	−570	1190	25.0	2799	163.2
TCIP	760 <-> 316	0.50	0.21	0.71	379	−583	962	−102.0	1940	137.8
CCOP	316 <-> 760	0.50	0.56	1.06	664	−299	963	182.5	647	133.9
TMIP	760 <-> 316	0.70	0.46	1.16	543	−794	1337	−125.5	829	41.1
TMOP	316 <-> 760	0.56	0.27	0.83	661	−415	1076	123.0	720	36.0

[a] Data from Ref *5*.
[b] For bithermal tests, temperatures under tensile and compressive loads, respectively. For thermomechanical tests, temperatures at peak tensile and compressive strains, respectively.
[c] For bithermal tests, t_f values include the temperature cycling time of 180 s/cycle.

loading is more damaging than out-of-phase bithermal loading for Haynes 188 (Fig. 3). The HRIP life curve more or less serves as the upper bound for the bithermal and thermomechanical in-phase fatigue data. Likewise, the HROP life curve forms an upper bound to the out-of-phase bithermal and thermomechanical fatigue data. Comparison of bithermal TCIP and CCOP data shows that at higher strain ranges TCIP loading is more damaging than the CCOP loading. However, this difference appears to diminish at lower strain ranges. The TMF data clearly show that at higher strain ranges TMIP loading is more damaging than TMOP loading and that at lower strain ranges TMOP loading is more damaging than TcIP loading.

In this study, one failed fatigue specimen from each of the four types of BF tests (HRIP, HROP, TCIP, and CCOP) and the two types of TMF tests (TMIP and TMOP) was selected to conduct fractographic and metallographic investigations. Samples with a total strain range, $\Delta\epsilon_t$ of approximately 1% were selected for these investigations. In the case of TCIP tests, the fracture surface of the specimen tested at a total strain range of 1.08% was damaged due to arcing after failure of the specimen. Therefore, another TCIP test specimen with a $\Delta\epsilon_t$ of 0.71% (next closest $\Delta\epsilon_t$ to 1%) was used in this study. The test type, test parameters, and fatigue life corresponding to each of the selected specimens are listed in Table 1. Cycle times of the TCIP and CCOP tests were longer than the cycle times of the HRIP, HROP, TMIP, and TMOP tests. In Table 1, the reported strain ranges are longitudinal mechanical strain ranges.

Fractography

From each fatigue specimen listed in Table 1, one half was used for fractography and the other for metallographic investigation of a longitudinal section. The fractographic samples were sputter-coated with palladium to improve the fracture surface images in a scanning electron microscope. Fatigue cracks that caused failure initiated at the surface in all BF and TMF test specimens.

In the bithermal HRIP and HROP tests (Table 1), as in the case of an isothermal high rate strain cycling fatigue test, transgranular cracking and fatigue striations were observed (Fig. 4). A larger number of striations was visible in the HROP test than in the HRIP test. This is possibly due to the higher cyclic life of the HROP test (Table 1). In both cases the observed striations exhibited nonuniform spacing and direction, which indicated an irregular crack front movement compared to classic fatigue striations. These striations are referred to as "distorted" striations. The "distorted" striations indicate the presence of secondary grain boundary activ-

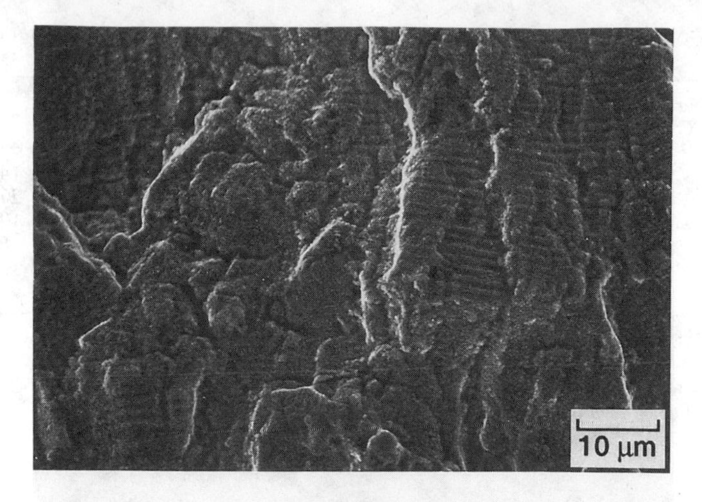

(a) HRIP test.

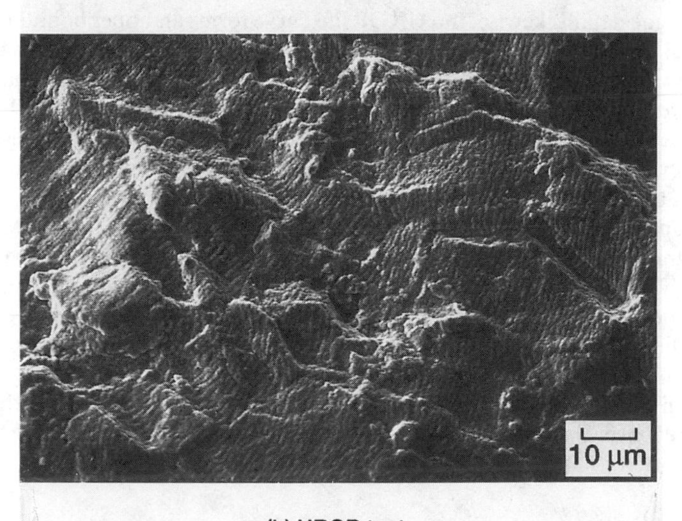

(b) HROP test.

FIG. 4—*Transgranular fracture and striations in bithermal fatigue tests:* (a) *HRIP test and* (b) *HROP test.*

ity (for example, grain boundary sliding) possibly caused by crystallographic slip in these types of bithermal tests.

In the bithermal TCIP test, fracture surface topography indicated intergranular cracking and no striations were observed (Fig. 5). Multiple regions of intergranular cracking were observed and the area between these regions failed in a ductile, tearing mode. On the contrary, the fracture surface of the bithermal CCOP test exhibited a transgranular mode of failure and heavily "distorted" striations (Fig. 6), possibly due to grain boundary sliding in compressive creep and the subsequent tensile plasticity. Similar failure modes were observed for an austenitic stainless steel under isothermal, tensile creep reversed by compressive plasticity, and compressive creep reversed by tensile plasticity, cyclic loading conditions [*13,14*].

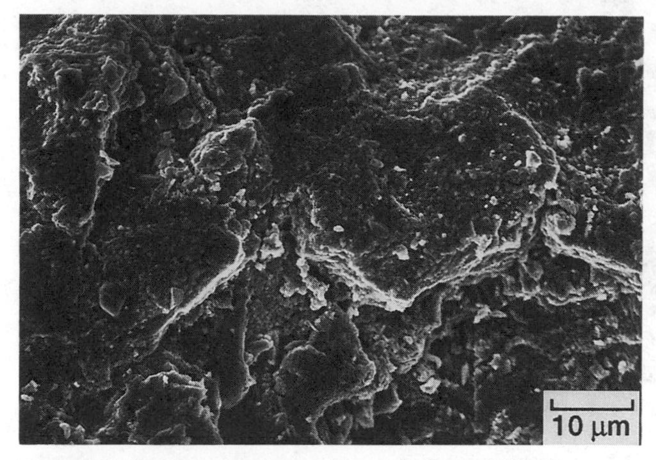

FIG. 5—*Intergranular cracking in bithermal TCIP test.*

Fracture surface topography for the thermomechanical TMIP test was distinctly intergranular with no striations (Fig.7). These fracture surface features are similar to those observed in the bithermal TCIP test (Fig. 5). In the case of the TMOP test the fracture surface was transgranular with mildly "distorted" striations (Fig. 8). However, occasional regions of intergranular fracture were also observed. The mildly "distorted" striations indicate a lower level of grain boundary activity (for example, grain boundary sliding) in the TMOP test as compared to the CCOP test (Fig. 6).

Metallography

Oxidation

Specimens were longitudinally sectioned and mounted in bakelite to preserve the oxide laden edges during sample preparation. The polished specimens were then photographed before etching (to preserve the oxide layer) with an optical microscope to document the oxi-

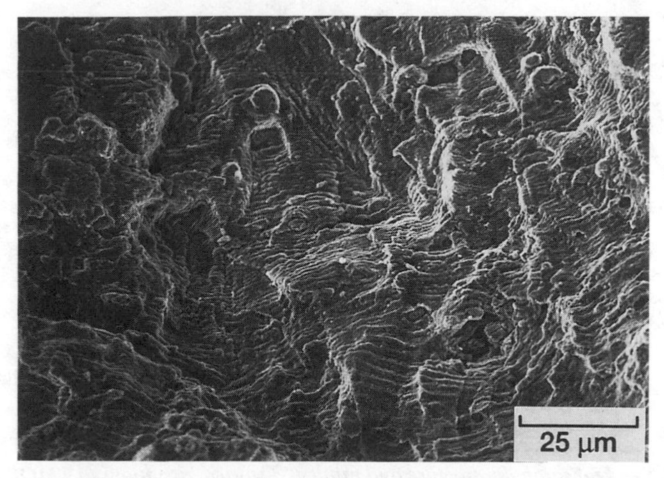

FIG. 6—*Transgranular cracking and heavily "distorted" striations in bithermal CCOP test.*

FIG. 7—*Intergranular fracture in TMIP test.*

dation patterns and to compare qualitatively the thicknesses of the oxide layers formed in the bithermal and thermomechanical fatigue tests. For the metallographic photographs presented in this paper (Figs. 9 to 16), vertical direction of the page corresponds to the mechanical loading direction of the hourglass fatigue specimen. No discernible oxide formed in the bithermal HRIP test on the external surface of the specimen. In the case of the bithermal HROP test very small amounts of oxide were noticed near the opened fracture surface. In the bithermal TCIP test oxide formed along the grain boundaries and secondary cracks (Fig. 9*a*; gray-colored regions are oxides). Extensive formation of oxide was observed on the fracture and external surfaces of the bithermal CCOP test specimen as shown in Fig. 9*b*. In this test the specimen was subjected to the lowest creep stress at high temperature for the longest amount of time among the four bithermal tests. In the TMIP test, formation of oxide was observed on the fracture surface and in the intergranular cracks, see Fig. 10*a*. The oxidation pattern in the TMIP test is similar to that in the TCIP test (Fig. 9*a*). A large amount of oxide was observed on the side of the specimen in the TMOP test as shown in Fig. 10*b*. A moderate amount of oxide was also observed along the main crack that caused failure in the TMOP test. The thickness of oxide scale formed in the TMOP test was less than that formed in the CCOP test (Fig. 9*b*), which had a longer time to failure.

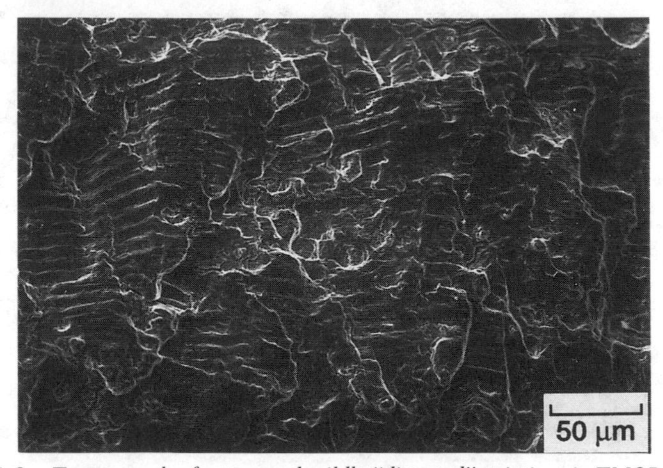

FIG. 8—*Transgranular fracture and mildly "distorted" striations in TMOP test.*

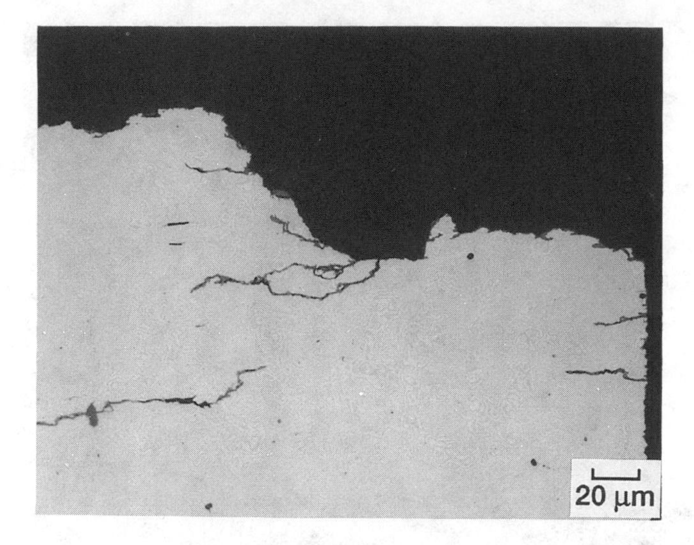

(a) TCIP test.

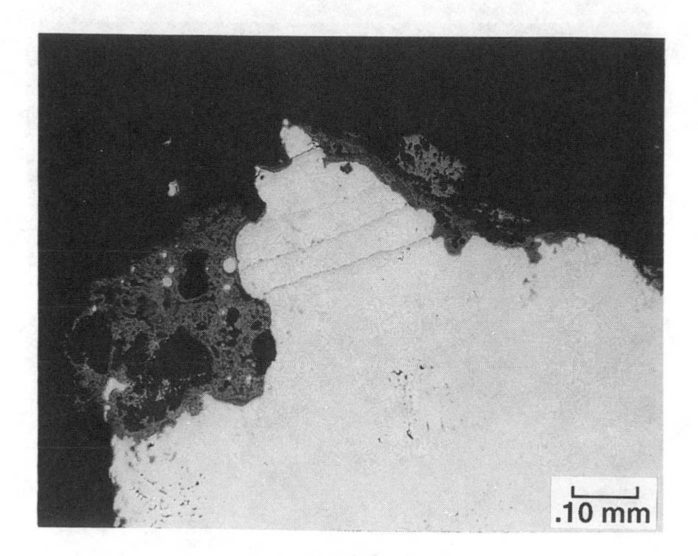

(b) CCOP test.

FIG. 9—*Oxide formation in bithermal tests:* (a) *TCIP test and* (b) *CCOP test.*

Microstructure

After examining all metallographic samples for oxidation, they were electrolytically etched in 5% hydrochloric acid solution with 4 V − 0.5 A current for 1 to 5 s. The metallographic samples were also sputter-coated with palladium to improve the images in a scanning electron microscope. The microstructures of all specimens tested under bithermal and thermomechanical fatigue displayed precipitation of carbides, most likely $M_{23}C_6$, predominantly along

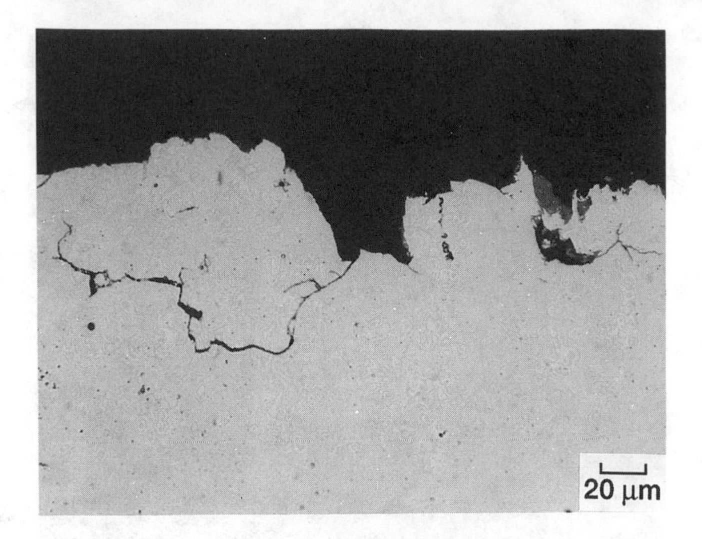

(a) TMIP test.

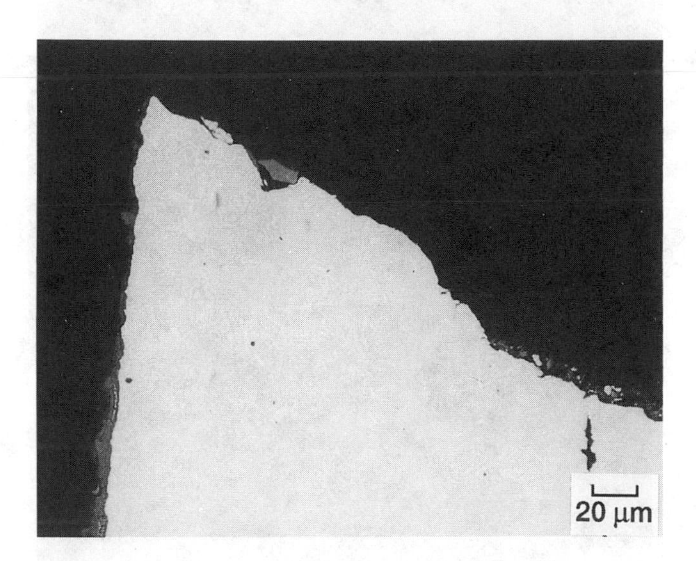

(b) TMOP test.

FIG. 10—*Oxide formation in thermomechanical tests:* (a) *TMIP test* (b) *TMOP test.*

the grain boundaries and sometimes along the twin boundaries as reported in Refs *6* to *8*. The testing times involved in these fatigue experiments were less than 200 hours which precluded the formation of Laves phase [*6*,*7*].

Grain boundary carbide precipitation near a secondary crack that was close to the fracture surface in the bithermal HRIP test is shown in Fig. 11. In the bithermal HROP test (Fig. 12), the following were observed: Small amounts of $M_{23}C_6$ carbides within the grains (in addition to the precipitation along the grain and twin boundaries) and crystallographic slip that pro-

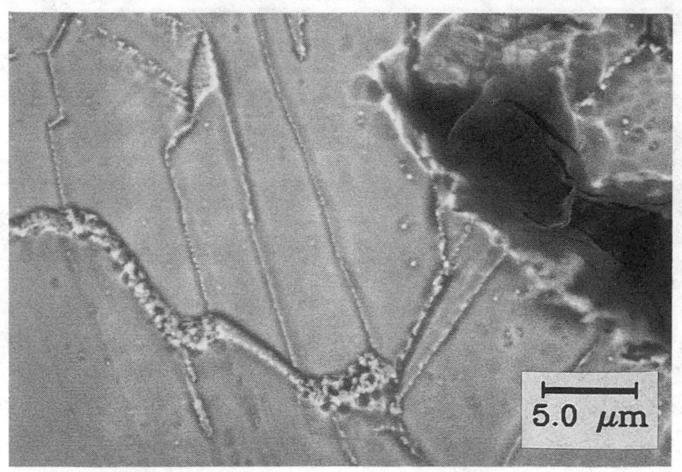

FIG. 11—*Grain boundary carbide precipitation near a secondary crack in the bithermal HRIP test.*

duced distortion of the twin boundaries. The microstructures of bithermal HRIP and HROP tests, while exhibiting subtle differences, were basically similar. In the TCIP test, precipitation and accumulation of $M_{23}C_6$ carbides along the grain boundaries and cracking of transversely oriented grain boundaries were noticed (Fig. 13). In the CCOP test, distortion of grains and twins due to plastic deformation (probably due to tensile overload at the end of the test) occurred in addition to the intergranular precipitation of carbides, see Fig. 14.

Microstructure of the TMIP test (Fig. 15) indicated small amounts of intergranular $M_{23}C_6$ carbide precipitation and an accumulation of these carbides in some locations. Intergranular fracture along nearly transversely oriented grain boundaries is distinctly visible in Fig. 15. Intergranular carbide precipitation and cracking within the longitudinal grain boundaries were observed in the TMOP test (Fig. 16). However, the main crack that caused failure in the

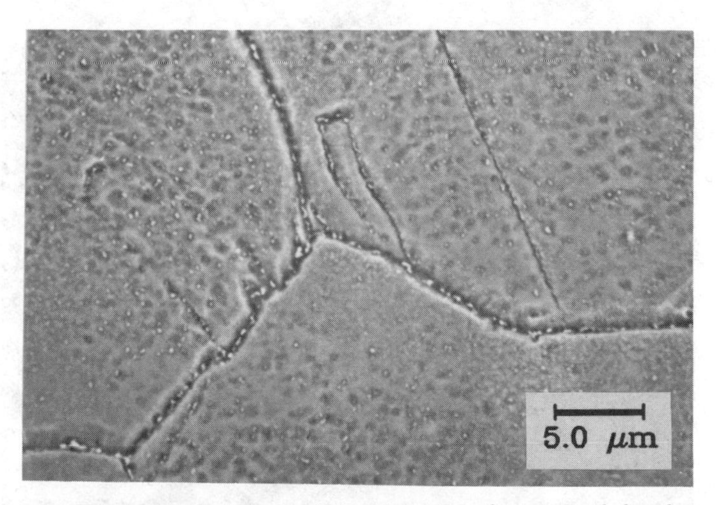

FIG. 12—*Intragranular precipitation of carbides, distortion of a twin, and slip plane activity in the bithermal HROP test.*

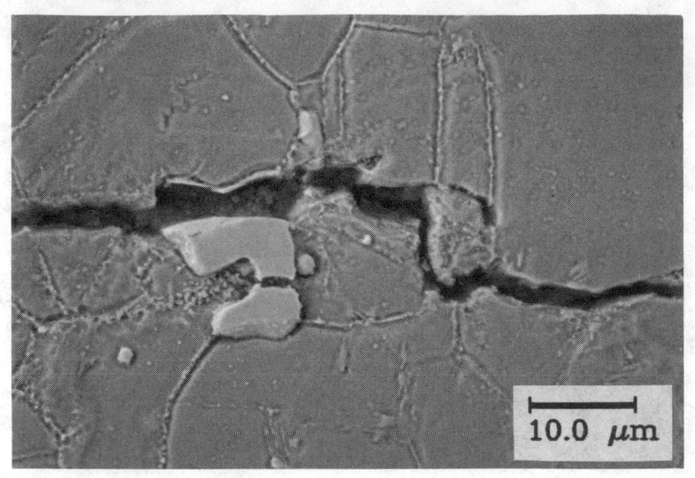

FIG. 13—*Grain boundary carbide precipitation and cracking of transverse grain boundaries in the bithermal TCIP test.*

TMOP test was transgranular. Table 2 summarizes the main fractographic and metallographic observations.

Discussion

Topographies of Fracture Surfaces

The bithermal HRIP and HROP tests, due to their high rate of loading, excluded time-dependent creep deformation. The fracture surfaces exhibited transgranular cracking and slightly "distorted" striations (Fig. 4). Among the bithermal tests that contained creep, the TCIP test exhibited intergranular cracking and no striations (Fig. 5) whereas the CCOP test exhibited transgranular fracture and highly "distorted" striations, see Fig. 6. These results indi-

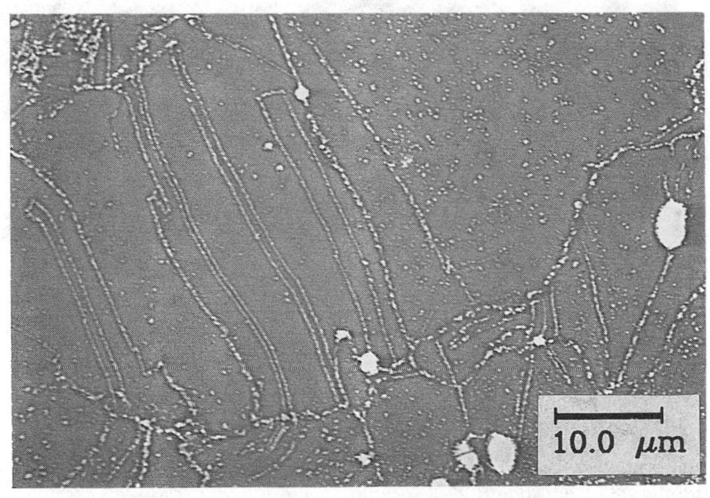

FIG. 14—*Distortion of twins and precipitation of carbides in the bithermal CCOP test.*

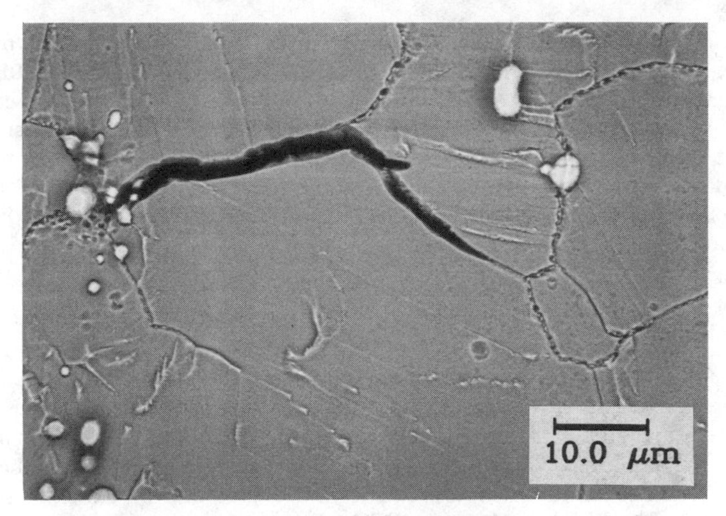

FIG. 15—*Intergranular carbide precipitation and fracture in TMIP test.*

cate that formation of striations occurred both at 316 and 760°C in BF tests whenever tensile time-independent plastic deformation was applied. The highly "distorted" striations in the bithermal out-of-phase CCOP test indicate a higher level of interaction between the striations and the grain boundaries they intersect than that encountered in the bithermal out-of-phase HROP test. Since the CCOP test involved compressive creep instead of compressive plasticity at 760°C, it is inferred that this creep deformation contributed to the interaction between the grain boundaries and striation formation. Potential mechanisms for such an interaction might be grain boundary sliding, intergranular carbide precipitation, or a combination of both.

It is indicated earlier in the paper that most of the fracture surface features of the TCIP and TMIP tests (Figs. 5 and 7) as well as the CCOP and TMOP tests (Figs. 6 and 8) are similar. Such a resemblance for Haynes 188 indicates the absence of synergistic damage mechanisms

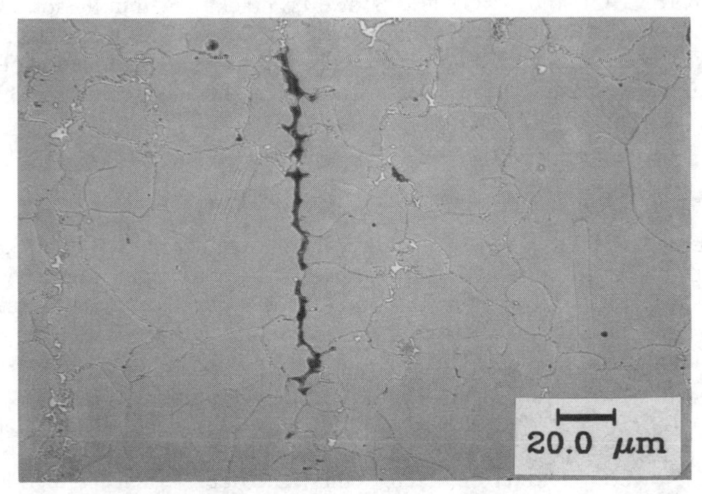

FIG. 16—*Intergranular carbide precipitation and cracking of longitudinal grain boundaries in TMOP test.*

in TMF due to simultaneous application of temperature and mechanical loads within the temperature range of 316 to 760°C. The resemblance between the fracture surfaces of BF and TMF tests is not unexpected because the TMF tests, both in-phase and out-of-phase, embody the same deformation modes as those in the BF tests. For instance, in the TMIP test, creep and plasticity are applied in tension with plasticity in compression, which is similar to the combination of deformations in TCIP and HRIP tests. Likewise, in a TMOP test, creep and plasticity are applied in compression with plasticity in tension, which is similar to the deformation in a combination of CCOP and HROP tests.

Oxidation Patterns

In both the BF and TMF tests (except in the CCOP test) the thicknesses of the oxide layers are small (Figs. 9 and 10). This is a testimony to the oxidation resistance of Haynes 188. However, at temperatures greater than 760°C, oxidation may play an important role in limiting the fatigue life of Haynes 188. Oxides formed mainly along the grain boundaries in the bithermal TCIP and thermomechanical TMIP tests as shown in Figs. 9 and 10. No oxidation of grain boundaries was noticed in the CCOP and TMOP tests. Instead, in these tests oxides formed on the outer specimen surfaces and along the main transgranular cracks that caused fatigue failure. These observations indicate that the formation of oxide along the grain boundaries is prevalent in the presence of tensile creep, which is common for both the TCIP and TMIP tests.

Microstructural Changes and Intergranular Cracking

In the BF as well as TMF tests, the microstructure of the failed samples revealed precipitation of $M_{23}C_6$ carbides mainly along the grain boundaries and occasionally along the twin boundaries and within the grains. Precipitation and accumulation of carbides along the grain boundaries and cracking of the transversely oriented grain boundaries was noticed in the TCIP and TMIP tests (Figs. 13 and 15). Tensile creep, which is present in both the TCIP and TMIP tests, appears to induce cracking along the grain boundaries that are nearly perpendicular to the loading direction. Grain boundary carbide precipitation and accumulation were also observed in both CCOP and TMOP tests. However, cracking within the longitudinally oriented grain boundaries was observed in some areas only in the TMOP test (Fig. 16). The significance of this observation is unknown. It is believed that the accumulation of carbide precipitates in the grain boundaries caused embrittlement and made them susceptible to fracture at prolonged exposure times.

Fatigue Lives and Damage Mechanisms

The bithermal HRIP fatigue lives were lower than the bithermal HROP fatigue lives (Fig. 3). The slightly higher amounts of fracture surface oxidation observed in the HRIP test specimen (Table 2), possibly due to the application of tensile plasticity at 760°C, may have contributed to the reduction in the HRIP fatigue lives compared to the HROP fatigue lives. It is not clear at this time whether the intragranular carbide precipitation observed only in the HROP test has any beneficial effect on fatigue life (Fig. 12). At higher strain ranges, the fatigue lives of TCIP and TMIP tests are lower than those of CCOP and TMOP tests (Fig. 3) because the former tests exhibited intergranular cracking, whereas the latter tests exhibited transgranular cracking. However, at lower strain ranges this trend in bithermal and thermomechanical fatigue lives appears to reverse itself, especially in the case of TMIP and TMOP tests (Fig. 3). To explain this trend reversal in fatigue lives, the following points should be considered. The

TABLE 2—*Summary of damage mechanisms in bithermal and thermomechanical fatigue of Haynes 188.*

Test Type	Grain Boundary Carbide Precipitation	Intergranular Cracking	Transgranular Cracking	Striation Formation	Location and Extent of Oxidation
HRIP	✓		✓	✓	Small amounts on fracture surface
HROP	✓		✓	✓	Very small amounts near the opened fracture surface
TCIP	✓	✓			Medium amounts on the fracture surface and within the secondary cracks along the grain boundaries
TMIP	✓	✓			"
CCOP	✓		✓	✓	Large amounts on the fracture and external surfaces
TMOP	✓	✓[a]	✓	✓	Medium amount on the fracture surface and large amount on the external surface

[a] Observed in some areas along the longitudinal grain boundaries.

out-of-phase CCOP and TMOP tests generated tensile mean stresses (Table 1) in the hysteresis loops (as opposed to the compressive mean stresses in the in-phase TCIP and TMIP hysteresis loops). The effect of these tensile mean stresses on fatigue life typically increases as the total strain range decreases [*15*]. In general, the CCOP and TMOP test specimens exhibited more oxidation on the external and fracture surfaces than the TCIP and TMIP test specimens (Table 2). These two observations seem to suggest that at lower strain ranges in the CCOP and TMOP tests, which exhibited transgranular cracking (with the exception of the longitudinal grain boundary cracking observed in the TMOP test), the combined effect of tensile mean stress and oxidation may be more detrimental to fatigue lfie than the intergranular cracking observed in the in-phase TCIP and TMIP tests. Note that the bithermal HROP tests also generated tensile mean stresses in the hysteresis loops (Table 1). However, the magnitudes of these mean stresses are relatively small in comparison to the magnitudes of those generated in the CCOP and TMOP tests and therefore may not have a significant detrimental effect on the HROP fatigue lives.

Several similarities were observed to exist in the modes of cracking, formation of oxide layers, and metallurgical microstructures among the in-phase and out-of-phase BF and TMF fatigue tests of Haynes 188. These similarities provide phenomenological bases for the use of bithermal fatigue data in thermomechanical fatigue life prediction. Indeed, Halford et al. [*16*] successfully predicted the in- and out-of-phase TMF fatigue lives of Haynes 188 to within nearly a factor of 2 by using the BF data shown in Fig. 3 to characterize the constants in the total strain version of the Strain Range Partitioning method. However, it must be pointed out that the similarities among the damage mechanisms of BF and TMF tests, which are observed in this study for Haynes 188 in the temperature regime of 316 to 760°C, may not exist for other materials and temperature regimes. Thus, it is the responsibility of the fatigue life estimator to ascertain that there are indeed similar damage mechanisms occurring in BF and TMF tests before attempting to use the BF data as the basis for TMF life prediction.

Conclusions

Fractographic and metallographic investigations were conducted on six specimens of Haynes 188 that were fatigued to failure under in-phase and out-of-phase bithermal and thermomechanical loading conditions between 316 and 760°C. These investigations identified the prevailing damaging mechanisms in both bithermal and thermomechanical fatigue tests.

Bithermal fatigue specimens tested under high strain rates (in-phase as well as out-of-phase) exhibited mildly "distorted" striations, transgranular cracking, and small amounts of fracture surface oxidation. Fatigue specimens of in-phase bithermal and thermomechanical tests involving tensile creep cracked in intergranular fashion with oxidation primarily along the grain boundaries. In out-of-phase bithermal and thermomechanical fatigue test specimens subjected to compressive creep, transgranular cracking and surface oxidation were observed. In all the specimens examined, precipitation of carbides were noticed primarily along the grain boundaries and occasionally along the twin boundaries and within the grains. Among the tests that imposed creep loading on the specimens, cracking of transverse grain boundaries was observed in in-phase bithermal and thermomechanical fatigue tests, whereas longitudinal grain boundary cracking was observed in some locations only in the out-of-phase thermomechanical fatigue test.

Based on the similarities that exist among the damage mechanisms of bithermal and thermomechanical fatigue tests, it was concluded that for Haynes 188 between 316 and 760°C, bithermal fatigue tests could be used as the basis for thermomechanical fatigue life prediction. However, this conclusion needs to be generalized for other materials, temperature regimes, and environmental conditions.

Acknowledgment

Mr. Ralph Corner and Mr. Todd Leonhardt meticulously prepared and photographed the samples for the fractographic and metallographic studies.

References

[*1*] Halford, G. R., "Low-Cycle Thermal Fatigue," *Thermal Stresses II,* R. B. Hetnarski, Ed., Elsevier Science Publishers, B. V., Amsterdam, The Netherlands, 1987, pp. 330–428.
[*2*] Manson, S. S., Halford, G. R., and Hirschberg, M. H., "Creep-Fatigue Analysis by Strain-Range Partitioning," *Symposium on Design for Elevated Temperature Environment,* American Society for Mechanical Engineers, New York, 1971, pp. 12–28.
[*3*] Halford, G. R., Hirschberg, M. H., and Manson, S. S., "Temperature Effects on the Strainrange Partitioning Approach for Creep Fatigue Analysis," *Fatigue at Elevated Temperatures, ASTM STP 520,* American Society for Testing and Materials, Philadelphia, 1973, pp. 658–667.
[*4*] Halford, G. R., McGaw, M. A., Bill, R. C., and Fanti, P. D., "Bithermal Fatigue: A Link Between Isothermal and Thermomechanical Fatigue," *Low Cycle Fatigue, ASTM STP 942,* H. D. Soloman, G. R. Halford, L. R. Kaisand, and B. N. Leis, Eds., American Society for Testing and Materials, Philadelphia, 1988, pp. 625–637.
[*5*] Halford, G. R., Verrilli, M. J., Kalluri, S., Ritzert, F. J., Duckert, R. E., and Holland, F. A., "Thermomechanical and Bithermal Fatigue Behavior of Cast B1900+Hf and Wrought Haynes 188," *Advances in Fatigue Lifetime Predictive Techniques, ASTM STP 1122,* M. R. Mitchell and R. W. Landgraf, Eds., American Society for Testing and Materials, Philadelphia, 1992, pp. 120–142.
[*6*] Herchenroeder, R. B., "Haynes Alloy No. 188 Aging Characteristics," *Proceedings of the International Symposium on Structural Stability in Superalloys,* ASTM/ASME/AMS/AIME, Vol. 2, Sept. 1968, Seven Springs, PA, pp. 460–500.
[*7*] Herchenroeder, R. B. and Ebihara, W. T., "In-Process Metallurgy of Wrought Cobalt-Base Alloys," *Metals Engineering Quarterly,* American Society for Metals, Metals Park, OH, May 1969, pp. 30–41.
[*8*] Matthews, S. J., "Thermal Stability of Solid Solution Strengthened High Performance Alloys,"

Superalloys: Metallurgy and Manufacture, Proceedings of the Third International Symposium, B. H. Kear, D. R. Muzyka, J. K. Tien, and S. T. Wlodek, Eds., Claitor's Publishing Division, Baton Rouge, LA, 1976, pp. 215–226.

[*9*] *Nickel Base Alloys,* International Nickel Company, Inc., New York, 1977.

[*10*] Halford. G. R., Saltsman, J. F., and Kalluri, S., "High Temperature Fatigue Behavior of Haynes 188," *Advanced Earth-to-Orbit Propulsion Technology Conference,* Marshall Space Flight Center, Huntsville, AL, R. J. Richmond and S. T. Wu, Eds., NASA CP-3012, Vol. I, National Aeronautics and Space Administration, Washington, DC, 1988, pp. 497–509.

[*11*] Verrilli, M. J., "Bithermal Fatigue of a Nickel-Base Superalloy Single Crystal," NASA TM-100885, National Aeronautics and Space Administration, Washington, DC, May 1988, pp. 1-13.

[*12*] Hirschberg, M. H., "A Low Cycle Fatigue Testing Facility," *Manual on Low Cycle Fatigue Testing, ASTM STP 465,* American Society for Testing and Materials, Philadelphia, 1969, pp. 67–86.

[*13*] Kalluri, S., Manson, S. S., and Halford, G. R., "Exposure Time Considerations in High Temperature Low Cycle Fatigue," *Mechanical Behavior of Materials—V, Vol. 2,* M. G. Yan, S. H. Zhang, and Z. M. Zheng, Eds., Pergamon Press, Beijing, People's Republic of China, June 1987, pp. 1029–1036.

[*14*] Kalluri, S., Manson, S. S., and Halford, G. R., "Environmental Degradation of 316 Stainless Steel in High Temperature Low Cycle Fatigue," *Environmental Degradation of Engineering Materials III,* M. R. Louthan, Jr., R. P. McNitt, and R. D. Sisson, Jr., Eds., Pennsylvania State University, University Park, PA, 1987, pp. 503–519.

[*15*] Halford, G. R. and Nachtigall, A. J., "Strainrange Partitioning Behavior of an Advanced Gas Turbine Disk Alloy AF2-1DA," *Journal of Aircraft,* Vol. 17, No. 8, 1980, pp. 598–604.

[*16*] Halford, G. R., Saltsman, J. F., Verrilli, M. J., and Arya, V., "Application of a New Thermal Fatigue Life Prediction Model to Two High-Temperature Aerospace Alloys," *Advances in Fatigue Lifetime Predictive Techniques, ASTM STP 1122,* M. R. Mitchell and R. W. Landgraf, Eds., American Society for Testing and Materials, Philadelphia, 1992, pp. 107–119.

Michael A. McGaw[1]

Cumulative Damage Concepts in Thermomechanical Fatigue

REFERENCE: McGaw, M. A., **"Cumulative Damage Concepts in Thermomechanical Fatigue,"** *Thermomechanical Fatigue Behavior of Materials, ASTM STP 1186*, H. Sehitoglu, Ed., American Society for Testing and Materials, Philadelphia, 1993, pp. 144–156.

ABSTRACT: A recently developed cumulative creep-fatigue damage model is applied to the prediction of thermomechanical and isothermal cumulative creep-fatigue loading histories. The model utilizes damage curve expressions to describe cumulative damage evolution under creep-fatigue conditions, qualitatively grouped according to dominant damage mechanism or mode. The damage coupling concept of McGaw is employed to describe the cumulative interaction among different mechanisms and modes. In this study, experiments consisting of two-level loadings of thermomechanical fatigue (both in-phase and out-of-phase) followed by isothermal fatigue to failure, were conducted on 316 stainless steel. It was found that the model gave good predictions for the out-of-phase two-level tests and provided reasonable bounds for the in-phase two-level tests.

KEYWORDS: cumulative damage, thermomechanical fatigue, creep-fatigue, fatigue, creep, creep damage, fatigue damage, stainless steel, damage theory

Nomenclature

A_{ij}	SRP life relation coefficients
a_{ij}	SRP life relation exponents
α_{ij}	Damage curve equation exponents
D	General damage variable
D_{ij}	Specific damage variable
F_{ij}	SRP inelastic strain component fraction
g_{ij}	General damage coupling functions
h_{ij}	Specific damage coupling functions
n	Number of cycles
n_1	Number of cycles applied in first loading block
n_2	Number of cycles observed in second loading block
N_f	Number of cycles to failure
N_1	Number of cycles to failure corresponding to the applied mechanical loading in the first loading block
N_2	Number of cycles to failure corresponding to the applied mechanical loading in the second loading block
N_{ij}	SRP partitioned component life, cycles to failure
Q_{ij}, R_{ij}	Damage coupling material parameters
$\Delta\epsilon_{in}$	Inelastic strain range
$\Delta\epsilon_{ij}$	SRP partitioned inelastic strain range

[1] Research engineer, NASA Lewis Research Center, Mail Stop 49-7, 21000 Brookpart Road, Cleveland, OH 44135.

Introduction

Background

The problem of thermomechanical fatigue (TMF) life prediction has received considerable attention in recent years, with efforts principally concentrated on the prediction of TMF under uniformly repeated loading conditions. Several researchers have developed models to treat this problem, generally based on isothermal considerations, ranging from engineering approaches to more fundamentally-based mechanistic approaches. In [*1*], Halford treats this problem using several methods, while referencing several reviews of the field. However, most applications (aircraft gas turbines, resuable rocket engines, etc.) that involve TMF loadings also involve complex multi-level cyclic loading patterns, so that cumulative damage effects are also a concern. In this regard, most of the presently available TMF life prediction methods rely on the use of the linear damage accumulation criterion of Palmgren, Langer and Miner [*2–4*], treated either as an explicit feature of the TMF model (see [*5*], for example) or as an additional consideration [*6*], that is, a linear cumulative damage law is used in conjunction with the basic TMF failure model. Others employ nonlinear evolution laws for monotonic creep and pure fatigue [*7*] and use an isothermal-based interpolative approach [*8*] to TMF and cumulative damage modeling.

The Bithermal Approach to TMF

Among the many approaches to TMF life prediction is the method of strain range partitioning (SRP) [*9,10*], as utilized in conjunction with the bithermal testing concept [*11*]. For the purposes of this paper, it is of interest to briefly summarize the approach and describe relevant results for 316 stainless steel. The bithermal approach has been utilized in extending the total strain version of SRP to encompass TMF, but addresses the uniform loading case [*6*]. In the bithermal approach, the thermomechanical in-phase (IP) or out-of-phase (OP) cycle is approximated by a cycle in which the tensile portion of the loading is conducted isothermally at one temperature, while the compressive portion of the loading is performed isothermally at another temperature (Fig. 1). Although summarized elsewhere [*11*], the advantages of this approach include: (1) the testing requirements of bithermal fatigue can be simpler than those of TMF; (2) isothermal behavior may be related to bithermal behavior, provided no new mechanisms of deformation and damage are introduced by the change of temperature; (3) the micromechanisms of deformation and damage in the bithermal cycle should be easier to interpret and relate to isothermal behavior due to the increased contrast offered by the discrete nature of the bithermal cycle; and (4) considerably more creep can be imposed in a bithermal cycle than in a cycle-time equivalent TMF cycle, which can lead, depending on material, to the establishment of lower bounds on life for the TMF cycle being bithermally approximated. Although this approach offers several benefits in the treatment of TMF, several aspects require consideration when addressing other materials and materials systems, namely:

1. Some materials display substantially different stress-strain response under TMF cycling than under isothermal conditions [*12*]. The extent to which this behavior impacts the damage development under TMF must be considered.
2. For coated and composite materials, the thermal expansion mismatch effects under cyclic TMF conditions almost certainly will lead to substantial differences in TMF versus bithermal failure behavior. This may be a concern for monolithic materials that undergo substantial environmental interaction (oxide coatings which may present severe properties gradients between the parent material and the oxide layer, in the context of damage development and failure).

In [*13*], Halford et al. show the partitioned (SRP) strain range-life behavior of 316 stainless

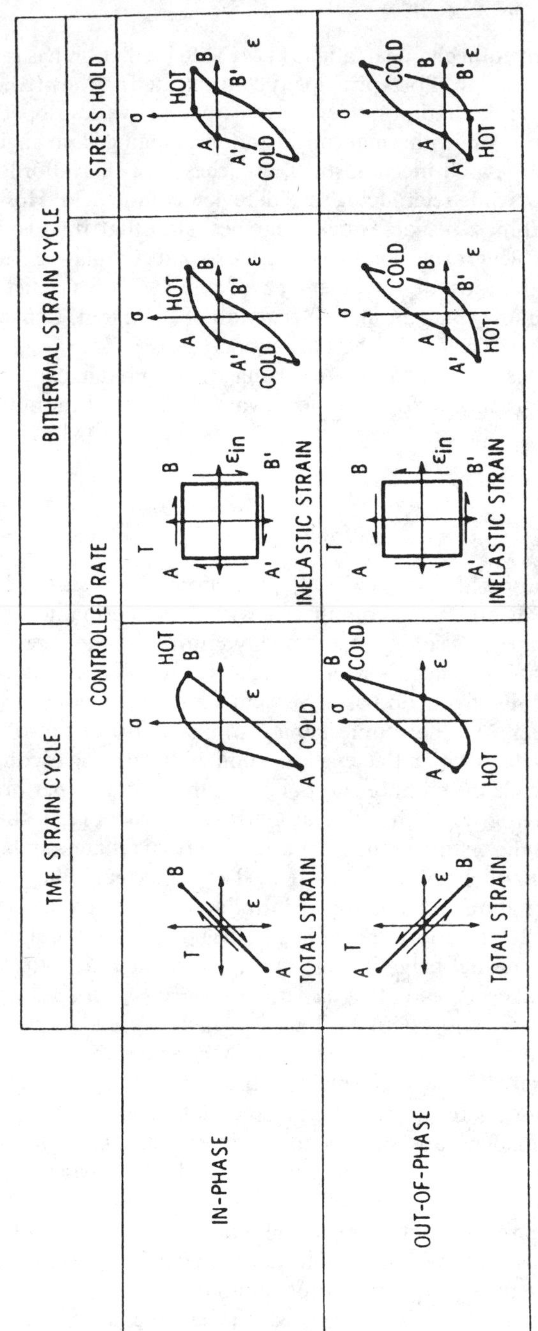

FIG. 1—*Schematic illustration of TMF and bithermal cycles (ε is mechanical strain).*

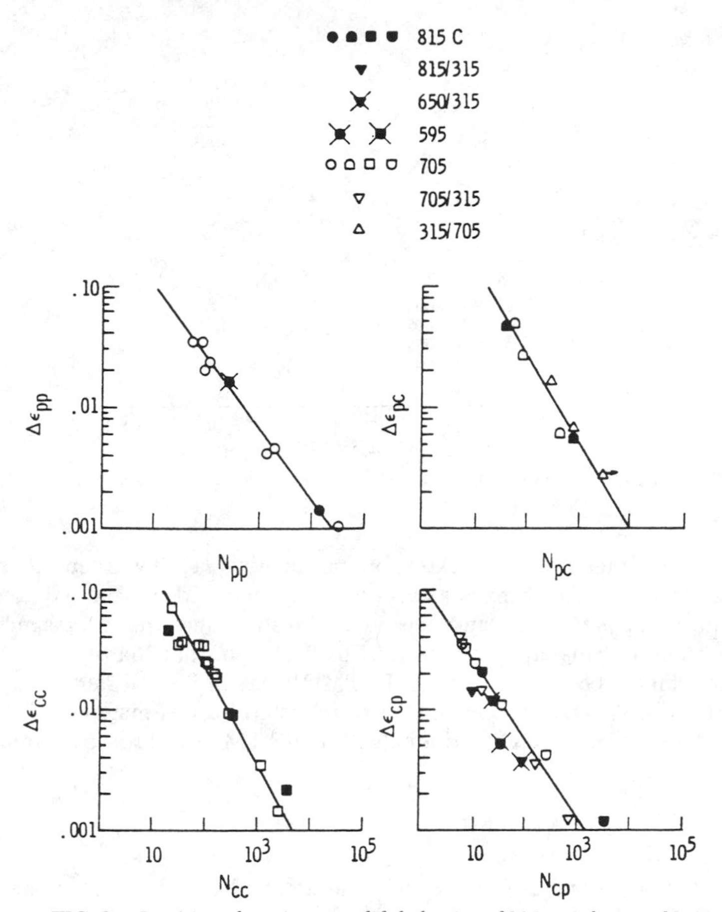

FIG. 2—*Partitioned strain range-life behavior of 316 stainless steel* [13].

steel, although a different heat from the material used in this study, for a variety of test temperatures, including results for what are now referred to as bithermal test conditions. These results are shown in Fig. 2. Although the deformation behavior of this material is strongly affected by temperature and cycle type (e.g., bithermal SRP versus analogous isothermal SRP cycle type), clearly the failure behavior of this material is largely independent of temperature and cycle type. This is corroborated by the fact that the creep and tensile ductilities for this material are themselves largely independent of temperature [*13*].

Halford et al. [*11*] augmented these results with additional, traditional TMF test results shown in Fig. 3. Even though this figure shows the results on an inelastic strain range-life basis (as opposed to a partitioned strainrange-life basis), we see that the behavior of this material is largely independent of cycle type. Isothermal SRP-type cycles lead to the same lives as bithermal SRP-type cycles as do the TMF IP and OP cycle types. As is pointed out in Ref *11*, the agreement of the TMF results with the isothermal and bithermal results could be improved if the Interaction Damage Rule (IDR) [*10*] was utilized in connection with the TMF results. This requires partitioned strain ranges for these cycles. These observations are consistent with the failure mechanisms found for the various test types. That is, IP TMF, IP bithermal and CP isothermal failure behaviors all were characterized by extensive intergranular cracking, whereas, the OP TMF, OP Bithermal and PC isothermal failure behaviors all were character-

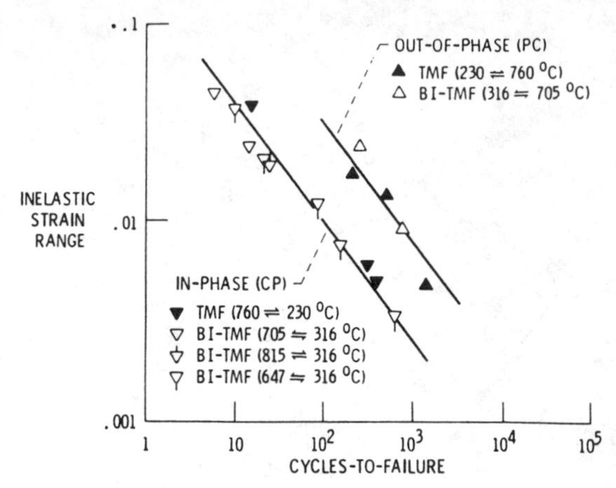

FIG. 3—*Comparison of bithermal and TMF LCF behavior of 316 stainless steel* [11].

ized by more benign transgranular cracking for this material [*11*]. It was concluded in Ref *11* that the damage accumulation behavior was reasonably independent of both the temperature and temperature time path taken, and depended instead on the type and magnitude of the inelastic deformation the material experienced. Finally, Halford and Manson [*14*] successfully used isothermal failure behavior to predict IP TMF behavior for 316 stainless steel. In this study, since a cyclic viscoplastic constitutive relationship for this material was not readily available, a detailed experimental partitioning of the IP TMF cycle was performed to make the life prediction.

Objective

A recently proposed life prediction model has addressed the need to capture the often nonlinear nature of cumulative creep-fatigue and fatigue damage under varying loading conditions [*15,16*]. In the present work, this model is used as a basis for addressing cumulative damage under TMF conditions. The resulting model is then utilized in the prediction of experiments consisting of two-level loadings of thermomechanical fatigue (IP and OP) followed by isothermal fatigue to failure conducted on 316 stainless steel.

Modeling Cumulative Creep-Fatigue Damage

Damage Curve Equations

The model of cumulative creep-fatigue damage proposed [*15,16*] rests on the notion that certain basic damaging modes of cyclic deformation characterize the aggregate failure behavior of a material, and utilizes the SRP method [*9,10*] as its basis. The dominant manifestation of damage is the formation of microcracks, their linkup to form a macrocrack, and the propagation of that macrocrack to fracture, regardless of whether such microcracks are the result of fatigue or creep-fatigue damage processes. A set of damage curve relations is proposed for the modeling of damage accumulation according to each of the four fundamental deformation and damage modes given in SRP

$$D_{ij} = \left(\frac{n}{N_{ij}}\right)^{(N_{ij})^{\alpha_{ij}}} \quad (1)$$

where α_{ij}, $i, j = P, C$ may be considered as material dependent parameters, and where the damage variable D possesses a range of 0 (undamaged condition) to 1 (failure), over the domain of the life fraction n/N_{ij} of 0 to 1.

Damage Metric

Remaining life n/N_f shall be used as a measure of the damage induced by a prior loading history, where N_f is the life corresponding to a given strain cycle, applied by itself, to failure, and n is the observed number of cycles to fracture for this strain cycle for a multiple strain or load level cycle history. This measure will be used in the context of the two-level fatigue test, where a certain number of cycles n_1 will be applied at a strain level corresponding to a life level N_1, and the balance of the test will be conducted at a different strain level (and corresponding life level N_2), to failure n_2.

Damage Coupling Equations

In a single damage mode any part of Eq 1 may be applied. When multiple damage modes are operative, Eq 1 cannot, in general, be utilized, since in addition to being mathematically inconsistent, doing so may also introduce under or over conservatism in the cumulative damage analysis, depending on the relative deleterious or benign effects of the damage modes involved. Therefore, an additional set of damage coupling relations is required

$$D = g_{ij}(D_{ij}) \tag{2}$$

where the functions g_{ij}, $i, j = P, C$, (where $g_{pp}(D_{pp}) \equiv D_{pp}$) must be determined by experiment. The coupling relations recognize that what is considered as *damage* in one context cannot necessarily be regarded as *damage* in another (e.g., PP-type cycling is associated with transgranular, classical fatigue-type damage, while CP-type cycling is often associated with intergranular, creep-type damage, [*10*]). However, the coupling relations do imply that the damage states are relatable. The coupling relations provide a mapping, or correspondence, between damage modes, that, together with the damage curve in Eq 1, permit an assessment of cyclic life for complex loading histories involving cycles of pure damaging modes. Physically, these equations may be expected to describe the interrelation between the processes of microcrack formation, nucleation, etc., for the basic damaging modes of CP and PP.

Synthesis

To model the more general problem wherein multiple damage modes may be present within a single cycle, one additional relationship must be established; namely, a description of how the various damaging contributions may be synthesized to provide a means of assessing the cumulative damage contribution. The following relationship is based on retaining consistency with trends demanded by the IDR [*10*] and provides this description [*15,16*]

$$\frac{n}{N_f} = F_{pp}D_{pp}^{N_{pp}^{-\alpha_{pp}}} + F_{cp}D_{cp}^{N_{cp}^{-\alpha_{cp}}} + F_{pc}D_{pc}^{N_{pc}^{-\alpha_{pc}}} + F_{cc}D_{cc}^{N_{cc}^{-\alpha_{cc}}} \tag{3}$$

where F_{pp}, F_{cp}, F_{pc}, and F_{cc} correspond to the relative amount of inelastic strain type present, e.g., $\Delta\epsilon_{pp}/\Delta\epsilon_{in}$, $\Delta\epsilon_{cp}/\Delta\epsilon_{in}$, $\Delta\epsilon_{pc}/\Delta\epsilon_{in}$, $\Delta\epsilon_{cc}/\Delta\epsilon_{in}$, and the terms, N_{pp}, N_{cp}, N_{pc}, and N_{cc} correspond to the life levels established from the SRP life relations, at the inelastic strain range, $\Delta\epsilon_{in}$. This equation describes the cumulative damage behavior of a complex cycle (with life N_f), as life fraction, expressed in terms of the four damage variables, D_{pp}, D_{cp}, D_{pc}, and D_{cc}.

Equation 3 may also be given in cyclic incremental form:

$$dn = \frac{F_{pp}N_{pp}^{-\alpha_{pp}}D_{pp}^{N_{pp}^{-\alpha_{pp}}-1}\,dD_{pp} + F_{cp}N_{cp}^{-\alpha_{cp}}D_{cp}^{N_{cp}^{-\alpha_{cp}}-1}\,dD_{cp} + F_{pc}N_{pc}^{-\alpha_{pc}}D_{pc}^{N_{pc}^{-\alpha_{pc}}-1}\,dD_{pc} + F_{cc}N^{-\alpha_{cc}}D_{cc}^{N_{cc}-1}\,dD_{cc}}{\frac{F_{pp}}{N_{pp}} + \frac{F_{cp}}{N_{cp}} + \frac{F_{pc}}{N_{pc}} + \frac{F_{cc}}{N_{cc}}} \quad (4)$$

In making application of Eqs 3 or 4, it is presumed that cyclic loading sequences that lead to the same *partitioned SRP result* produce the same micromechanistic damage, and lead to identical cyclic lives. This permits superposability of damage within a cycle, irrespective of the (nonlinear) cycle-to-cycle cumulative behavior. Equations 3 and 4 present numerous possibilities; many of these are described in [*15*]. Note that in general, the coupling relations must be invoked to recast Eqs 3 and 4 in terms of one effective damage variable *D*.

Treatment of Thermomechanical Fatigue

Basis

The model previously described can be readily extended to treat the case of TMF through the use of the bithermal fatigue approximation to TMF [*11*], described earlier. To address TMF, the life relations, damage curve equations and damage coupling relations can be directly replaced by bithermal counterparts. Note that in the bithermal case, two PP life relations are possible, namely IP and OP. Further, the bithermal CC cycle is of less interest since the CC component of inelastic strain in most TMF cycles of interest is small and nearly disappears as the total strain range becomes nominally elastic.

Experimental Program

Program

To explore the applicability of the proposed model to cumulative TMF, experiments were performed in two categories. In the first category, the baseline TMF behavior of 316 stainless steel was established for specific IP and OP cycles. The life level targeted was nominally 100 cycles to failure. In the second category, two-level loading experiments involving either IP TMF followed by isothermal fatigue (PP) to failure or OP TMF followed by isothermal fatigue (PP) to failure were performed to assess the cumulative damage aspects of TMF loading relative to its isothermal counterpart. In these experiments, the same strains were imposed for the TMF portion of the loading as had been imposed to obtain the baseline TMF behavior. The fatigue loading employed in the second portion of the block was done at an inelastic strain range level leading to a nominal 5000 cycle to failure life level, as observed in the baseline isothermal fatigue (PP) behavior.

Material and Procedure

The specific details concerning experimental techniques, equipment, etc. used in this study have been covered elsewhere [*15,16*]. Consequently, only those aspects that are new or different are covered in the present study. For the TMF testing, both the IP and OP tests were conducted using sinusoidally-varying temperature and strain-time (diametral displacement) waveforms, in strain control, under computer control, using the software described in [*17*]. Both TMF cycle types used 5 min cycle periods in conjunction with a maximum temperature of 816°C, and a minimum temperature of 316°C. In performing these experiments, each specimen was thermally cycled at zero load for at least ten cycles to obtain the free thermal expansion behavior (obtained for the tenth cycle). This was used, together with the desired (sinusoidal) strain-time waveform, to obtain the actual, experimentally enforced, strain-control waveform, and to obtain the true mechanical strains for the half-life cycles. This procedure was used for both the TMF baseline (constant amplitude) experiments and the TMF portions of the two-level block loading experiments.

In the two-level block loading experiments, the second loading was PP fatigue-type loading performed isothermally at 816°C. The process employed in transitioning from the TMF block to the isothermal block involved stopping the TMF cycling at a small positive strain to enable the unloading of the specimen to a nominally zero stress, zero strain state. The temperature was then changed to 816°C (under load control), the strain zero experimentally reestablished, and strain-controlled fatigue cycling commenced, at 0.2 Hz, until the specimen fractured into two halves.

Results

Refined Damage Coupling Relations

During the course of performing the experimental program as outlined above, additional isothermal fatigue testing was performed in an effort to gain a better understanding of the nature of the damage coupling behavior for 316 stainless steel, especially for the case of CP+PP loading. This was important, as the planned IP two-level block loading tests were conceived in relation to the isothermal observations for this loading case and because the cumulative damage behavior under this type of loading was observed to be substantially different from that observed for the cases of PP+PP, CC+PP and PC+PP [*15,16*].

In the set of two-level loading experiments involving creep-fatigue followed by fatigue loading to failure, the inelastic strain ranges obtained in the second (fatigue, PP) loading level were somewhat lower than those present in the baseline fatigue characterization data. To establish the life level corresponding to these inelastic strain ranges, a Manson-Coffin relation was employed [*15,16*]. Since the inelastic strain ranges were somewhat below the region where data had been obtained, the life levels calculated were extrapolations. Since the strains were still of the same order of magnitude, this was thought to be an adequate approximation of the fatigue life at this strain range. However, additional fatigue testing in this lower inelastic strain range regime revealed a substantial deviation away from the Manson-Coffin representation used (see Table 1, Fig. 4). The behavior is consistent with that found by Jaske [*18*], in which

TABLE 1—*Constant amplitude, inelastic strain-life fatigue behavior of 316 stainless steel at 816 °C.*

Specimen	Test	$\Delta\epsilon_{in}$	N_f
BY-126	PP	0.055 4	90
BY-114	PP	0.057 0	99
BY-202	PP	0.003 64	2 538
BY-115	PP	0.002 37	4 257
BY-104	PP	0.002 68	5 292
BY-128	PP	0.002 25	5 694
BY-203	PP	0.002 58	6 246
BY-204	PP	0.001 73	16 380
BY-206	PP	0.001 44	29 748
BY-210	PP	0.001 28	47 559
BY-201	PP	0.001 35	58 153
BY-209	PP	0.001 12	135 609

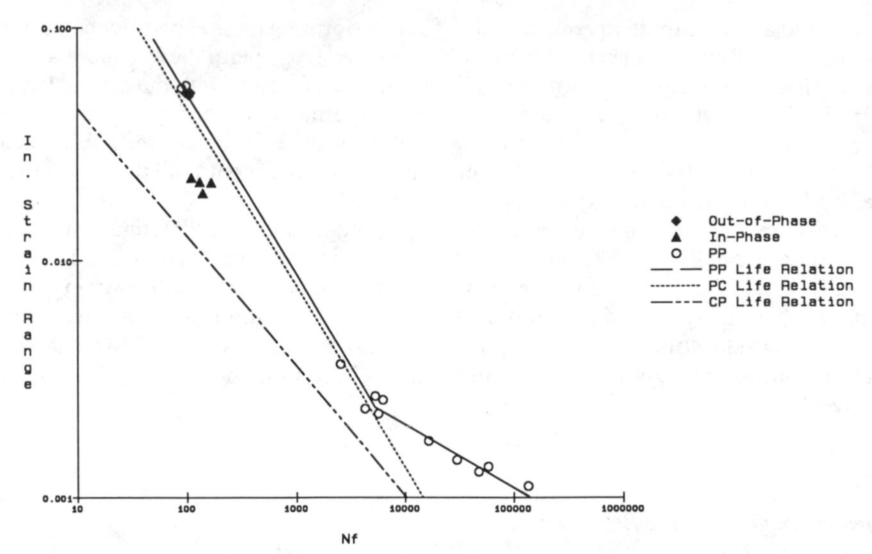

FIG. 4—*Constant amplitude, inelastic strain-life behavior of 316 stainless steel.*

long life experiments were performed on 316 stainless steel and a total strain range basis was employed for interpretation. Unusually large inelastic strains are present even at very high cyclic lives for this material, and it is reasonable that the inelastic strain-life data must tend away from the usual Manson-Coffin representation to maintain self-consistency with a strain-range (total, elastic and inelastic) -life interpretation. A bi-linear function has been used to represent these data (Table 1) and is shown in Fig. 4.

When this representation is used in calculating the life levels corresponding to the inelastic strains found in the second loading of the two-level loading experiments, the corresponding N_2 lives go up, bringing the life fraction, n_2/N_2, to lower values. This is true for all of the two-level creep-fatigue followed by fatigue results. Thus, the conclusions in [*15,16*] regarding *relative* behavioral differences between loading cases still hold, but the *absolute* nature of the two-level results must be revised to reflect these new findings. The net effect was a revision in the damage coupling relations to

$$D = D_{pp} = D_{pc} = D_{cc}$$
$$D = D_{cp}^{N_{cp}^{(Q_{cp} \ln N_{cp} + R_{cp})}} \tag{5}$$

where $Q_{cp} = -0.1235$ and $R_{cp} = 1.3443$, determined by regression analysis.

TMF IP and OP Results

The constant-amplitude failure behavior of each test type is shown in Fig. 4, where the data are from Table 2. Also shown on this figure are the updated PP, CP-type and PC-type isothermal life relationships obtained for this material in a previous study [*15,16*]. Because a cyclic viscoplastic constitutive relationship was not available for use in evaluating the deformation behavior to resolve nominally time-dependent and time-independent inelastic strains within the cycle, a true prediction of the TMF life behavior could not be performed. However, one may expect the IP failure response to be bounded by the CP and PP life relations and the OP

TABLE 2—*Constant amplitude, inelastic strain-life thermomechanical fatigue behavior of 316 stainless steel 316 to 816 °C.*

Specimen	Test	$\Delta\epsilon_{in}$	N_f
BY-225	IP	0.0194	139
BY-222	IP	0.0216	166
BY-220	IP	0.0217	131
BY-221	IP	0.0226	110
BY-226	OP	0.0515	105
BY-219	OP	0.0529	108
BY-218	OP	0.0530	97

failure response to lie between the PC and PP life relations. This is the case for the IP results, with the (log) average life level being 135 cycles to failure for the inelastic strain range experienced. The OP results are clustered tightly among the PC and PP life relationships, which, for the inelastic strain ranges considered, are very similar. The (log) average life level for this case was 103 cycles to failure.

TMF+PP Two-Level Block Loading Results

The results of the two-level block loading tests for IP-TMF followed by PP to failure, IP-TMF+PP, and OP-TMF followed by PP to failure, OP-TMF+PP, are presented in Table 3. The (log) average life levels obtained in the IP and OP TMF baseline testing correspond to the inelastic strain ranges imposed in the two-level block loading tests. The life levels corresponding to the PP (second) loading may be obtained from the bi-linear PP life relation given earlier. With these items, one may form the life fractions n_1/N_1 and n_2/N_2, and present the results on a remaining life plot, wherein the observed remaining life fraction in the second (PP, fatigue) loading n_2/N_2 is given as a function of the applied life fraction in the first (TMF) loading n_1/N_1. Figure 5 presents the results for the case of OP-TMF+PP and Fig. 6 presents the results of IP-TMF+PP.

Discussion

Clearly, the studies summarized earlier show that for 316 stainless steel, it is sufficient to obtain the isothermal failure behavior to predict the nonisothermal one. This was experimen-

TABLE 3—*Cumulative thermomechanical-fatigue interaction behavior of 316 stainless steel, two-level block loading.*

Specimen	Test Type	$\Delta\epsilon_{in}$, Block 1	n_1	$\Delta\epsilon_{in}$, Block 2	n_2
BY-227	OP + PP	0.0521	10	0.002 64	2561
BY-228	OP + PP	0.0528	20	0.002 50	2011
BY-229	OP + PP	0.0527	30	0.002 87	1595
BY-230	IP + PP	0.0233	10	0.002 68	1193
BY-231	IP + PP	0.0225	20	0.002 75	2200
BY-232	IP + PP	0.0222	30	0.002 75	1845
BY-233	IP + PP	0.0222	10	0.002 78	2408
BY-234	IP + PP	0.0227	20	0.002 47	2047
BY-235	IP + PP	0.0222	30	0.002 51	1480

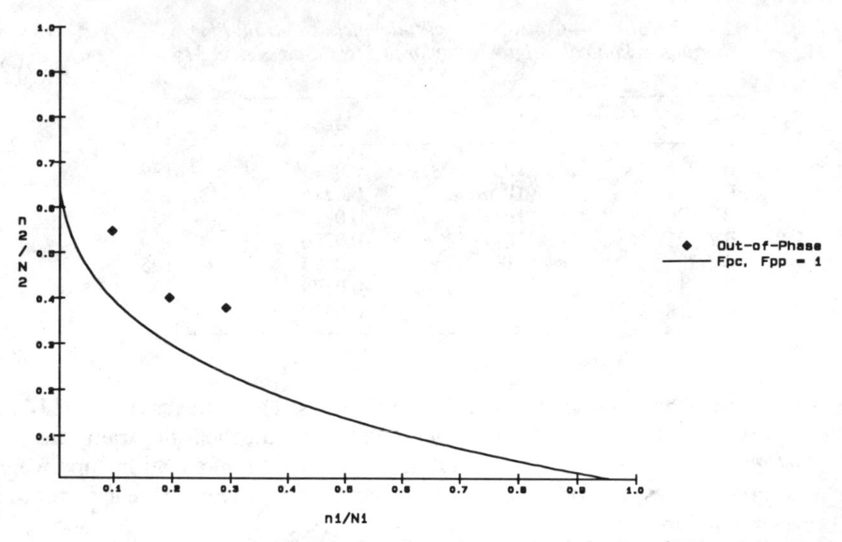

FIG. 5—*OP-TMF + PP two-level block straining interaction behavior of 316 stainless steel.*

tally established for several temperatures and cycle types, and was corroborated by detailed metallographic and fractographic findings with regard to operative failure mechanisms. Based on this, it may be presumed that the isothermal life relations determined for the present heat of material, at 816°C, may also be used to represent the bithermal life relationships with sufficient validity. Based on the consistency of failure mechanism among temperatures and non-isothermal cycle types, it may also be presumed that the bithermal damage curve equation constants and the damage coupling relations are reasonably approximated by their isothermal counterparts.

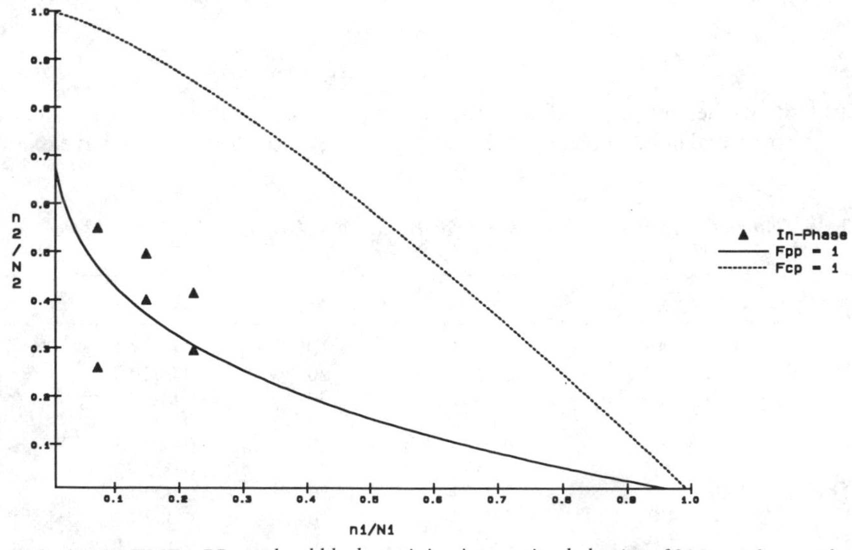

FIG. 6—*IP-TMF + PP two-level block straining interaction behavior of 316 stainless steel.*

As mentioned earlier, the lack of an available cyclic viscoplastic constitutive model precludes an analysis by which the time-dependent and time-independent inelastic strains may be partitioned, and hence, a direct application of either Eqs 3 or 4 solved for the two-level loading case herein was not possible. However, one may use either Eqs 3 or 4 to establish bounds on the expected results by considering that the basic life levels are held constant, while the partitioned inelastic strain fractions take on values of $F_{pp} = 1$ ($F_{pc} = 0$), and $F_{pp} = 0$ ($F_{pc} = 1$) (cases for OP-TMF+PP), and $F_{pp} = 1$ ($F_{cp} = 0$), and $F_{pp} = 0$ ($F_{cp} = 1$) (cases for IP-TMF+PP). The curves that result from these assumptions are also shown on Figs. 5 and 6.

Since there is little difference in the PP and PC life relations, and as the damage coupling relations used describe the coupling between PP and PC damage as being linear and directly additive, one expects to see minimal differences in the two curves shown in Fig. 5, as is the case. Although the experimental results are outside of the region circumscribed, one must consider the quality of the overall prediction to be quite good, as the qualitative behavior of the results are well described, while in quantitative terms, the accuracy of the prediction is well within a factor of two on life. Additional support of this conclusion is provided by limited results reported by Manson, et al. [*19*], wherein OP-TMF+PP testing was performed on 316 stainless steel, although over somewhat different temperatures (316 to 760°C for TMF, 760°C for the PP portion). In these results, the predicted behavior according to the Damage Curve Approach [*20*] was seen to be quite good. This is precisely the prediction that would be given using the model in the present study, since PP and PC damage types are considered to be interchangeable for this material.

The two curves that result for the IP-TMF+PP two-level loading case (Fig. 6) are much more distinct. This is due to the influence of the damage coupling term as one moves from $F_{pp} = 1$, wherein perfect coupling applies (e.g., PP couples linearly with itself) to $F_{pp} = 0$, where the nonlinear effects of the coupling are manifest. Note that the data are largely bounded by this set of curves. In fact, the data should be expected to lie closer to the $F_{pp} = 1$ lower curve, since, in the IP-TMF cycle employed, a 5 min strain-controlled cycle was used, and clearly does not provide the opportunity for creep to accumulate to the same extent that would in a stress-hold isothermal cycle. One would expect that as the level of creep strain increases in the IP-TMF+PP case, the more nonlinear interaction that was found in the isothermal case (where substantially more CP-type strains were accumulated) would be observed.

Conclusions

A recently-developed model of cumulative creep-fatigue damage was extended to address cumulative damage under thermomechanical fatigue conditions. This was accomplished in part, by the use of the bithermal fatigue approximation to thermomechanical fatigue. This model has been characterized in its isothermal form for 316 stainless steel. It was found that the model gave good predictions of the OP TMF block loading test results and provided reasonable bounds for the IP TMF block loading results. For the material and conditions considered in this study, the isothermal behaviors were seen to be excellent predictors of the TMF behavior, whether considering the problem of constant amplitude TMF loading or the analysis of complex TMF and isothermal multi-level histories. This, however, may not be true when other materials and conditions are considered.

References

[*1*] Halford, G. R., "Low-Cycle Thermal Fatigue," *Thermal Stresses II*, R. B. Hetnarski, Ed., Elsevier Science Publishers B.V., 1987.

[2] Palmgren, A., "Die Lebensdauer von Kugellargen," Verfahrenstechinik (Berlin), Vol. 68, 1924, pp. 339–341.

[3] Langer, B. F., "Fatigue Failure from Stress Cycles of Varying Amplitude," *Journal of Applied Mechanics*, Vol. 59, 1937, pp. A160–A162.

[4] Miner, M. A., "Cumulative Damage in Fatigue," *Journal of Applied Mechanics*, Vol. 67, 1945, pp. A159–A164.

[5] Neu, R. W. and Sehitoglu, H., "Thermomechanical Fatigue, Oxidation and Creep: Part II Life Prediction," *Metal Transactions*, Vol. 20A, 1989, pp. 1769–1783.

[6] Saltsman, J. F. and Halford, G. R., "Life Prediction of Thermomechanical Fatigue Using Total Strain Version of Strainrange Partitioning (SRP)—A Proposal," NASA TP-2779, 1988.

[7] LeMaitre, J. and Chaboche, J. L., "A Non-Linear Model of Creep-Fatigue Cumulation and Interaction," Office National D'Etudes et de Recherches Aerospatiales, T.P. No. 1934, 1974.

[8] Chaboche, J. L., "Thermodynamic and Phenomenological Description of Cyclic Viscoplasticity With Damage," ESA-TT-548, 1979.

[9] Manson, S. S., Halford, G. R., and Hirschberg, M. H.,"Creep Fatigue Analysis by StrainRange Partitioning," *Proceedings*, First Symposium on Design for Elevated Temperature Environment, San Francisco, CA, 1971, pp. 12–24, disc. pp. 25–28.

[10] Manson, S. S., "The Challenge to Unify Treatment of High Temperaturc Fatigue—A Partisan Proposal Based on StrainRange Partitioning," *Fatigue at Elevated Temperatures, ASTM STP 520*, American Society for Testing and Materials, 1973, pp. 744–782.

[11] Halford, G. R., McGaw, M. A., Bill, R. C., and Fanti, P. D., "Bithermal Fatigue: A Link Between Isothermal and Thermomechanical Fatigue," *Low Cycle Fatigue, ASTM STP 942*, H. D. Solomon, G. R. Halford, L. R. Kaisand, and B. N. Leis, Eds., American Society for Testing and Materials, Philadelphia, 1988, pp. 625–637.

[12] Castelli, M. G., Miner, R. V., and Robinson, D. N., "Thermomechanical Deformation Behavior of a Dynamic Strain Aging Alloy, Hastelloy X," appearing elsewhere in this volume.

[13] Halford, G. R., Hirschberg, M. H., and Manson, S. S., "Temperature Effects on the Strainrange Partitioning Approach for Creep Fatigue Analysis," *Fatigue at Elevated Temperatures, ASTM STP 520*, American Society for Testing and Materials, Philadelphia, 1973, pp. 658–669.

[14] Halford, G. R. and Manson, S. S., "Life Prediction of Thermal-Mechanical Fatigue Using Strainrange Partitioning," *Thermal Fatigue of Materials and Components, ASTM STP 612*, D. A. Spera and D. F. Mowbray, Eds., American Society for Testing and Materials, Philadelphia, 1976, pp. 239–254.

[15] McGaw, M. A., "Modeling and Identification of Cumulative Creep-Fatigue Damage in an Austenitic Stainless Steel," Ph.D. Dissertation, Case Western Reserve University, 1990.

[16] McGaw, M. A., "Cumulative Creep-Fatigue Damage Evolution in an Austenitic Stainless Steel," *Advances in Fatigue Lifetime Predictive Techniques, ASTM STP 1122*, American Society for Testing and Materials, Philadelphia, 1991.

[17] McGaw, M. A. and Bonacuse, P. J., "Automation Software for a Materials Testing Laboratory," *ASTM STP 1092*, American Society for Testing and Materials, Philadelphia, 1990, pp. 211–231.

[18] Jaske, C. E. and Frey, N. D., "Long-Life Fatigue of Type 316 Stainless Steel of Temperatures Up to 593°C," ASME 81-MAT-2, 1981.

[19] Manson, S. S., Halford, G. R., and Oldrieve, R. E., "Relation of Cyclic Loading Pattern to Microstructural Fracture in Creep Fatigue," Second International Conference on Fatigue and Fatigue Thresholds, Birmingham, England; also NASA TM-83473, 1984.

[20] Manson, S. S. and Halford, G. R., "Practical Implementation of the Double Linear Damage Rule and Damage Curve Approach for Treating Cumulative Fatigue Damage," *International Journal of Fracture*, Vol. 17, 1981, pp. 169–192.

Jean-Yves Guedou[1] *and Yves Honnorat*[1]

Thermomechanical Fatigue of Turbo-Engine Blade Superalloys

REFERENCE: Guedou, J.-Y. and Honnorat, Y. **"Thermomechanical Fatigue of Turbo-Engine Blade Superalloys,"** *Thermomechanical Fatigue Behavior of Materials, ASTM STP 1186,* H. Sehitoglu, Ed., American Society for Testing and Materials, Philadelphia, 1993, pp. 157–175.

ABSTRACT: A thermomechanical fatigue (TMF) facility has been developed in the Snecma Materials Laboratory to characterize the mechanical behavior and the damaging processes of blade superalloys under realistic loading conditions. Three nickel-base superalloys have been investigated using a reference TMF cycle representative of blade leading edge. Stabilized stress-strain loops exhibit greater plastic strains and higher stress amplitudes for an equiaxed alloy IN-100, mainly due to elastic anisotropy of the directionally solidifed DS-200 and AM1 single crystal alloys. The application of a viscoplastic constitutive model shows a good correlation between calculation and experiment for DS-200 and AM1 but indicates that the IN-100 model coefficients need improvement.

The TMF lives are much more important for the two oriented alloys (that is, AM1 and DS-200) as compared with the equiaxed IN-100, as in isothermal LCF conditions, but a correlation between TMF and LCF lives cannot be easily proposed and depends upon the material considered. The damaging mechanisms have been investigated on single crystal failed specimens, and compressive loads at high temperature play an important role in the damaging process. A continuous damage model delivered good predictions for TMF lives on equiaxed alloy specimens. Nevertheless, the application of such a model to structure calculations must take into account possible effects of crystallographic anisotropy in the case of directionally solidified (DS) and single crystal alloys.

KEYWORDS: cast superalloys, equiaxed, directionally solidified (DS), single crystal, constitutive law, damage

Interest of Thermomechanical Fatigue Tests

Stressing of turbine blades operating in quite stringent thermomechanical environment requires two mandatory conditions: first, an acute knowledge of high temperature resistant materials mechanical properties, especially in the case of new, high technology materials that may exhibit peculiar responses to in-service loadings, due to heterogeneity, anisotropy, and so on. The modelling of the mechanical behavior and damage processes can then be performed using sophisticated constitutive laws that can be introduced in computer codes developed in parallel. For both aspects, realistic lab tests on specimens need to be performed to take into account all the relevant parameters: changes in mechanical properties with temperature, phase transformations, environment effects, influence of loading history, and others.

Therefore, mechanical testing methodologies have improved over the years to allow the materials mechanics engineer more accurate information on material response. Classical isothermal tests (either monotonic or cyclic) are insufficient because they do not discriminate enough for behavior validation, and even, sometimes, they appear unrealistic for identifying

[1] Material structure, and mechanics manager and head, respectively, Materials and Processes Department, respectively, Snecma, BP 81, 91003 Evry Cedex, France.

damage mechanisms. Thermal fatigue tests [*1*] are more accurate for this purpose, however, several parameters cannot be easily controlled in these tests. Finally, thermomechanical fatigue (TMF) tests in which a volume element is submitted to simultaneous controlled load (or strain) and temperature, have been developed for superalloys over the last 15 years [*2–7*], and more recently, for metallic matrix composites [*8*].

Since 1985, in France [*9–11*] and in Western Europe [*12–14*], TMF methodologies have been implemented in structural life prediction. In Snecma, a TMF facility has been operative for two years, mainly devoted to turbine blade alloy studies because this type of loading appears to be highly representative of loading conditions in actual turbo-engine parts. The TMF tests provide a lot of valuable information on the material mechanical responses, but these tests are rather complicated, long, and very expensive: consequently, they are used in SNECMA Materials Laboratory to assess (and improve) constitutive models and to identify damage processes, enabling the determination of damage laws on a limited number of well-instrumented specimens. Those two specific items will be described subsequently.

Experimental Procedure

Testing Device and Specimen

The TMF facility uses a servohydraulic 100-kN testing machine that is controlled with a microcomputer. The closed loop control is ensured for each signal by a temperature controller and feedback signal from the testing machine itself. The imposed total strain ϵ_t is the sum of the mechanical strain ϵ_m and the thermal strain ϵ_{th}. The free thermal expansion is memorized by the computer at zero-imposed load, and next, slight variations of thermal stresses are corrected by the testing machine at zero-imposed strain. Then, the signal of synchronization between the thermal cycle and imposed mechanical strain is delivered by the testing machine, and the TMF test begins. Heating is ensured by a radiation furnace fitted with four, 1.5-KW quartz lamps. No forced cooling appears to be necessary for minimum temperatures above 300°C.

The temperature is controlled by a thermocouple wired to the specimen in the gage-length section. The imposed mechanical strain is monitored by a 12-mm gage length extensometer with alumina rods, which can be used up to temperatures of 1100°C. The specimen is similar to that used in other studies [*15*]: it consists of a hollow cylindrical bar (9-mm internal diameter, 1-mm wall thickness) with threaded heads. It can be achieved either by machining bulk castings or by casting direct on-size specimens. In the first case (the case of this study) both the inner and outer surfaces are polished and possibly coated.

In addition to stress-strain homogeneity in the effective volume (due to specimen geometry), negligible temperature difference across the specimen section and along the gage length are recorded simultaneously (less than 5°C): therefore, the TMF test in the Snecma laboratory is quite representative of the behavior of a volume element, which is the main purpose of such a test. The analysis of TMF tests is frequently carried on with reference to isothermal LCF tests, which have been performed on cylindrical bars (5.64-mm diameter), machined in cast bulks, and polished.

TMF Cycles

The TMF equipment allows several types of TMF cycles with any desired phase shift between strain and temperature. Nevertheless in order to develop a consistent database for various materials, a basic TMF cycle which was the same as used in Ecole des Mines de Paris laboratory [*9*] was selected. Computations of temperature stress-strain histories in actual

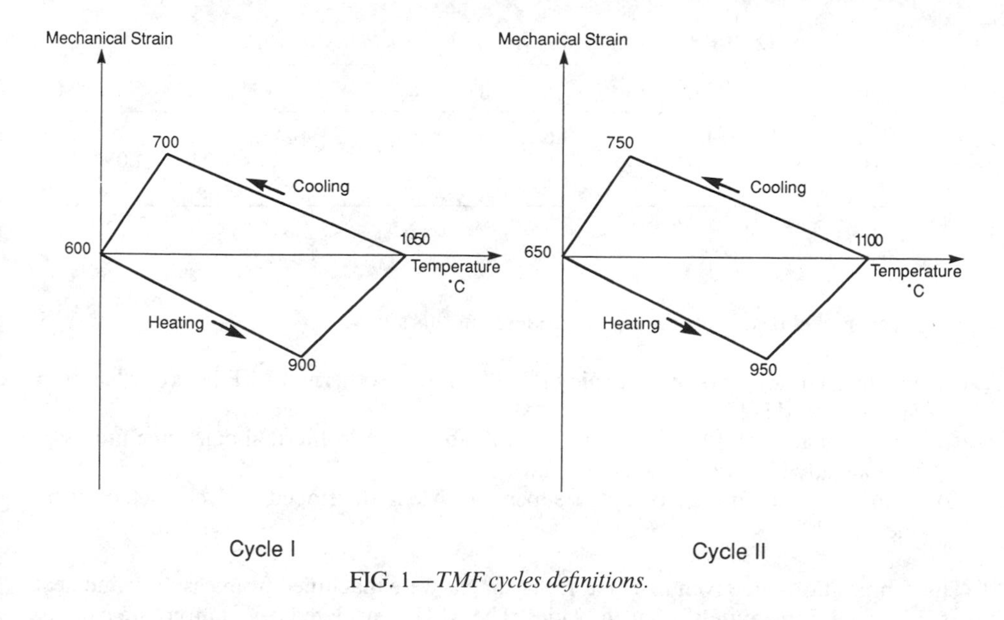

FIG. 1—*TMF cycles definitions.*

cooled turbine blades have led to identification of a realistic cycle, simulating the leading edge region of the blade. It consists of a triangular temperature cycle between 600 or 650 and 1050 or 1100°C associated with an alternate four slopes strain wave ($R_\epsilon = -1$). The frequency of those cycles is 5.6×10^{-3} Hz (180 s per cycle). The extreme temperatures (600 to 1050 or 650 to 1100°C) are chosen according to the material resistance, but they are always coupled with zero strain while maximum tensile and compressive strains occur at mid-temperatures. Individual plots of the TMF cycles, which present an intermediate shape between pure classical in-phase and out-of-phase cycles [*4,7*] are shown in Fig. 1. Comparison with other TMF cycles has been performed [*15*], and more damaging thermal mechanical loading could be easily found to validate a specific model, if necessary. The experimental shapes of both cycle types, associated with recorded temperatures, are shown in Fig. 2.

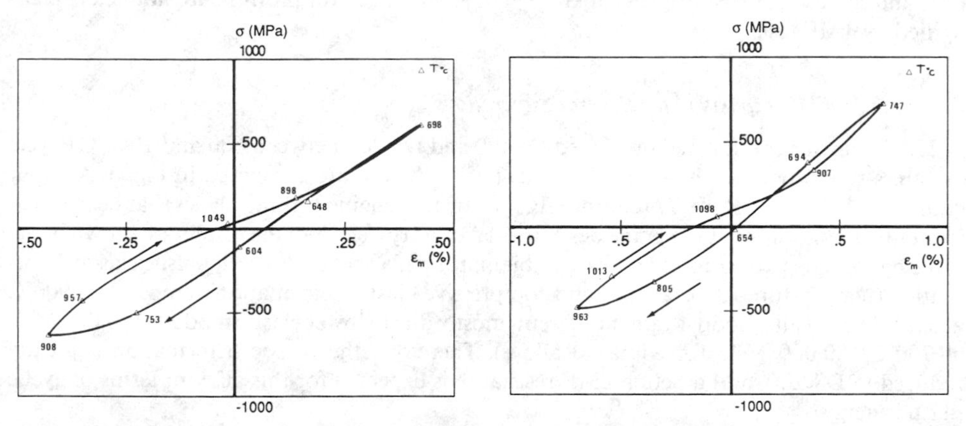

FIG. 2—*Experimental stress-strain TMF loops on IN 100 (Cycle I) and AM1 (Cycle II).*

TABLE 1—*Compositions (in weight %) of alloys studied.*

Alloys	Ni	Cr	Co	Mo	Ti	Al	Ta	W	V	Hf	Nb	C
IN100	Bal.	10.3	14.7	3.2	4.6	5.7	...	...	1.0	...	...	0.18
DS200	Bal.	8.4	10.0	...	1.8	5.0	...	12.5	...	2.2	1.0	0.13
AM1	Bal.	7.8	6.5	2.0	1.1	5.2	7.9	5.7	...	...	...	0.01

Materials

Three cast nickel base materials are considered in this study:

(1) an equiaxed well known superalloy IN-100, whose isothermal LCF behavior has been widely studied [*16*];
(2) a directionally solidification alloy, DS-200, for which isothermal damaging processes have already been quantified [*17*]; and
(3) a single crystal French patented superalloy AM1, developed and characterized by Snecma [*18*].

The compositions are given in Table 1. Specimens were machined from cast bars and heat treated in conditions quite similar to blades. The AM1 considered orientations are close to ⟨001⟩ within a ±6°C scatter range. IN-100 and AM1 were investigated in two surface states: bare or coated, that is, aluminum coating for IN-100 [*19*] and chromium-aluminum for AM1 [*20*]. DS-200 specimens were aluminum coated (same process as IN-100). In all cases, the coating was applied on internal and external cylindrical surfaces of the specimens.

Mechanical Behavior Characterization of Blades Alloys

Scope

As previously mentioned, TMF tests are first devoted to assess the stress-strain response, that is, the mechanical behavior of materials loaded in realistic conditions. The same approach has been conducted by other turboengine manufacturers [*21*]. Therefore, those responses have been studied for different alloys, as the influence of coatings: those investigations ought to allow the validation (or not) of the constitutive models which are identified from isothermal tests and, if necessary, to improve them by taking into account prominent parameters highlighted by TMF tests.

Nonisothermal Comparison for Several Superalloys

TMF tests were performed on coated IN 100 and DS 200 between 600 and 1050°C (Cycle I). Stress-strain loops at the same imposed strain (±0.4%) are compared in Fig. 3. Another equiaxed blade alloy, René 77 (commonly used in civil engine turbine blades) has been tested in identical conditions at the Ecole des Mines laboratory [*22*], and the results are shown in the same figure. The two equiaxed alloys exhibit quite similar behavior, with tensile elastic deformation (temperature 700°C at σ_{max}) and compressive plastic deformation (temperature 900°C at σ_{min}). The DS alloy loop is rather different, mostly due to lower elastic moduli (100 000 MPa at 700°C/180 000 MPa for equiaxed alloys). Therefore, the plastic deformation is greatly reduced in DS-200, and a better TMF resistance is expected for this alloy in terms of cyclic plastic energy.

Such effects are emphasized when considering equiaxed, DS and single crystal superalloys. The diagram in Fig. 4 illustrates that DS-200 and AM1, for the same total TMF strain ampli-

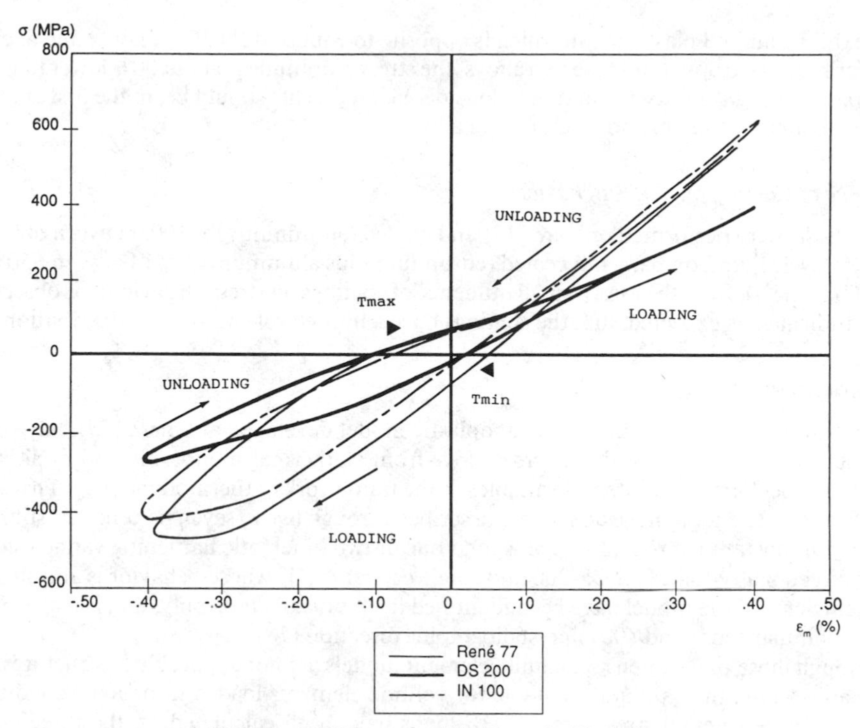

FIG. 3—*TMF loops for* ±*0.4% strain (Cycle I).*

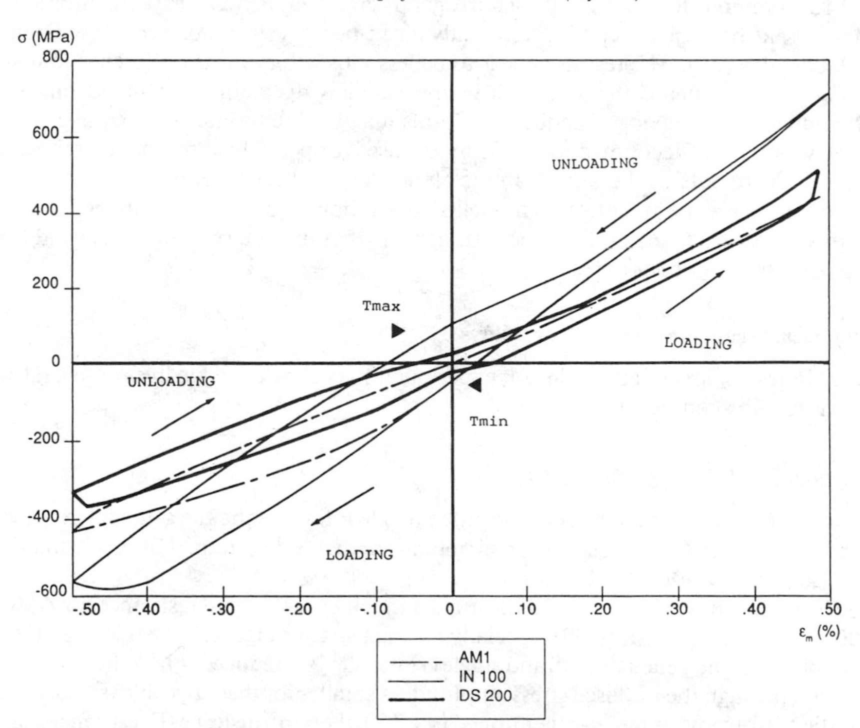

FIG. 4—*TMF loops for* ±*0.5% strain (Cycle I).*

tude, exhibits limited plastic strain which is opposite to equiaxed IN-100, mainly due to crystallographic anisotropy. For those two alloys, the stress amplitude is about 30% lower than for IN-100: whether a $\Delta\sigma$ or $\Delta\epsilon p$ criterion is considered, fatigue life should be, in the first approximation, better for the DS and single crystal alloys.

Influence of Coating on TMF Behavior

TMF tests were performed on bare [*15*] and coated (aluminum) IN-100, between 600 and 1050°C (Cycle I) and on bare and coated (chromium plus aluminum) AM1 between 650 and 1100°C (Cycle II). In both cases, a small influence of coatings on stress-strain loops is observed, which indicates, as expected, that the coatings have a limited role in the load distribution.

Behavior Modelling

Constitutive laws, using the cyclic viscoplastic model developed at Onera [*23*], have been identified on the three considered superalloys from isothermal monotonic and cyclic tests [*24*]. The model included internal variables in the framework of thermodynamics of irreversible processes. The total inelastic strain is described through five to seven parameters: stiffness, viscosity, a constant isotropic hardening, and one or two kinematic hardening variables.

For directionally solidified DS-200 and single crystal AM1, whose behavior is anisotropic, the coefficients of the model have been identified from uniaxial tests only, that is respectively in the columnar sense and $\langle 001 \rangle$ crystallographic direction [*18*].

Although those one-dimensional finite element models are not applicable to structures due to 3D anisotropy, they are quite suitable for volume elements loaded in uniaxial condition. Therefore, experimental stress-strain TMF loops have been calculated on the three above alloys. From previous tests on IN-100, a fairly good agreement between experimental and predicted loops has been observed [*25*], especially for in-phase tests performed at the Ecole Des Mines laboratory [*15*]. The results appear to be less satisfactory for the reference four slopes wave (Fig. 5). This means the identified viscoplastic law does not simulate adequately the actual compressive behavior at about 900°C: this may be due to microstructural changes in the alloy (which is not accounted for in the model used) or possibly to the fact that those loading conditions are outside the validity limits of the identified coefficients.

On the contrary, experimental and modelled loops show a good correlation on DS-200 and AM1 single crystal (Fig. 6). In these cases, it appears that the determined law is valid for the range of deformations considered.

Thermomechanical Damage in Blade Alloys

The TMF test is, first of all, a relevant experiment to compare fatigue lives of several alloys in realistic loading conditions.

Blade Alloys TMF Lives Comparison

The three coated alloys have been submitted to TMF tests, using the same reference Cycle I (600 to 1050°C), up to macroscopic crack initiation, that is determined by a 20% maximum tensile load decrease after stabilization.

The results are shown in Fig. 7 and illustrate a much better fatigue resistance for AM1 and DS-200 as compared with IN-100, especially for high strain ranges (above 0.7%). In terms of stress amplitudes, the benefit for DS and single crystal alloys is reduced (Fig. 8) but still remains noticeable: although the imposed stress amplitude is smaller for these two alloys (due to favorable elastic anisotropy) as already mentioned (see 3.2), their intrinsic TMF resistance is appreciably higher.

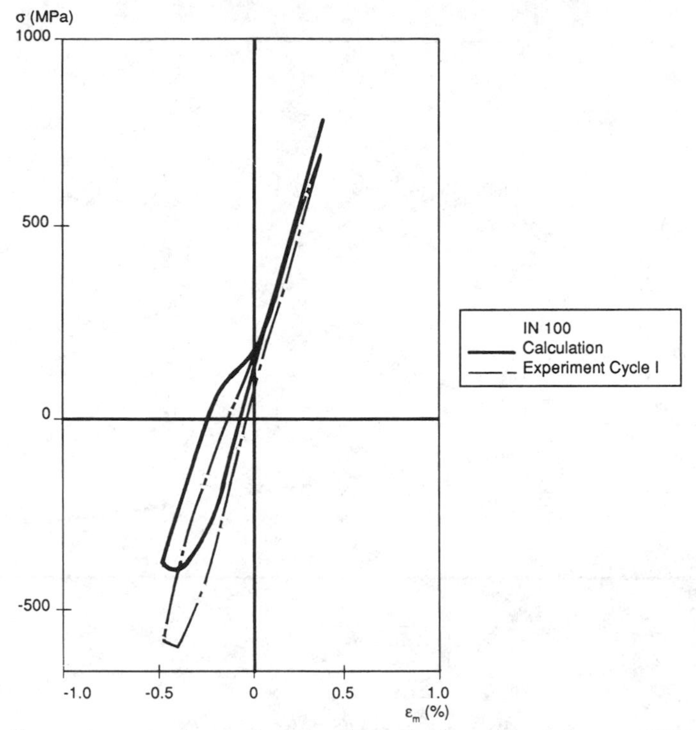

FIG. 5—*Comparison experimental/calculated stabilized stress strain loops on IN100 (Cycle I).*

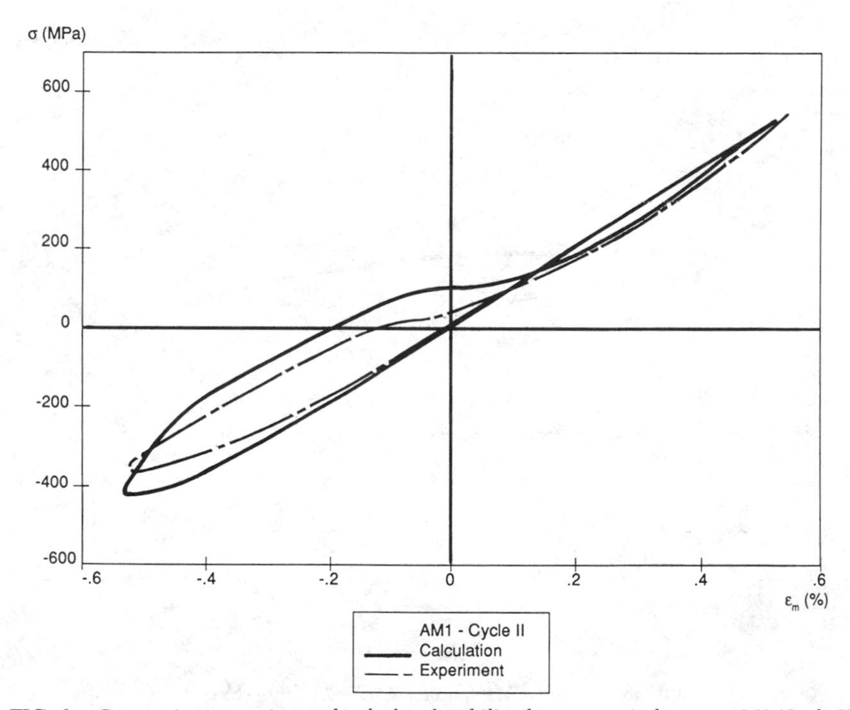

FIG. 6—*Comparison experimental/calculated stabilized stress-strain loop on AM1 (Cycle II).*

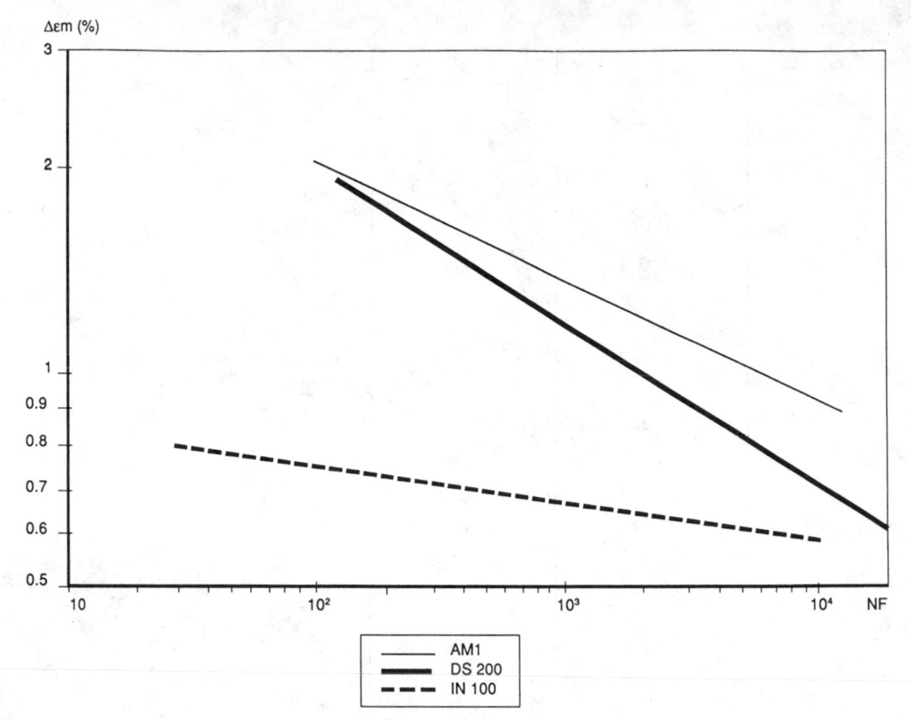

FIG. 7—$\Delta\epsilon$(N) *TMF lives for AM1—DS200—IN100.*

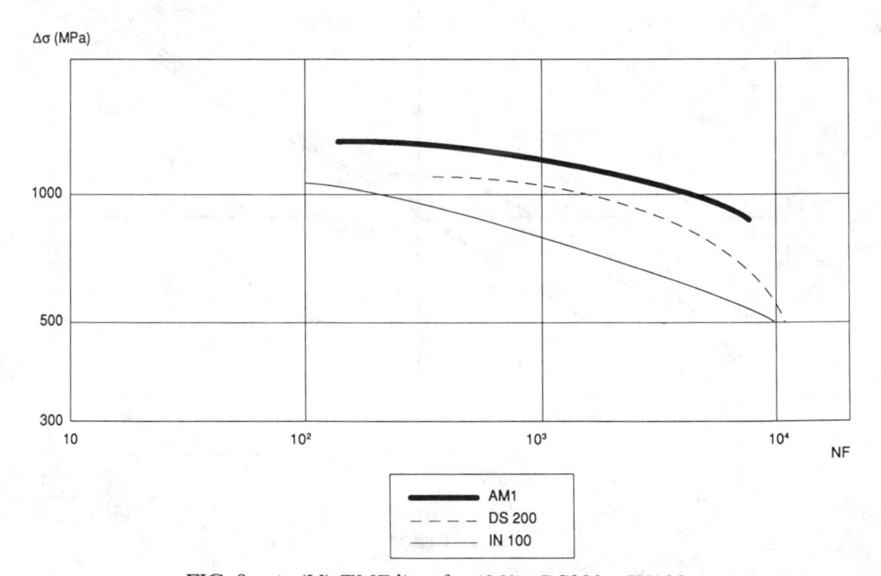

FIG. 8—$\Delta\sigma$(N) *TMF lives for AM1—DS200—IN100.*

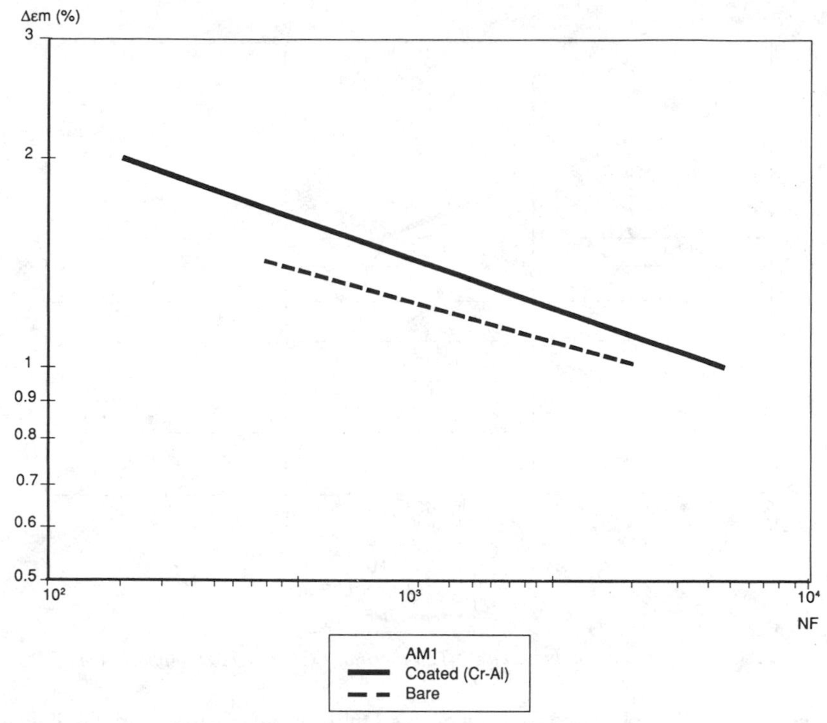

FIG. 9—*Influence of Cr-Al coating on TMF life for AM1.*

Coating Effect on TMF Durability

Aluminide coating has been shown to improve the TMF life on IN100 [*26*], especially at low strain levels and consequently long lives. TMF type II cycles (650 to 1100°C) have been performed on AM1 bare and chromium-aluminum coated specimens. Figure 9 clearly illustrates the beneficial effect of coating on TMF durability. Whatever the strain level, a factor of approximately 2 is observed for lives in favor of coated materials.

Isothermal alternating stress control tests were performed at 950°C, that is, the temperature at maximum compressive load in TMF tests (where the larger amount of plasticity is observed). The specimens were AM1 cylindrical bars (Fig. 2*b*) with and without Cr-Al coating. In that case, no beneficial effect of coating is observed. Micrographics investigations may contribute to explain those results. From those observations, it appears that for TMF tests, maximum stresses associated with intermediate temperatures (750°C in tension; 950°C in compression) are not the only damaging factors. The high temperature levels, which are known to be effective on life reduction for uncoated materials (due to environment effects) play an important role in the damage process, even associated with very low load in TMF tests.

Influence of TMF Cycle

Experimental results on AM1 single crystal (uncoated) tested according to Cycle II (Θ_{min} = 650°C, Θ_{max} = 1100°C with $\sigma = 0$ in both cases — σ_{max} tension at Θ = 750°C and σ_{min} compression at Θ = 950°C) have been compared to tests [*27*] on the same material, using a similar TMF test but shifted 50°C downwards: Θ_{min} = 600°C — Θ_{max} = 1050°C — σ_{max} tension at Θ = 700°C — σ_{min} compression at Θ = 900°C.

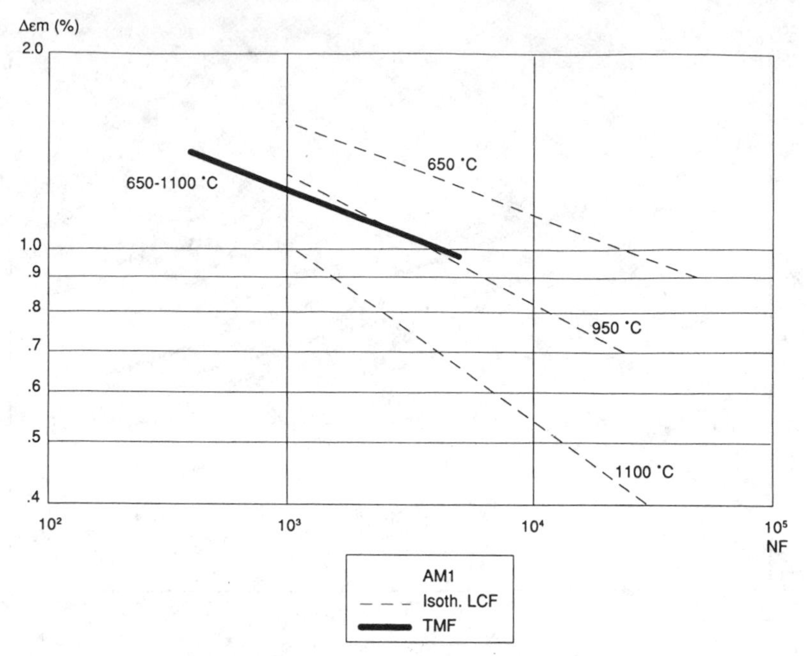

FIG. 10—*Comparison between* ε(N) *isothermal LCF and TMF on bare AM1.*

Quite similar lives are observed for both cycles. On the contrary, isothermal fatigue tests on bare cylindrical specimens, under alternate strain control ($R_\epsilon = -1$) indicate an important fatigue resistance decreasing when the temperature raises (Fig. 10), especially at high temperature. The fatigue life is decreased by a factor of about 5 when the temperature increases from 950 to 1100°C. In that diagram, the TMF life curve is close to the 950°C curve: it must be noticed that this temperature is precisely coupled with the maximum compressive load and the maximum plastic strain range (see Fig. 2). For IN-100 loaded according to Cycle I, it was observed that TMF life was close to 1000°C LCF life [*9*] that is 100°C above compressive peak stress temperature. Another study on AM1 [*28*] revealed that the TMF curve crosses the various LCF curves at 650 and 1100°C with decreasing mechanical strain range. In the latter case, the failure criterion was different (0.3-mm crack detection) from that of the present study and could explain this discrepancy because in our tests, a larger part of crack propagation is taken into account, mostly at low strain levels.

The results are different when plotted in stress amplitudes (Fig. 11). In the present study, the TMF curve is lower than isothermal LCF curves up to 950°C. A 1100°C curve obtained at Ecole des Mines Lab [*28*], under loading conditions differing from the above ones, appears too to be higher than our TMF curve.

So, the comparison between the results described previously and those of other laboratories on the same single crystal alloy, points out that the analysis of TMF experiments is complex based on observations of stress strain behavior. Microstructural investigations were conducted to enlighten us on the damage processes in AM1.

Micrographic Investigations on AM1 Failed Specimens

SEM fractography was performed on TMF and isothermal LCF specimens. On uncoated TMF specimens, slip lines are revealed on the outer surface (Fig. 12*a*): they correspond to

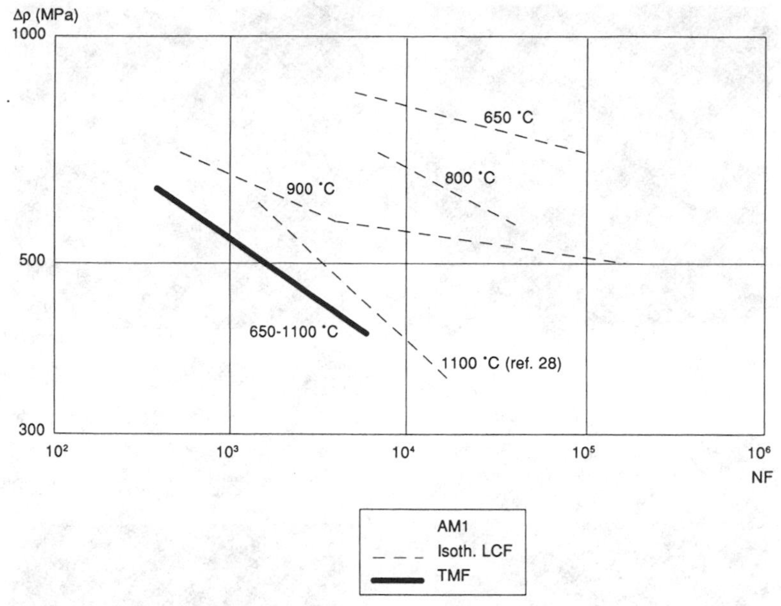

FIG. 11—*Comparison between σ*(N) *isothermal LCF and TMF on bare AM1.*

classical octahedral slip systems, that is, ⟨111⟩ planes, as expected for fatigue: those systems are activated at 650°C on AM1 loaded in ⟨001⟩ direction [*29*]. The rupture surface investigation (Fig. 12*b*) reveals oxides alignments, cracked perpendicularly to the loading direction and also crack initiation sites around microshrinkages (Fig. 12*c*). An examination of the γ' precipitates' morphology (Fig. 12*d*) shows that rafts are grown in a direction parallel to the load. Such a microstructure has been associated with high temperature compressive creep [*30,31*]: although the maximum temperature is combined in our TMF cycle, with near zero stress, the total duration of the considered test (3858 cycles, that is, about 200 h between 650 and 1100°C) with maximum compressive stress at 900°C appears to be sufficient to allow γ' coalescence, as in a pure creep test. This observation clearly shows the role of compression in the involved damage mechanisms for such a TMF test.

On coated TMF specimens, crack initiation sites were determined at the interface between the Cr-Al coating and the single crystal material for low strain amplitudes and long lives (Fig. 13*a*). In this case, porosities or other defects are not noticed on failed surface. However, for high strain amplitude (and long lives), crack initiation sites around casting defects are clearly evidenced (Fig. 13*b*). On isothermal LCF specimens, surface failure investigations indicate that fatigue cracks are always nucleated from internal casting defects (porosity or microshrinkage), whatever the considered test temperature between 650 and 1100°C (Fig. 14*a*, *b*, and *c*). The defect may sometimes be localized in subsurface (Fig. 14*d*). The presence of coating does not change the crack initiation process (Fig. 14*e*), which is again associated with casting defects.

In brief, microstructural investigations reveal that the damaging mechanisms may be different between LCF and TMF. The environment plays an important role in TMF, although casting defects are always the most detrimental nucleation sites for isothermal LCF. These concluding remarks explain the difficulty in correlating the results of both types of fatigue tests, and to establish a criterion to predict TMF life from LCF data.

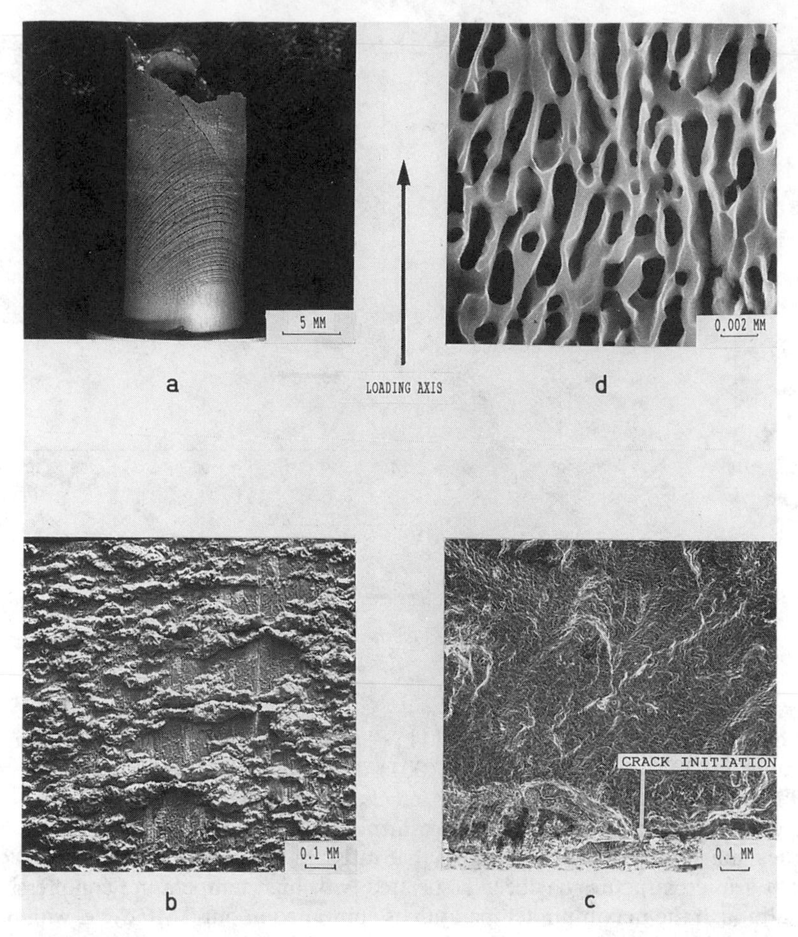

FIG. 12—*SEM investigations of AM1 uncoated TMF specimens:* (a) *slip lines,* (b) *oxides cracking,* (c) *crack initiation around microshrinkage, and* (d) γ' *rafting.*

Damage models are therefore needed to calculate adequately the lives for specimens, and beyond, for blades.

Damage Models

For nearly 20 years, damage models derived from isothermal LCF such as strain range partitioning [*32*] or thermal fatigue concepts [*33*] have been developed for TMF with some success. Other approaches based on crack propagation or energetic concepts [*34–36*] have been proposed, but their identification appears complex and they are rather inadequate for structure calculations.

A simple relationship between TMF and isothermal LCF lives, using harmonic or geometric means of number of cycles to crack initiation (50 μm) at $T(\epsilon_{max})$ and $T(\epsilon_{min})$ has been applied to a cobalt alloy and a superalloy single crystal [*37,38*]. The results are scattered, which prevents generalizing this method. In the case of our TMF AM1 results (Fig. 11), it is obvious that this approach will lead to nonconservative predictions due to the fact that both $T(\epsilon_{max})$ and T

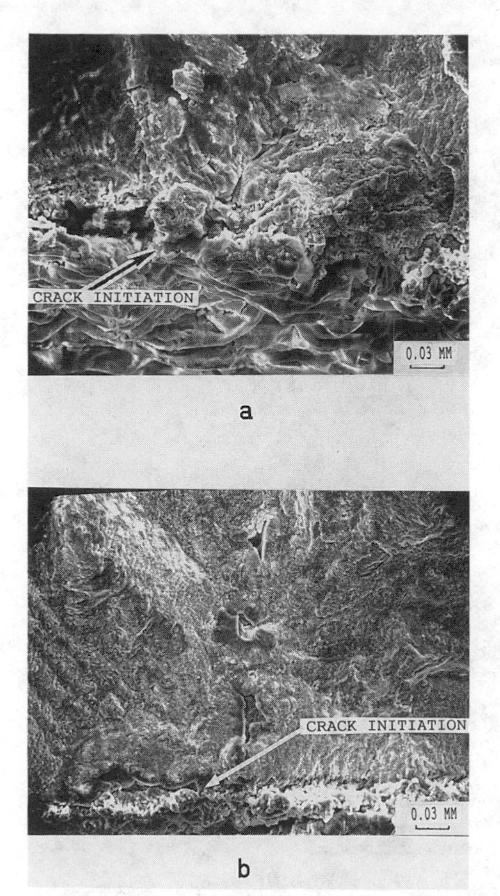

FIG. 13—*SEM investigations on AM1 coated TMF specimens:* (a) *coating-basic alloy interface rupture and* (b) *casting defects in the failed surface.*

(ϵ_{min}) LCF curves are above the TMF data. The development of a fatigue-oxidation base model [*39*], which gives quite good predictions for LCF and thermal fatigue in equiaxed alloys, seems more promising.

In SNECMA, the continuous damage model, developed by ONERA [*40*] for high temperature LCF, has been applied to TMF tests on IN 100, using first a simplified code with original fatigue damage law [*40*], and second, a more sophisticated derived model with a modified fatigue law [*41*] (see Table 2). For this application, the total fatigue damage per cycle has been determined using coefficients that are independent of temperature, and stress standardization by the ultimate stress at the considered temperature. The results in Fig. 15 show that both models are suitable. Nevertheless, the first one overpredicts TMF lives at high strain amplitudes and tends to be conservative for long lives. The second improved model is quite fitted with experimental data, except for low strain amplitudes, where its prediction is too optimistic. So, within a factor of about two, it seems to be suitable to predict fatigue life of structures under realistic loading conditions. However, it must be remarked that the extrapolation of this method to directionally solidified (DS) and single crystal alloys needs to be assessed cautiously according to possible effects of crystallographic anisotropy. Snecma has initiated an important research program in this direction in cooperation with other French laboratories.

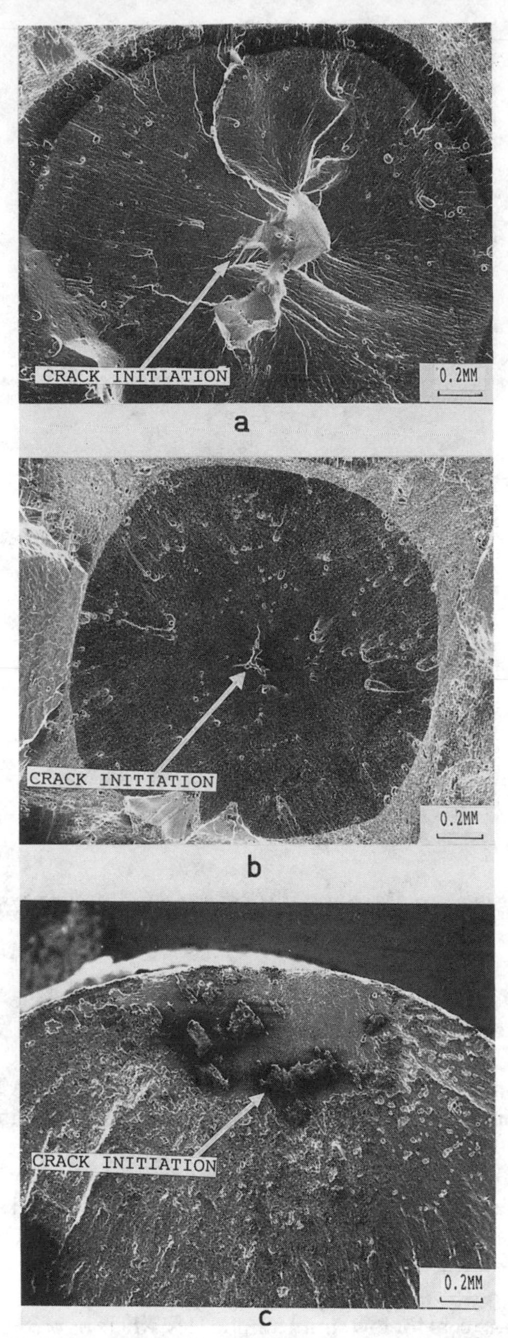

FIG. 14—*SEM investigations on AM1 isothermal LCF specimens:* (a) *650°C uncoated-crack initiation from internal casting defect,* (b) *950°C-uncoated-crack initiation from internal casting defect,* (c) *1100°C-uncoated-crack initiation from internal casting defect,* (d) *subsurface defect crack initiation (800°C), and* (e) *950°C coated-crack initiation from internal casting defect.*

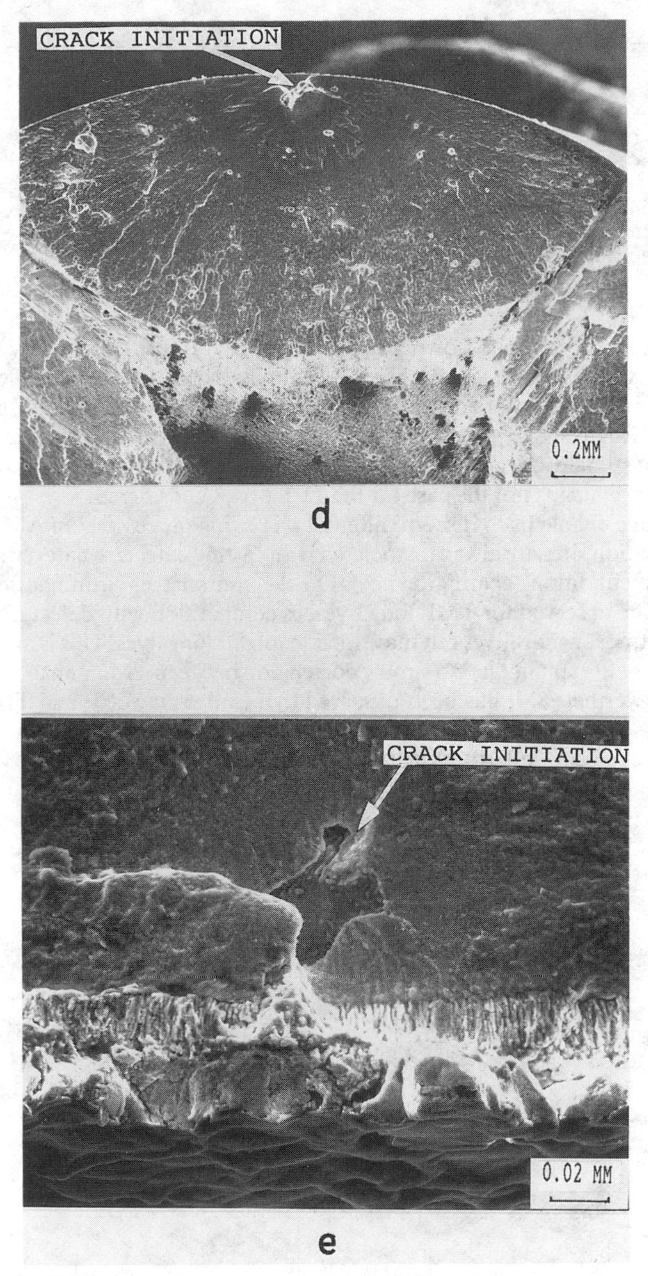

FIG. 14—*Continued.*

Discussion and Prospects

TMF tests are discriminatory experiments to validate constitutive law for material behavior. Direct comparison, in the same loading conditions, of stress-strain loops has highlighted the behavior differences for equiaxed and anisotropic alloys (DS and single crystal), conveyed in terms of lives. So, the stress amplitude value appears to be an important parameter for life

TABLE 2—*Continuous damage models.*

Damage	Pure Fatigue	Pure Creep	Creep-Fatigue Cumulation
Model I	Chaboche original law [*40*]	Kachanov[b] [*40*]	linear
Model II	Chaboche derived[a] law [*41*]	Rabotnov-Kachanov[c] [*40*]	nonlinear

[a] Same as original plus non-damaging stress limit.
[b] Two coefficients.
[c] Same as above plus 1 stress depending coefficient.

prediction. This is corroborated owing to the adequate results delivered by the continuous damage model, which is founded on the stress amplitude concept. Nevertheless, such a thorough approach is necessary for accurate predictions since simplified relationships, based only on experimental analysis, are not always as fruitful. In fact, correlating isothermal fatigue and TMF data is not straightforward. It must be noticed that the TMF cycle employed favors compressive damage, which is not the case for the LCF cycles considered.

TMF tests have enable us to discern different damaging processes. On AM1 single crystal, LCF crack initiation sites are always associated with casting defects, whatever the temperature and the surface condition (coating or not). On the contrary, environment effects on crack nucleation can be observed for TMF long lives, in competition with defect nocivity, which is always dominating for short lives but may operate too for long lives. This is another important factor that may help explain that no simple correlation between TMF and isothermal LCF has been found. Nevertheless, it has been observed that isothermal LCF and TMF induced the same damaging effects in equiaxed IN100 [*9,15*] and in that case, an equivalence between TMF and LCF can be found experimentally [*16*]. This is not the case for AM1 single crystal, where damaging processes are more complex under realistic conditions than under isothermal

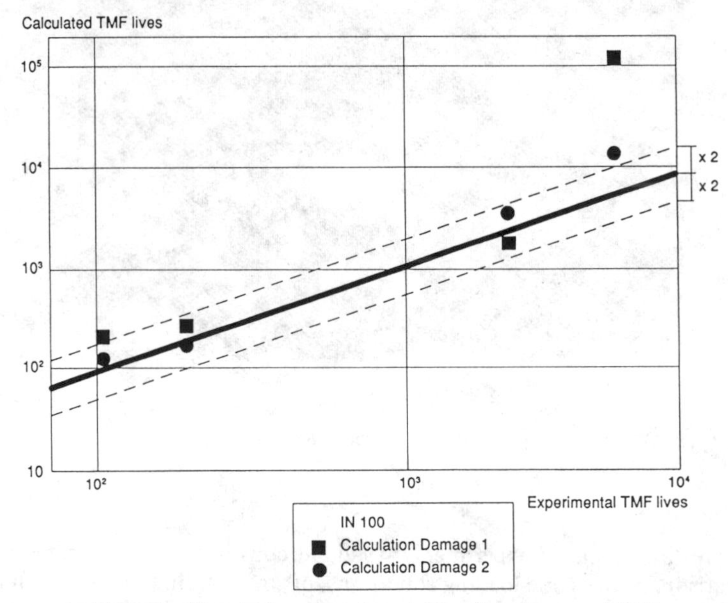

FIG. 15—*Comparison experimental/predicted lives on IN100.*

LCF conditions: TMF tests tend to simulate those complex conditions and the knowledge of damaging processes is very important to direct damage models development, especially for alloys that exhibit metallurgical peculiarities such as anisotropic materials.

Considering a two-materials approach representative of the base superalloy and associated coating, the conclusions need to be detailed since the effect of coating is not the same under isothermal and TMF loading conditions. It must be considered that the coating exhibit two quite different behaviors according to the temperature [*26*]: brittle at low and intermediate temperature, and ductile at high temperature. TMF tests according to various cycles (ϵ_{max} and ϵ_{min} at different temperatures above or under the brittle to ductile transition) are necessary to characterize the observed phenomenon and to separate individual effects of base alloy and coating.

However, it must be pointed out that the effect of coatings on fatigue lives of blade alloys, both isothermal and TMF, is generally not so clear. As isothermal fatigue is concerned, it has been observed that coating may have a beneficial (or at least neutral) effect [*42*] or detrimental [*43*] on life. For TMF, no definitive conclusion can be drawn since coating may improve the lives [*26*] or not [*44*]. More exhaustive studies are needed to account for these observations.

In a turbo engine, apart from turbine blades, other critical parts are submitted to variable temperatures under high stresses. Disks now work for definite periods at high temperature, where viscoplastic effects become dominant. Moreover, fatigue crack propagation mechanisms change within this temperature range. Therefore, TMF tests on disk alloys, between low and intermediate temperature are of particular interest [*45*] and SNECMA is adapting its TMF device to conduct such tests on disk superalloys, mainly powder metallurgy (PM) materials.

Conclusion

TMF tests have been performed on three blade alloys under a reference cycle representative of actual loading conditions. These experiments allowed the validation of behavior-constitutive laws for two of these alloys, but they have also highlighted the inadequacy of the model for the third material.

The damage processes have been investigated in the single crystal superalloy. Also, the beneficial influence of coating on the TMF life has been shown, opposite to isothermal LCF tests. Different damage processes may be involved in TMF loading, thus a unique relationship between isothermal LCF and TMF data cannot generally be made. A damage model based on stress amplitude variations seems adequately predict TMF lives for equiaxed superalloys but improvements are needed to take into account anisotropic effects for DS and single crystal alloys.

TMF tests are found to be very attractive for material mechanics studies. Beyond high temperature blade cast alloys, they are now developed for intermediate temperature disk superalloys, where viscoplastic behavior and crack propagation regimes need to be considered in realistic conditions.

References

[*1*] Rémy, L., *Fatigue at High Temperature,* Société Française de Métallurgie, Paris, 1986, pp. 252–274.

[*2*] Jaske, C. E., *Thermal Fatigue of Materials and Components, ASTM STP 612,* American Society for Testing and Materials, Philadelphia, 1976, pp. 170–198.

[*3*] Taira, S., Motoaki, F., and Takashi, H., *Symposium on Mechanical Behavior of Materials,* Society of Materials Science, Kyoto, Japan, 1974, pp. 257–264.

[*4*] Marchand, N., L'Espérance, G., and Pelloux, R. M., *Low Cycle Fatigue, ASTM STP 942,* American Society for Testing and Materials, Philadelphia, 1988, pp. 638–656.

[5] Hopkins, S. W., *Thermal Fatigue of Materials and Components, ASTM STP 612,* American Society for Testing and Materials, Philadelphia, 1976, pp. 157–169.
[6] Rau, C. A., Gemma, A. E., and Leverant, G. R., *Fatigue at Elevated Temperature, ASTM STP 520,* American Society for Testing and Materials, Philadelphia, 1973, pp. 166–178.
[7] Cook, T. S., Kim, K. S., and McKnight, R. L., *Low Cycle Fatigue, ASTM STP 942,* American Society for Testing and Materials, Philadelphia, 1988, pp. 692–708.
[8] Karayaka, M. and Sehitoglu, H., *Metallurgical Transactions A,* Vol. 22, 1991, pp. 697–706.
[9] Malpertu, J. L. and Rémy, L., *Low Cycle Fatigue, ASTM STP 942,* 1988, pp. 657–671.
[10] Rémy, L., *High Temperature Fracture Mechanisms and Mechanics,* Mechanical Engineering Publication, EGF 6, Dourdan, France, 1990, pp. 353–378.
[11] Korn, C. and Pluvinage, G., *Fatigue des Structures Industrielles,* Lieurade Editor, IITT Paris, 1989, pp. 74–104.
[12] Baumgärtner, T., Bothe, K., Hurta, S., and Gerold, V., *High Temperature Materials for Power Engineering,* Kluwer Academic Publishers, Liège, Belgium, 1990, pp. 1087–1096.
[13] Henderson, P., Lindblom, J., and Ivarsson, B., *High Temperature Materials for Power Engineering,* Kluwer Academic Publishers, Liège, Belgium, 1990, pp. 1151–1162.
[14] Joos, R., Elzey, D. M., and Artz, E., *High Temperature Materials for Power Engineering,* Kluwer Academics Publishers, Liège, Belgium, 1990, pp. 1173–1184.
[15] Malpertu, J. L. and Rémy, L., *Metallurgical Transactions A,* Vol. 21, 1990, pp. 389–399.
[16] Reger, M. and Rémy, L., *Metallurgical Transactions A,* Vol. 19, 1988, pp. 2259–2268.
[17] Rongvaux, J. M. and Guédou, J. Y., *High Temperature Fracture Mechanisms and Mechanics,* Mechanical Engineering Publication, EGF 6, Dourdan, France, 1990, pp. 95–118.
[18] Guédou, J. Y. and Honnorat, Y., *Constitutive Laws of Plastic Deformation and Fracture,* Krauz et al., Eds., Kluwer Academic Publishers, Ottawa, Canada, 1990, pp. 35–42.
[19] Gaugé, G. and Morbioli, R., *High Temperature Protective Coatings, Proceedings of the 112th American Institute of Mechanical Engineers Annual Meeting,* E. Singhal, Ed., 1983, pp. 13–26.
[20] Morbioli, R. and Honnorat, Y., *Coating for Heat Engines, Proceedings of 1st NATO Advanced Workshop,* J. Clark, Ed., Acquafredda, Italy, 1984, pp. 179–191.
[21] Moreno, V. and Jordan, E. H., *International Journal of Plasticity,* Vol. 2, 1986, pp. 223–245.
[22] Vincent, J. N. and Rémy, L., *Thermal Mechanical Fatigue on René 77,* Internal Report, EMP/SNECMA, Evry, France, 1987.
[23] Chaboche, J. L., *International Journal of Plasticity,* Vol. 5, 1989, pp. 247–302.
[24] Dambrine, B. and Mascarell, J. P., *Recherche Aerospatiale,* Vol. 1, 1988, pp. 35–45.
[25] Dambrine, B. and Mascarell, J. P., *High Temperature Fracture Mechanisms and Mechanics,* Mechanical Engineering Publication, EGF 6, Dourdan, France, 1990, pp. 195–210.
[26] Bernard, H. and Rémy, L., *High Temperature Materials of Power Engineering,* Kluwer Academic Publishers, Liège, Belgium, 1990, pp. 1185–1194.
[27] Vincent, J. N. and Rémy, L., *Thermomechanical Fatigue on Single Crystal Turbine Blades Alloys,* (DRET/ARMINES) Report 84-221, French Ministry of Defense/Ecole des Mines de Paris, 1986.
[28] Fleury, E. and Rémy, L., *High Temperature Materials for Power Engineering,* Kluwer Academic Publishers, Liège, Belgium, 1990, pp. 1007–1016.
[29] Hanriot, F., Fleury, E., and Rémy, L., *High Temperature Materials for Power Engineering,* Kluwer Academic Publishers, Liège, Belgium, 1990, pp. 997–1006.
[30] Fredholm, A. and Strudel, J. L., *Superalloys 84, 15th International Symposium on Superalloys,* Gell et al., Ed., The Metals Society of American Institute of Mechanical Engineers, 1984, pp. 211–223.
[31] Ayrault, D., *Creep at High Temperature in Nickel-Base Single Crystal Superalloys,* Thesis, Ecole des Mines de Paris, France, 1989.
[32] Halford, G. R. and Manson, S. S., *Thermal Fatigue of Materials and Components, ASTM STP 612,* American Society for Testing and Materials, Philadelphia, 1976, pp. 239–254.
[33] Taira, S., *Fatigue at Elevated Temperature, ASTM STP 520,* American Society for Testing and Materials, Philadelphia, 1973, pp. 80–101.
[34] Antolovitch, S. D., Liu, S., and Baur, R., *Metallurgical Transactions A,* Vol. 12, 1981, pp. 473–481.
[35] Degallaix, G., Korn, C., and Pluvinage, G., *Fatigue and Fracture for Engineering Materials and Structures,* Vol. 13, No. 5, May 1990, pp. 473–485.
[36] Pejsa, P. N. and Cowles, B. A., *ASME Journal of Engineering for Gas Turbine and Power,* No. 86-GT-123, 1986.
[37] François, M. and Rémy, L., *Fatigue and Fracture for Engineering Materials and Structures,* Vol. 14, No. 1, Jan. 1991, pp. 115–129.
[38] Fleury, E. and Rémy, L., *Advance in Fracture Research,* ICF 7, Pergamon Press, Houston, TX, 1989, pp. 1133–1140.

[*39*] Rezai-Aria, F. and Rémy, L., *Engineering Fracture Mechanics,* Vol. 34, No. 2, 1989, pp. 283–294.
[*40*] Chaboche, J. L., *Journal of Applied Mechanics,* Vol. 55, 1988, pp. 65–72.
[*41*] Chaboche, J. L. and Lesne, P. M., *Fatigue and Fracture for Engineering Materials and Structures,* Vol. 11, No. 1, 1988, pp. 1–17.
[*42*] Strangman, T. E., Fujii, M., and NGuyen-Dinh, X., *Superalloys 84,* M. Gell et al., Eds., The Metals Society of the American Institute of Mechanical Engineers, 1984, pp. 795–804.
[*43*] Wood, M. I. and Restall, J. E., *High Temperature Alloys for Gas Turbines and other Applications,* Reidel Publishing Co., Liège, Belgium, 1986, pp. 1215–1226.
[*44*] Bain, K. R., *The Effect of Coatings on the Thermomechanical Fatigue Life of a Single Crystal Turbine Blade Material,* AIAA, 1985.
[*45*] Heil, M. L., Nicholas, T., and Haritos, G. K., *Thermal Stresses, Material Deformation and Thermomechanical Fatigue,* ASME Pressure Vessel Piping, Vol. 123, H. Sehitoglu and S. Zamrik, Eds., American Society for Mechanical Engineers, New York, 1987, pp. 23–29.

Gary R. Halford,[1] *Bradley A. Lerch,*[1] *James F. Saltsman,*[1] *and Vinod K. Arya*[2]

Proposed Framework for Thermomechanical Fatigue (TMF) Life Prediction of Metal Matrix Composites (MMCs)

REFERENCE: Halford, G. R., Lerch, B. A., Saltsman, J. F., and Arya, V. K., **"Proposed Framework for Thermomechanical Fatigue (TMF) Life Prediction of Metal Matrix Composites (MMCs),"** *Thermomechanical Fatigue Behavior of Materials, ASTM STP 1186,* H. Sehitoglu, Ed., American Society for Testing and Materials, Philadelphia, 1993, pp. 176–194.

ABSTRACT: The framework of a mechanics of materials model is proposed for thermomechanical fatigue (TMF) life prediction of unidirectional, continuous-fiber metal matrix composites (MMCs). Axially loaded MMC test samples are analyzed as structural components whose fatigue lives are governed by local stress-strain conditions resulting from combined interactions of the matrix, interfacial layer, and fiber constituents. The metallic matrix is identified as the vehicle for tracking fatigue crack initiation and propagation. The proposed framework has three major elements. First, TMF flow and failure characteristics of in situ matrix material are approximated from tests of unreinforced matrix material, and matrix TMF life prediction equations are numerically calibrated. The macrocrack initiation fatigue life of the matrix material is divided into microcrack initiation and microcrack propagation phases. Second, the influencing factors created by the presence of fibers and interfaces are analyzed, characterized, and documented in equation form. Some of the influences act on the microcrack initiation portion of the matrix fatigue life, others on the microcrack propagation life, while some affect both. Influencing factors include coefficient of thermal expansion mismatch strains, residual (mean) stresses, multiaxial stress states, off-axis fibers, internal stress concentrations, multiple initiation sites, nonuniform fiber spacing, fiber debonding, interfacial layers and cracking, fractured fibers, fiber deflections of crack fronts, fiber bridging of matrix cracks, and internal oxidation along internal interfaces. Equations exist for some, but not all, of the currently identified influencing factors. The third element is the inclusion of overriding influences such as maximum tensile strain limits of brittle fibers that could cause local fractures and ensuing catastrophic failure of surrounding matrix material. Some experimental data exist for assessing the veracity of the proposed framework.

KEYWORDS: metal matrix composites, fatigue (metal), thermal fatigue, thermomechanical fatigue, bithermal fatigue, low cycle fatigue, high temperature fatigue, creep fatigue, life prediction, strainrange partitioning, crack initiation, crack propagation, thermal expansion

Nomenclature

Symbols

B	Intercept of elastic strainrange—life relations
C	Intercept of inelastic strainrange—life relations
C'	Intercept of inelastic line for combined creep-fatigue cycles

[1] Senior scientific technologist and research engineers, respectively, NASA Lewis Research Center, Mail Stop 49-7, Cleveland, Ohio 44135-5000.

[3] Resident research associate, University of Toledo, NASA Lewis Research Center, Mail Stop 49-7, Cleveland, OH 44135-5000.

CC	Creep strain in tension, creep strain in compression
CP	Creep strain in tension, plastic strain in compression
CTE	Coefficient of thermal expansion
F	Inelastic strain fraction
K	Cyclic strain-hardening coefficient
N	Number of applied cycles
PC	Plastic strain in tension, creep in compression
PP	Plastic strain in tension, plastic strain in compression
R	Ratio, algebraic minimum to maximum
V	Ratio, mean to alternating
MF	Multiaxiality factor
TF	Triaxiality factor
Δ	Range of variable
ϵ	Strain
Γ	Ratio, absolute value of the compressive to tensile flow strengths evaluated at their respective temperatures and strain rates, preferably from cyclic isothermal or bithermal experiments
σ	Stress
Σ	Summation

Subscripts

cc	Creep strain in tension, creep strain in compression
cp	Creep strain in tension, plastic strain in compression
el	Elastic
f	Failure with zero mean stress
fm	Failure with mean stress
ij	pp, pc, cp, or cc
in	Inelastic
i	Microcrack initiation with zero mean stress
L	Load
p	Microcrack propagation with zero mean stress
pc	Plastic strain in tension, creep in compression
pp	Plastic strain in tension, plastic strain in compression
t	Total
y	Yield (0.2% offset)
ϵ	Strain
σ	Stress

Superscripts

b	Power for elastic strainrange—life relations
c	Power for inelastic strainrange—life relations
n	Cyclic strain-hardening exponent ($\approx b/c$)

Introduction

Low resistance to thermomechanical fatigue (TMF) is expected to be a limitation to the use of continuous filament, metal matrix composites (MMCs) material systems in high temperature applications. This is because an MMC will experience cyclic stresses and strains from three sources (Fig. 1) rather than only two for conventional monolithic alloys. The potential

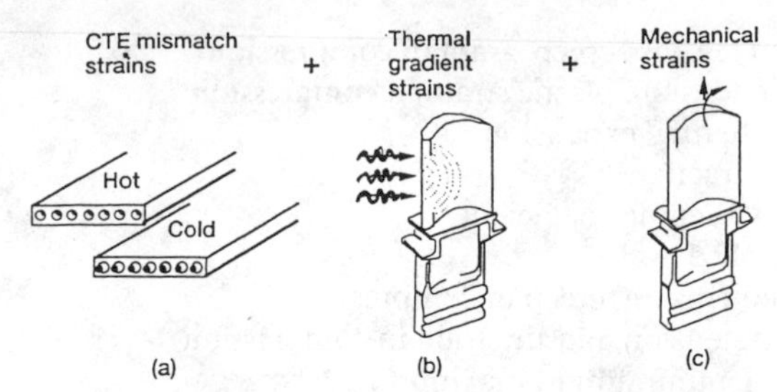

FIG. 1—*Triple contributions to stresses and strains in thermal fatigue of MMCs;* (a) *Coefficient of thermal expansion mismatch induced thermal stresses and strains,* (b) *Thermal gradients producing thermal stresses and strains,* (c) *Applied mechanical loading producing active stresses and strains.*

TMF degradation is particularly severe when the phasing of thermal and mechanical loadings cause the three strain contributors to have the same algebraic sign.

Typically, the coefficient of thermal expansion of the fibers is considerably different (lower) than that of the matrices. Thus, when the temperature of an MMC material is changed, opposing internal thermal stresses and strains are distributed between the constituents in a manner consistent with equilibrium, compatibility, and constitutive stress-strain relations of fiber, matrix, and interfacial material.

The thermal expansion mismatch imposes deleterious stresses and strains within the matrix, fiber, and interface, even in the absence of mechanical loads or conventional thermal strain-inducing temperature gradients. Thus, structural components made of an MMC with a sizable CTE-mismatch and operating over a large temperature range will experience a potentially large range of internal cyclic mechanical strain.

Consequently, component design procedures will mandate use of engineering structural analyses and durability models in attempts to mitigate effects of potential TMF degradation in MMCs.

This manuscript presents a framework for performing fatigue life analyses of MMCs for thermal and thermomechanical loading.

Proposed Framework for TMF Life Prediction Model

General

The framework of an engineering model that addresses the key durability issues involved in the TMF resistance of continuous fiber, [0°] MMCs for high temperature applications. Axially loaded MMC material specimens are treated as conventional ministructures (Fig. 2) whose weakest, most highly stressed links are the origins of fatigue cracking. As with conventional structures, the TMF durability of an MMC is expected to be governed by localized stress,

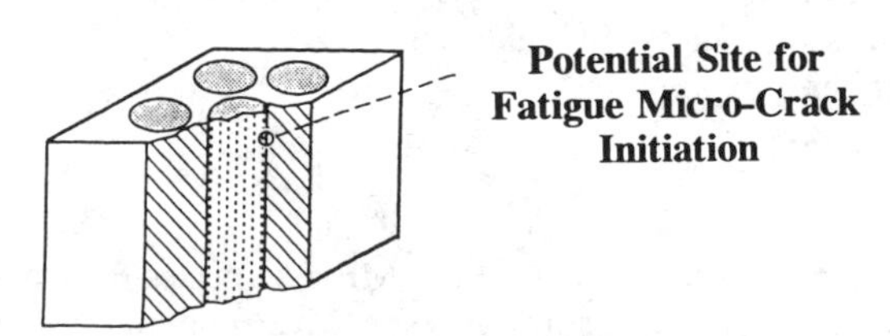

FIG. 2—*A composite material is a mini-structure.*

strain, and temperature conditions, and their cyclic fluctuation with time. Linear elastic and nonlinear viscoplastic structural analysis techniques applicable to thermally cycled MMCs are under development [*1–4*].

If the reinforcing fibers are brittle ceramic, they would not be expected to be prone to cycle-dependent fatigue or time-dependent creep. Nor would one expect brittle interfacial layer material to undergo these classical mechanisms of degradation. Cracking occurs in these constituents, but the cracks initiate, grow, and arrest according to brittle fracture criteria. Thus, insofar as modeling damage initiation and propagation by cycle-dependent fatigue and time-dependent creep are concerned, we focus our attention on the ductile metallic matrix material. The matrix is used as the vehicle for tracking fatigue damage evolution.

The proposed framework consists of three major elements.

1. The TMF equations that approximate the in situ matrix material are calibrated from test results of unreinforced matrix material. The measured macrocrack initiation fatigue life (failure of coupon-sized axial specimen) of the matrix material is partitioned into microcrack initiation and microcrack propagation phases in accordance with proposed equations.
2. The influencing factors created by the presence of fibers and interfaces are analyzed, characterized, and documented in equation form. Some of the influences act on the microcrack initiation portion of the matrix fatigue life, others on the microcrack propagation life, while some affect both. Influencing factors will be discussed in an ensuing section.
3. There is the recognition of overriding influences such as a maximum tensile strain limit of brittle fibers that could cause individual fiber fractures and ensuing catastrophic failure of the surrounding matrix.

Some experimental data are available to assess the potential capability of the proposed framework.

TMF Failure Behavior of the Matrix

Since isothermal fatigue and creep-fatigue behavior alone may be insufficient for estimating fatigue behavior under thermal cycling conditions [*5*], it is desirable that the matrix material's baseline failure behavior be characterized through thermal, rather than isothermal, cycling experiments. We recommend using bithermal-cycling experiments as proposed in Ref *6* and adopted in the thermomechanical fatigue (TMF) life prediction model [*7*] based on the concepts of the total strain version of strainrange Partitioning (TS-SRP) [*8,9*]. The applicability of the TMF/TS-SRP approach has been verified recently for two high-temperature aerospace superalloys, cast B-1900 + Hf and wrought Haynes 188 [*10*].

Bithermal test results are used to establish the constants in the matrix TMF life prediction equation. Bithermal fatigue (in-phase and out-of-phase PP) and bithermal creep-fatigue (in-phase CP and out-of-phase PC) experiments [*6*] produce hysteresis loops of the types shown in Figs. 3*a* to *d*. The term creep-fatigue is used because the cycles emphasize creep deformation superimposed on strain-limited fatigue cycling. As the tests typically are performed on specimens in a laboratory air atmosphere, they are also subject to oxidation. For present purposes, it is to be understood that creep-fatigue experiments in air produce results that reflect the interactions of all three important failure behavior factors: fatigue, creep, and oxidation. Failure behavior is not particularly sensitive to the exact details of the wave shape of the cycle, provided the partitioning of the inelastic strains is similar.

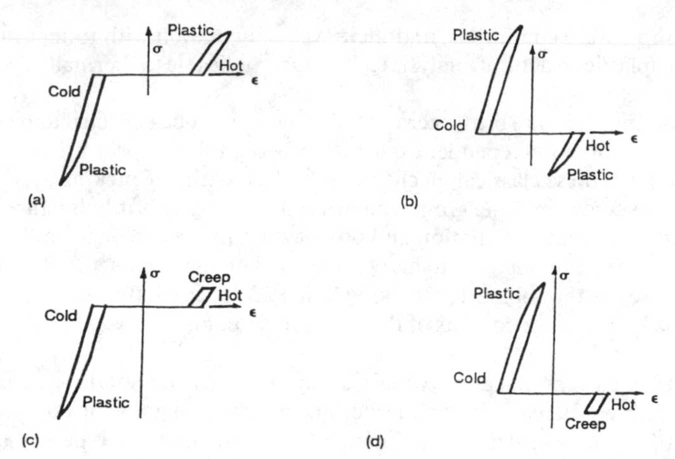

FIG. 3—*Schematic stress-strain hysteresis loops for bithermal fatigue and creep-fatigue;* (a) *bithermal, high-rate, in-phase, continuous strain cycle producing 100% PP strain range,* (b) *bithermal, high-rate, out-of-phase, continuous strain cycle producing 100% PP strain range,* (c) *bithermal, tensile creep, in-phase, compressive plasticity cycle producing CP and PP strain ranges,* (d) *bithermal, compressive creep, out-of-phase, tensile plasticity cycle producing PC and PP strain ranges.*

TMF Constitutive Flow Behavior of the Matrix

In addition to documentation of bithermal fatigue and creep-fatigue failure characteristics, it is also necessary to establish the cyclic viscoplastic flow behavior of the matrix. A particularly advantageous feature of the currently proposed TMF life prediction approach is its ability to be applied to any generalized time- and temperature-dependent TMF cycle. This is achieved through the use of any one of numerous unified viscoplastic models that have been proposed over the last decade. Evaluation of the constants in these frequently complex evolutionary-type constitutive equations remains a serious problem that has limited their widespread use. The procedures are not always straightforward. Furthermore, the precisely appropriate data may not be readily available without performing additional experiments. Nonetheless, considerable cyclic flow results are automatically made available during the performance of the bithermal failure behavior tests. These results and supplemental test results, as necessary, depending upon the selected flow model, are used to evaluate the constants in a cyclic viscoplastic model. Among the unified viscoplastic constitutive models used successfully in various NASA Lewis Research Center programs are the models proposed by Bodner and Partom [*11*], Walker [*12*], Robinson and Swindeman [*13*], and Freed [*14*]. Considerable research remains to develop reliably accurate models that are easy to calibrate and implement into finite element structural analysis and life prediction codes. On occasion, we have resorted to simpler, less general, empirical, power-law relations to document the required cyclic flow behavior for use with the current TMF life prediction method [*7,8*]. In these instances, because of the sensitivity of the flow behavior to details of the wave shape, it is necessary to employ a wave shape for flow behavior documentation that is similar to the mission cycle being analyzed.

Failure Behavior Equations for the Matrix

Macrocrack Initiation—Once the flow and failure characteristics of the matrix material have been established, the total mechanical strain range versus fatigue life (macrocrack initiation life ≡ fatigue life of coupon-sized axial specimen) equation can be written for any arbitrary in-phase or out-of-phase TMF cycle.

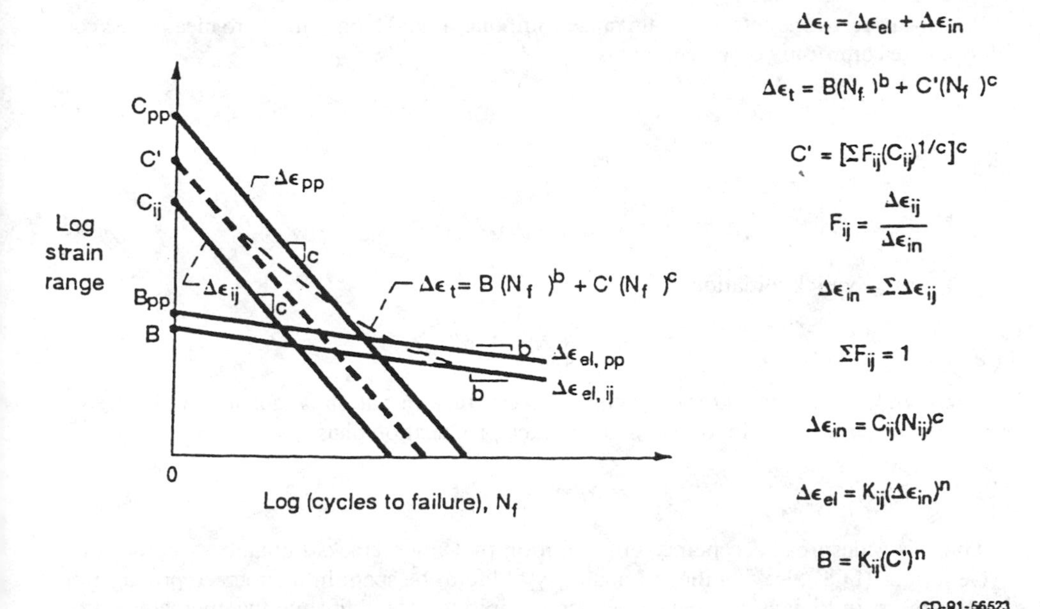

FIG. 4—*TMF fatigue life equations and low cycle fatigue curve for matrix material. Condensed from Ref* 7, (ij = pp, pc, cp, cc).

Figure 4 graphically displays the total mechanical strain range versus fatigue life relation for matrix material based on the TS-SRP representation. Pertinent equations are shown in the figure. Step-by-step procedures for the calibration of the constants in the subject equations for TMF are given in Ref 7. It is important to realize that the resultant TMF life prediction equation is for the matrix material and is applicable for the specifically stated loading condition, that is, the exact temperature and strain versus time history, or a reasonable approximation, is prescribed. Should a different set of loading conditions be imposed, a new set of equation constants are to be calculated from the calibrated flow and failure characteristics.

The TMF life prediction equation shown at this stage represents conditions involving no fibers, no geometric discontinuities (that is, no stress concentrations), zero mean stress, a uniaxial stress-strain state, and continuously repeated loadings of constant total strain range, frequency, temperature range, and so on. Modifications of the equation to deal with specific influencing factors associated with a composite are discussed in the subsequent section. Some of the influences can be described analytically and quantitative computations of the expected effect on TMF life are possible. Other influences are not as yet at the quantifiable stage, and for the meantime will require empiricism and calibration of computations with experimental TMF results.

Some of the influencing factors of the fibers will affect microcrack initiation more than microcrack propagation, and vice versa. It is convenient therefore to distinguish, analytically, these two important components of the total macrocrack initiation life.

As a starting point, we will assume that the relative proportions of microcrack initiation and propagation in TMF are similar to isothermal fatigue. Approximate equations for the microcrack initiation and propagation portions of the total fatigue life are suggested below. First, the total strain range can be decomposed into its elastic and inelastic strain range components

$$\Delta\epsilon_t = \Delta\epsilon_{el} + \Delta\epsilon_{in} \tag{1}$$

Power-law relations between strain range components and fatigue life (zero mean stress conditions) are commonly observed, that is

$$\Delta\epsilon_{el} = B(N_f)^b \quad (2)$$

and

$$\Delta\epsilon_{in} = C(N_f)^c \quad (3)$$

thus, for macrocrack initiation

$$\Delta\epsilon_t = B(N_f)^b + C(N_f)^c \quad (4)$$

Microcrack Initiation and Propagation—Macrocrack initiation fatigue life can be decomposed into microcrack initiation and microcrack propagation phases

$$N_f = N_i + N_p \quad (5)$$

There is widespread acceptance of the notion that microcracks initiate very early in low cycle fatigue (LCF), leaving the vast majority of life to be spent in microcrack propagation. Furthermore, in high cycle fatigue, microcracks initiate late in life and the microcrack propagation phase is a small fraction of the total life. Although somewhat arbitrary, it is not unreasonable to assume the following. At the lowest possible life of $N_f = 10$

$$N_i \approx 0.1N_f \quad (6a)$$

$$N_p \approx 0.9N_f \quad (6b)$$

and, at an arbitrarily high cyclic life of $N_f = 10^7$

$$N_i \approx 0.9N_f \quad (6c)$$

$$N_p \approx 0.1N_f \quad (6d)$$

Simple power-law equations can be written between N_i and N_f and N_p and N_f by using the respective sets of coordinates of Eqs 6a, b, c, and d. The resultant equations, when substituted in Eq 4 with $b = -0.12$ and $c = -0.60$ [*15*], yield the following relations between total mechanical strain range and microcrack initiation and propagation lives.

For microcrack initiation

$$\Delta\epsilon_t \approx 0.76B(N_i)^{-0.10} + 0.25C(N_i)^{-0.50} \quad (7)$$

For microcrack propagation

$$\Delta\epsilon_t \approx 1.04B(N_p)^{-0.14} + 1.20C(N_p)^{-0.70} \quad (8)$$

Influences of Fibers on Matrix Properties and Behavior

The fact that fibers are present in a composite imparts changes in both the flow and failure response of the matrix. A fully developed MMC life-prediction method must deal directly with

TABLE 1—*Factors associated with fibers mechanically influencing matrix fatigue response.*

Factor	N_i	N_p
CTE mismatch strains	Yes	Yes
Residual (mean) stresses	Yes	Yes
Multiaxial stress state	Yes	Yes
Off-axis fibers	Yes	Yes
Internal stress concentrations	Yes	No
Multiple initiation sites	Yes	Yes
Nonuniform spacing	Yes	Yes
Interfacial layers	Yes	No
Fractured fibers	Yes	No
Fiber debonding	No	Yes
Fiber crack retardation	No	Yes
Fiber bridging	No	Yes

these induced changes. A listing of the most significant mechanical influences on the surrounding matrix is given in Table 1. Each factor is identified as to whether it influences the microcrack initiation or microcrack propagation phases of the macrocrack initiation life. A few of the influences currently can be handled analytically; others require development. Currently identified influences are discussed in the following paragraphs.

Effects of Fibers and Interfaces on Matrix Flow and Failure

CTE Mismatch Strains—One of the largest influences on the cyclic strains induced in the matrix of an MMC under TMF loading is the mismatch between the fiber and matrix coefficients of thermal expansion. As shown in Table 2 for three common MMC systems, the matrix CTE is larger than the fiber CTE by as much as two to four times. Based upon the mechanics concepts of equilibrium, compatibility, and constitutive stress-strain response, Garmong [*16*] has presented analyses of the constituent stresses and strains in thermally loaded composites. Although Garmong dealt with eutectic composites, his analyses are directly applicable to [0] MMCs.

Slow, uniform temperature cycling (with no externally applied load) of a typical [0] MMC will induce a cyclic strain range in the matrix. Since the matrix is mechanically strained by the fibers as the temperature changes, the matrix undergoes a TMF cycle. The matrix TMF cycle is out-of-phase. Should the MMC also be loaded in an out-of-phase TMF cycle, the externally applied mechanical strain will add directly to the internally induced mechanical CTE mismatch strain. The resultant increased strain range in the matrix will lower the macrocrack initiation life of the matrix in accordance with Eq 4. On the other hand, if the MMC is subjected to in-phase TMF loading, the two matrix strain contributions will tend to subtract from one another, depending upon the exact details of the phasing of temperature and strain. For iso-

TABLE 2—*Approximate values of coefficients of thermal expansion (CTE) at room temperature of matrix and fiber for three common high temperature composite systems.*[a]

Composite System	W/Cu	SCS-6/Ti-15-3	SCS-6/Ti-24Al-11Nb
Matrix	16.0	9	12
Fiber	4.4	4.9	4.9

[a] CTE ($\times$ 10^{-6}°C^{-1}).

thermal fatigue at room temperature, the CTE mismatch strain caused by cooling from a fabrication/heat treatment temperature usually produces a tensile residual stress in the matrix that will act as a mean stress in subsequent fatigue loading provided the residual stress does not relax due to inelasticity in the matrix.

Mean Stress Effects—Procedures for dealing with mean stresses (due to residual stresses or actively applied mean loads) in strain-based fatigue life models are covered in Ref *7*. The currently adopted mean stress model is based on a modification [*17*] of the Morrow mean stress approach [*18*]. The following equation can be derived from Morrow's approach. It has direct applicability to isothermal, nominally elastic cyclic loading conditions

$$(N_{fm})^{-b} = (N_f)^{-b} - V_\sigma \tag{9}$$

V_σ is the ratio of mean to alternating stress and is equal to the inverse of the classical fatigue A ratio. For strain-controlled cycles involving detectable amounts of inelasticity, any initially present mean stresses will tend to cyclically relax, numerically reducing the value of V_σ. During creep-fatigue loading, the numerical value of V_σ may be non-zero, yet it should not be expected to affect cyclic life in a conventional mean stress manner. Consequently, an effective V_σ was defined in Ref *17*, $V_{eff} = k \times V_\sigma$, where, $k \rightarrow 0$ for $\Delta\epsilon_{in}/\Delta\epsilon_{el} \rightarrow 0.2$ and larger, and $k \rightarrow 1$ for $\Delta\epsilon_{in}/\Delta\epsilon_{el} \rightarrow 0$.

The effective mean stress ratio, V_{eff}, for nonisothermal conditions has been proposed [5] to take the form

$$V_{eff} = (1 + R_\sigma/\Gamma)/(1 - R_\sigma/\Gamma) \tag{10}$$

where Γ is a strength ratio defined in the nomenclature.

Accurate calculation of matrix mean stresses requires accurate nonlinear structural analysis procedures and viscoplastic constitutive modeling of the matrix material. Mean stress can affect the fatigue macrocrack initiation resistance of MMCs by more than an order of magnitude in cyclic life. For example, for an isothermal, nominally elastic case with $b = -0.12$, $V_\sigma = 1.0$ (that is, zero-to-max stresses), and a corresponding life with zero mean stress, $N_f =$ 100 000, the fatigue life, N_{fm}, with a sustained mean stress is calculated from Eq 9 to be less than 9000 cycles to failure.

Multiaxiality of Stress—Fibers cause multiaxial stress-strain states within the matrix (even though the loading on the unidirectional MMC is uniaxial) as a result of their differing elastic and plastic stress-strain properties, and their differing thermal conductivities and expansion coefficients. Any deviation from uniaxial loading can be handled by using any of a number of multiaxiality rules proposed in the literature. A relatively simple procedure for calculating the life influencing effects of multiaxial stress states is given by Manson and Halford [*19,20*]. The approach is based on von Mises effective stress-strain and the multiaxiality factor, *MF* (a measure of the degree of hydrostatic stress normalized by the corresponding von Mises effective stress). The equation for the multiaxial factor is

$$MF = TF \quad TF \geq 1 \tag{11a}$$

$$MF = 1/(2 - TF) \quad TF \leq 1 \tag{11b}$$

where

$$TF = (\sigma_1 + \sigma_2 + \sigma_3)/(1/\sqrt{2})\sqrt{(\sigma_1 - \sigma_2)^2 + (\sigma_1 - \sigma_3)^2 + (\sigma_2 - \sigma_3)^2} \tag{11c}$$

and, σ_i = principal stresses ($i = 1, 2, 3$). While multiaxial stress-strain states are always present in thermally cycled MMCs, their effect on cyclic durability of axially loaded [0] MMCs does not appear to be exceptionally great, at least for the degrees of multiaxiality that are self-induced (as opposed to externally imposed multiaxial loading). The multiaxiality factor for unidirectional loading of a unidirectional fiber layup for the SCS-6/Ti-15-3 MMC system was determined to be only 1.04 [*4*]. Multiplying 1.04 times the computed effective strain range yields the strain range for entering the matrix total strain range versus macrocrack initiation life equation. This would give rise to a calculated decrease in fatigue life of only about 10%.

Off-Axis Fibers—Using data for an *E*-glass fiber/polymeric matrix composite, Hashin and Rotem [*21*] analyzed the influence of off-axis fibers. As an example of their analytical and experimental findings, a shift from a 5 to a 10° off-axis loading showed a loss of a factor of approximately two in isothermal fatigue strength. The corresponding loss in cyclic lifetime was measurable in terms of multiple orders of magnitude. A 60° off-axis loading resulted in a loss of nearly an order of magnitude in fatigue strength. Relatively small deviations from [0] can be responsible for large losses in fatigue resistance. Although the example was for a polymeric matrix composite, MMC fatigue behavior would be expected to be comparable. Modifications to Hashin and Rotem's equations will be necessary to make their analysis compatible with the matrix strain-based approach under development herein.

Internal Stress Concentrations/Multiple Initiation Sites—An internal microstress concentration factor produces higher local internal stresses and strains and promotes earlier microcrack initiation. Once the microcrack grew away from the local concentration, however, the concentration effect would diminish and the microcrack propagation portion of life would be relatively unaffected. An internal stress-strain concentration factor can be multiplied times the calculated nominal stresses and strains, or the nominal strain range can be entered into a modification of Eq 7 (wherein the coefficients 0.8 and 0.34 have been divided by the value of the concentration factor).

Internal crack initiation at local stress-strain concentrations can occur in MMCs at multiple initiation sites, thus leading to shorter paths for cracks to follow prior to linking together and hastening the macrofracture of the composite. If microcrack growth paths are decreased by an average of a factor of two, it would be expected that the micropropagation phase of life would also decrease by a factor of two.

Nonuniform Spacing—The recent work of Bigelow [*22*] represents an excellent example of how structural analysis is used to determine the influence of nonuniform fiber spacing on the stress-strain response behavior of the in situ matrix. Using finite element structural analysis modeling of fabrication cool-down stresses in an MMC with uneven fiber spacing, Bigelow has been able to demonstrate that increased matrix stresses (longitudinal, radial, and hoop directions relative to fiber) result from decreasing local fiber spacing. Greater local stresses translate into increased ranges of local strain for cyclic temperatures. Hence, Eq 7 would be used to predict the expected lower microcrack initiation life. Little effect on the microcrack propagation life would be expected for this influencing factor.

Interfacial Layers—Additional examples of using structural analyses to ascertain effects of fibers and interfaces on matrix stress-strain response is found in the work of Jansson and Leckie [*23*] and Arnold *et al.* [*24*] for assessing the influence of compliant interfacial layers between fibers and matrix.

Additional Influencing Factors Associated with Cracks or Debonding—Further analytic development of the proposed framework will be required to quantitatively model the impact on microcrack propagation life of some of the influencing factors not discussed above. Of particular interest are those factors directly associated with microcracks within an MMC, that is, fractured fibers and fiber debonding (the concepts proposed by Chen and Young [*25*], among others, offers a promising approach), fiber crack retardation, and fiber bridging (see applicable work of Ghosn et al. [*26*]).

Once determined, the induced stresses and strains from any of the influences discussed above are expected to alter the matrix failure behavior in accordance with experience on monolithic metallic materials, that is, local increases in strain range will reduce N_i, higher tensile mean stresses will decrease N_i, a higher multiaxiality factor will reduce N_i, etc. The more highly localized is the stress and strain, the less likely a significant influence will be expected on N_p.

Metallurgical Interactions—The very presence of distributed nonmetallic fibers can also influence the metallurgical state of the matrix material. Heat treating of the matrix in the presence of fibers can result in a somewhat different microstructure, yield strength, ultimate tensile strength, ductility, and hardness than obtained by heat treating unreinforced matrix metal [*27*]. This influence will compound the problem of isolating the true cyclic flow and failure behavior of the in situ matrix material.

Another crucial factor is the potential for increased internal oxidation of the MMC made possible by the interfacial layers acting as diffusional pipelines within the interior of the MMC. Internal oxidation will dramatically reduce an MMC's TMF resistance by promoting early microcrack initiation and faster microcrack propagation.

TMF models with separate terms for cyclic oxidation damage interaction with creep and fatigue have been proposed recently by Nissley, Meyer, and Walker [*28*], Neu and Sehitoglu [*29,30*], and Miller et al. [*31*]. Perhaps features of these models can be adapted to the prediction of internal oxidation effects in high temperature MMCs.

Mechanical Effects of Matrix on Fibers

It is also necessary to examine the mechanical influence of the matrix on the response characteristics of the reinforcing fibers. Fibers of primary concern are elastic and brittle. They fail progressively throughout the fatigue life of the MMC as a result of the continual shedding of tensile stresses from the matrix material as it cyclically deforms. Cyclic relaxation of mean (residual) stresses and strain hardening or softening of the matrix results in various scenarios of behavior depending upon the combination of operative conditions.

For illustrative purposes, it is informative to select the specific condition of zero to maximum tensile load control to aid in a qualitative micromechanics analysis of how the fibers are influenced by the matrix behavior. Cyclic tensile mean stress relaxation as well as cyclic strain softening of the matrix causes a shift of additional tensile stress (and hence, strain) to the fibers, pushing them closer to their critical fracture strain. If the matrix cyclically strain hardens, there will be a tendency for the matrix to carry a larger portion of the total peak tensile load. This is counteracted, and possibly overshadowed, by the cyclic relaxation of any initial tensile residual stresses. For all other things being equal, the case of cyclic strain softening of the matrix will tend to strain the fibers in tension to a greater extent than for cyclic strain hardening.

In addition to cyclic hardening, softening, and relaxation of mean stresses (either by cyclic or time-dependent means), fibers are forced to carry a greater portion of the imposed tensile load due to matrix cracking perpendicular to the fibers. As the matrix fatigues due to microcrack initiation and microcrack propagation, more and more of the applied load is shed to the fibers in the plane of the cracks. Furthermore, as fibers begin to crack, the load they carried is transferred to the remaining unbroken fibers, thus increasing their peak tensile stresses and strains and pushing them closer to ultimate fracture. Failure of the composite into two pieces occurs at the point wherein the remaining axial fibers are being asked to carry stresses and strains in excess of their critical value. It is this later important fact that forms the basis for the "fiber-dominated" approach to composite life prediction proposed by Johnson [*32,33*]. An

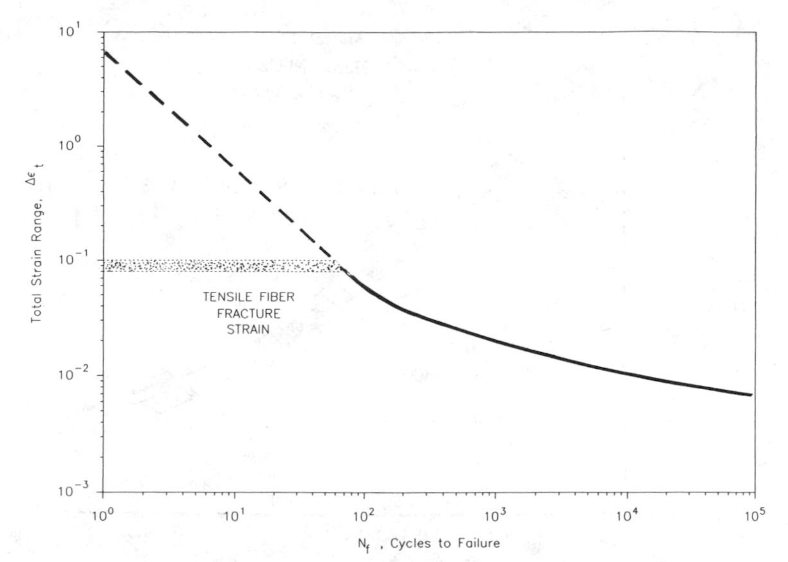

FIG. 5—*Strain range versus fatigue life curve for MMC with brittle fiber.*

integral part of the currently proposed TMF MMC life prediction method is the inclusion of an upper strain limit that will cover those conditions wherein the life of the material is clearly governed by the tensile strain capacity of the fibers. This is represented in Fig. 5 where a horizontal line is drawn at the value of the critical tensile fracture strain of the fiber. To properly place this strain value on a strain-range axis requires a micromechanics strain analysis of the MMC to be able to determine the initial mean strain in the fibers.

Comparison with Experimental Results

To date, there has been no direct experimental evaluation or verification of the proposed MMC modeling efforts. While TMF data for MMCs exist, the necessary matrix material property data required by the proposed TMF life prediction method have as yet to be generated for these systems. Nevertheless, some isothermal fatigue results from the open literature for MMCs and their corresponding matrices have been examined in light of the currently adopted points of view. Examples of published isothermal results are available from NASA and Air Force programs for three commonly available materials: W/Cu [*34*], SiC/Ti-15-3-3-3 [*35*], and SiC/Ti-24Al-11Nb [*37,37*]. The first example represents a ductile/ductile MMC system while the latter two represent brittle/ductile MMC and IMC (intermetallic matrix composite) systems, respectively.

Tungsten/Copper MMC

Strain-controlled, completely-reversed, low cycle fatigue test results have been reported by Verrilli and Gabb [*34*] for 9 and 36 vol% W/Cu at 260 and 560°C. They compared their 560°C results to 538°C isothermal data for both annealed and hardened copper generated by Conway et al. [*38*]. Figure 6 is taken from Ref *34*. It is seen that there are only slight differences in the strain-cycling fatigue resistance of the copper and the W/Cu MMC. This would imply that the numerous potentially detrimental effects on the matrix due to the presence of fibers are not realized under the current circumstances. This observation is consistent with several impor-

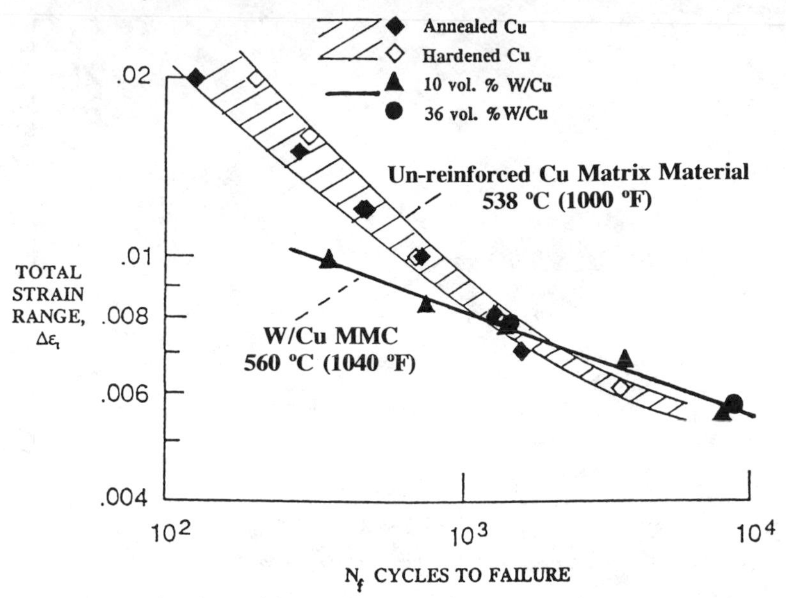

FIG. 6—*Comparison of isothermal fatigue resistance of W/Cu MMC (Verrilli and Gabb* [34]*) with Cu unreinforced matrix material (Conway et al.* [38]*).*

tant aspects of this MMC system and its constituents. Because of the low strength and high ductility of the matrix, any detrimental original tensile residual stresses and any applied tensile mean stresses (MMC loading was zero to maximum tension) would have cyclically relaxed very rapidly at the strain levels of the experiments. The degree of multiaxiality in this system during isothermal loading is also quite small, and hence the MF may be too low to have a discernible effect on fatigue life. Large internal stress concentrations were not present, although 0.4-μm-sized voids apparently were introduced into the matrix during fabrication of the MMC. Fiber alignment and spacing were well controlled during the fabrication process. In addition, the fiber and matrix do not have a brittle interfacial reaction layer as the two elements chemically "wet" each other and form excellent bonds. Fibers are not brittle at elevated temperature and hence cracked fibers are not a source of microcrack initiation. Furthermore, fibers do not debond from the matrix during axial loading, and fibers appear to prolong the life to a small degree at the longer life levels, perhaps due to the beneficial effects of crack retardation and crack bridging brought about by the excellent interfacial bond.

Silicon Carbide/Titanium MMC

Unpublished isothermal, load-controlled, low cycle fatigue results for SCS-6/Ti-15-3 MMC have been generated in conjunction with a cooperative research program between NASA Lewis and Pratt and Whitney. These results, shown in Fig. 7, are for 427°C (800°F) with $R_L \approx 0$ and cyclic frequencies ranging between about 0.03 and 0.3 Hz, depending upon strain range and the testing laboratory. Fiber volume fractions range from 34 to 38%. The MMC results are compared to isothermal, strain-controlled, $R_\epsilon = -1$, data reported [*35*] for matrix material processed in the same fashion as the composite. Frequencies ranged between 0.03 and 1.0 Hz, and were the highest at the lowest strain ranges. Substantially reduced strain fatigue resistance is exhibited by the MMC when compared to the unreinforced matrix. The reduction is a factor of approximately 2 in strain range for a fixed life, or a factor of about 40 in life for a

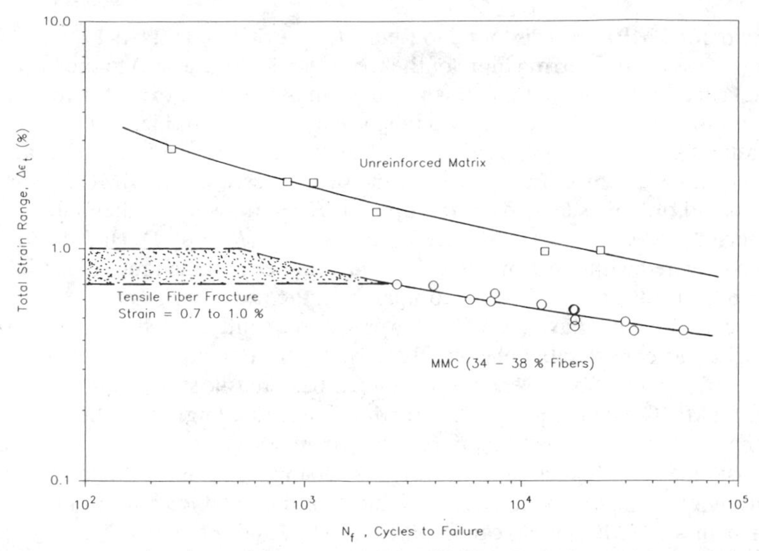

FIG. 7—*Comparison of isothermal fatigue resistance of SCS-6/Ti-15-3 MMC (previously unpublished NASA and Pratt & Whitney data) with Ti-15-3 unreinforced matrix material (Gayda et al.* [35]*).*

given strain range. Among the dozen influential factors previously discussed, few could be responsible for the significantly large loss in fatigue resistance caused by the presence of fibers in the Ti-15-3 matrix. The most pertinent influences are discussed in the following paragraphs.

Any mean stresses in the matrix that may be present due either to the nature of loading or to residual fabrication stresses are expected to cyclically relax to zero for Ti-15-3 at 427°C (800°F). Experimental results supporting this argument have been reported by Gayda et al. [*35*] for axially loaded specimens of a 33 vol% $[0]_4$ SCS-6/Ti-25-3 MMC. They also generated cyclic strain-controlled data for unreinforced specimens of the matrix material and showed that, for $R_\epsilon = 0$ tests, the initially large tensile mean stress quickly relaxed to zero, that is, $V_\sigma \rightarrow 0$, resulting in hysteresis loops identical to those for $R_\epsilon = -1$ tests. Thus, from the standpoint of the effect of mean stresses in the matrix, it is not expected that the matrix material would initiate and propagate microcracks earlier in the MMC than if fatigued alone.

However, a larger peak tensile stress is experienced by the fibers as they carry a greater portion of the tensile load following relaxation of the matrix tensile mean stresses. Hence, the fibers are forced to operate closer to their peak tensile strain capacity and the MMC becomes less tolerant of subsequent matrix cracking that will further erode load-bearing area and transfer more and more tensile stress (and hence, strain) to the fibers. The cyclic durability resistance of the MMC thusly has been compromised, not by fatigue cracking per se, but by the cyclic flow response characteristics of the matrix and the limited strain capacity of the brittle fibers.

A somewhat similar situation can occur during $R_L = 0$, in-phase, TMF cycling of an MMC in the high strain fatigue regime. Because of the low flow strength of the matrix material at the maximum TMF temperature, the matrix undergoes inelasticity. In turn, the matrix is driven into compressive stresses (at zero external applied load to the MMC) during the low temperature excursion of the TMF cycle. The end result of the matrix flow is to cause the fibers to carry a larger and larger share of the peak applied tensile loads. If these loads are high enough, fibers begin to fracture in a runaway sequence that feeds upon itself, and the MMC fails as a result of fiber overload before any fatigue cracking can occur in the matrix. An example of such behavior for TMF cycling is found in Ref *37* for the SCS-6/Ti-24Al-11Nb MMC system. While the process of transferring more of the tensile stress (and strain) to the fiber will have a detrimental effect on cyclic lifetime, this mechanism alone is not expected to be responsible for the large discrepancy between the two fatigue curves shown in Fig. 7.

Explanation for an additional portion of the discrepancy can be found in examination of the differences in the control modes used in the two test series. The MMC fatigue tests used load control and only tensile loads were imposed ($R_L = 0$), whereas the matrix fatigue tests were conducted under completely reversed strain control ($R_\epsilon = -1$). Had the MMC fatigue tests been conducted with the same range of load, and hence same range of strain with $R_L = -1$ (to keep the results of Fig. 7 more comparable), the peak tensile stress (or strain) experienced by the fibers would have been 50% lower. Consequently, much more extensive matrix fatigue cracks and considerable fiber cracking would have had to have occurred before the remaining load carrying fibers were pushed to their peak tensile strain limit. Using this line of reasoning, the MMC specimens could have endured more loading cycles, thus resulting in lives closer to those exhibited by the unreinforced matrix material.

Additional qualitative explanation for the discrepancy is found in the microcrack propagation portion of cyclic life, which can be reduced in the MMC as a result of two important factors. First, in the MMC, cracks can initiate internally and subsequently link, thus reducing cyclic life. Secondly, internal oxidation at 427°C can occur along the interfaces and any interior cracks. A loss of a factor of four in microcrack propagation life by these means is certainly within the realm of possibility. Furthermore, the microcrack initiation portion of specimen fatigue life has been observed via fractography [*35*] to be essentially bypassed in this MMC system. This is likely due to the nature of the interfacial material and its relatively poor bonding to both fiber and matrix. At a cyclic strain range giving a life of 10^7 cycles to failure for the unreinforced matrix material, the microcrack initiation portion is expected to be 90% of the total macrocrack initiation life in accordance with Eq 6c. Thus an additional factor of ten loss in life can be attributed to the loss of microcrack initiation life. The two life losses are multiplicative, that is, $4 \times 10 = 40$, which is in general agreement with the losses observed in Fig. 7. For the time being, it is not possible to quantify the degree of loss with any confidence in accuracy. A series of strain-controlled, $R_\epsilon = -1$ low cycle fatigue tests of MMC coupons is planned. The results are expected to provide a better understanding of the causes of the large observed life losses in this MMC material for $R_L = 0$.

Silicon Carbide/Titanium Aluminide IMC

Isothermal, load-controlled, low cycle fatigue results for coupons of SCS-6/Ti-24Al-11Nb have been reported by Brindley et al. [*36*], and by Russ et al. [*37*]. Fiber volume percentages range from 27 to 33 vol%. In both cases, the only strain data reported was the maximum tensile cyclic strain imposed on the MMC specimens. Both sets of investigators have since supplied the current authors with the average cyclic strain ranges for each fatigue test analyzed herein. Isothermal tensile and fatigue data on Ti-24Al-11Nb unreinforced matrix material have been reported by DeLuca et al. [*39*]. The matrix fatigue data are from completely reversed, strain-cycling ($R_\epsilon = 1$) axial experiments with a frequency of 0.17 Hz, whereas the MMC data are for unidirectional coupons subjected to $R_L = 0$ [*36*] or 0.1 [*37*] with frequencies of about 0.2 to 0.3 Hz [*36*] and 3.0 [*37*]. The basic mechanical loading conditions were similar to those for the SCS-6/Ti-15-3 system analyzed in the previous section.

Figures 8*a* and *b* show comparisons between the fatigue resistances of the unreinforced matrix material and MMC coupons at 427 and 650°C, respectively. There is remarkable similarity between the sets of curves in Figs. 8*a* and *b* and those of Fig. 7. The MMC's resistance

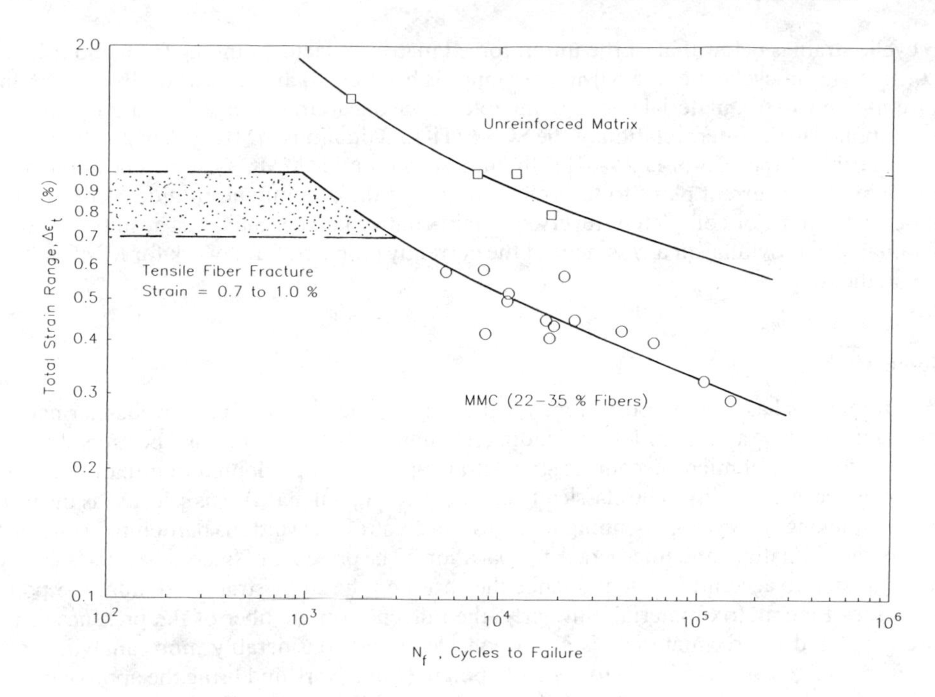

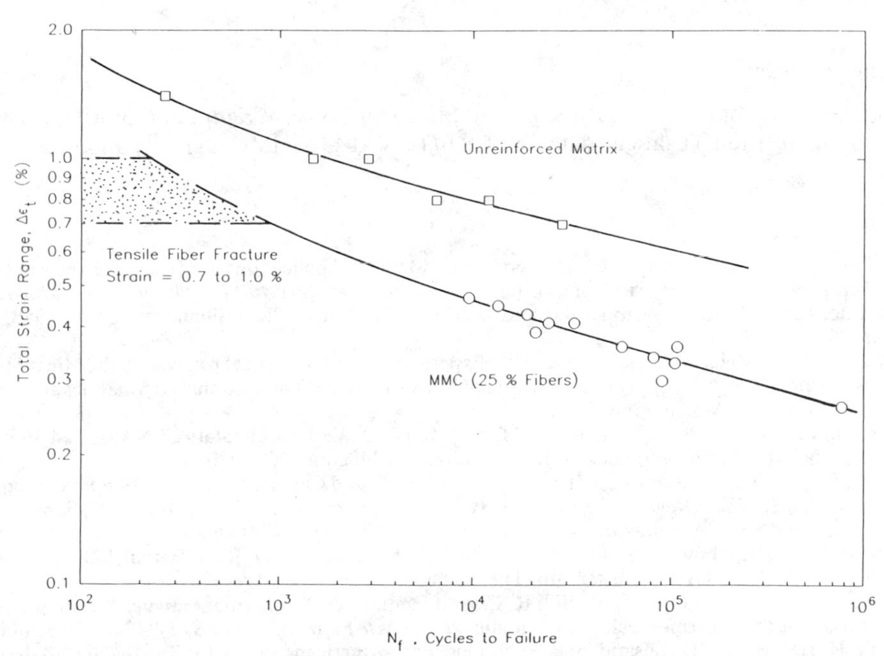

FIG. 8—*Comparison of isothermal fatigue resistance of SCS-6/Ti-24Al-11Nb MMC with Ti-24Al-11Nb unreinforced matrix material (DeLuca et al.* [39]*)*. (a) *427°C (800°F) data from Brindley et al.* [36] *and* (b) *650°C (1200°F) data from Russ et al.* [37].

to cyclic strain is below that of the unreinforced matrix by factors ranging from about 1.7 to 2.0. In terms of cyclic life, the MMC resistance is between 10 and 100 times lower than the unreinforced matrix material for the same total mechanical strain range. The arguments that were applied to the interpretation of the SCS-6/Ti-15-3 titanium MMC system results are just as applicable to the SCS-6/Ti-24Al-11Nb titanium aluminide MMC system. The authors are unaware of any current plans to test thick samples of the titanium aluminide MMC system under conditions of completely reversed, strain-control fatigue. Such test results would be invaluable in providing an assessment of the currently proposed framework for MMC fatigue life prediction.

Summary

A framework has been proposed for a life prediction system applicable to cyclic thermal and mechanical fatigue and creep-fatigue loading of unidirectional, continuous-fiber, metal matrix composites. A local micromechanics stress-strain approach was adopted to impart the maximum degree of generality. The classically fatigue-prone metal matrix was selected as the vehicle for tracking the cycle-depending changes (mean stress relaxation, hardening, softening, microcrack initiation, and microcrack propagation). The presence of fibers and interfacial layers is taken into account by the influence they exert on the stress-strain and failure response behavior of the matrix material. Similarly, the influence on the fiber of the presence of the interfacial and matrix materials is given consideration. Considerably more analytical and experimental research is needed to expand upon the framework and bring the approach to its full potential. Examples of isothermal fatigue data for MMCs taken from the literature illustrate important aspects of the modeling framework.

Acknowledgment

The assistance of Dr. S. M. Russ, United States Air Force, Wright Laboratories, Dayton, OH, in sharing valuable fatigue data for SCS-6/Ti-24Al-11Nb IMC is greatly appreciated.

References

[*1*] Caruso, J. J. and Chamis, C. C., "Superelement Methods Applications to Micromechanics of High Temperature Metal Matrix Composites," *29th Structures, Structural Dynamics, and Materials,* American Institute of Aeronautics and Astronautics (AIAA) '88, Williamsburg, VA, 1988, pp. 1388–1400.

[*2*] Lerch, B. A., Melis, M. E., and Tong, M., "Experimental and Analytical Analysis of the Stress-Strain Behavior in a $[90/0]_{2s}$, SiC/Ti-15-3 Laminate," NASA TM 104470, National Aeronautics and Space Administration, Washington, DC, July 1991.

[*3*] Chamis, C. C., Caruso, J. J., and Lee, H.-J., "METCAN Verification Status," NASA TM-103119, National Aeronautics and Space Administration, Washington, DC, 1990.

[*4*] Halford, G. R., Lerch, B. A., Saltsman, J. F., Arya, V., and Caruso, J. J., "LCF Life Prediction for MMCs," *HITEMP Review 1989,* (ITAR Restricted Document), NASA CP-10039, National Aeronautics and Space Administration, Washington, DC, 1989, pp. 64-1 to 64-9.

[*5*] Halford, G. R., "Low-Cycle Thermal Fatigue," *Thermal Stresses II,* R. B. Hetnarski, Ed., Elsevier Science Publishers B.V., Amsterdam, The Netherlands, 1987, pp. 329–428.

[*6*] Halford, G. R., McGaw, M. A., Bill, R. C., and Fanti, P. D., "Bithermal Fatigue: A Link Between Isothermal and Thermomechanical Fatigue," *Low Cycle Fatigue, ASTM STP 942,* H. D. Solomon, G. R. Halford, L. R. Kaisand, and B. N. Leis, Eds., American Society for Testing and Materials, Philadelphia, 1988, pp. 625–637.

[*7*] Saltsman, J. F. and Halford, G. R., "Life Prediction of Thermomechanical Fatigue Using the Total Strain Version of Strainrange Partitioning (SRP)—A Proposal," NASA TP-2779, National Aeronautics and Space Administration, Washington, DC, Feb. 1988.

[8] Saltsman, J. F. and Halford, G. R., "An Update on the Total Strain Version of SRP," *Low Cycle Fatigue, ASTM STP 942,* H. D. Solomon, G. R. Halford, L. R. Kaisand, and B. N. Leis, Eds., American Society for Testing and Materials, Philadelphia, 1988, pp. 329–341.
[9] Halford, G. R., Verrilli, M. J., Kalluri, S., Ritzert, F. J., Duckert, R. E., and Holland, F., "Thermomechanical and Bithermal Fatigue Behavior of Cast B1900 + Hf and Wrought Haynes 188," *Advances in Fatigue Lifetime Predictive Techniques, ASTM STP 1122,* M. R. Mitchell and R. W. Landgraf, Eds., American Society for Testing and Materials, Philadelphia, 1991, pp. 120–142.
[10] Halford, G. R., Saltsman, J. F., Verrilli, M. J., and Ayra, V., "Application of a New Thermal Fatigue Life Prediction Model to Two High-Temperature Aerospace Alloys," *Advances in Fatigue Lifetime Predictive Techniques, ASTM STP 1122,* M. R. Mitchell and R. W. Landgraf, Eds., American Society for Testing and Materials, Philadelphia, 1991, pp. 107–119.
[11] Bodner, S. R. and Partom, Y., "Constitutive Equations for Elastic-Viscoplastic Strain-Hardening Materials," *Journal of Applied Mechanics,* Vol. 42, No. 2, June 1975, pp. 385–389.
[12] Walker, K. P., "Research and Development Program for Nonlinear Structural Modeling With Advanced Time-Temperature Dependent Constitutive Relations," NASA CR-165533, National Aeronautics and Space Administration, Washington, DC, 1981.
[13] Robinson, D. A. and Swindeman, R. W., "Unified Creep-Plasticity Constitutive Equations for 2-1/4Cr-1Mo Steel at Elevated Temperature," ORNL Report TM-8444, Oak Ridge National Laboratories, Oak Ridge, TN, 1982.
[14] Freed, A., "Thermoviscoplastic Model With Application to Copper," NASA Report TP 2845, National Aeronautics and Space Administration, Washington, DC, Dec. 1988.
[15] Manson, S. S., "Fatigue: A Complex Subject—Some Simple Approximations," *Experimental Mechanics,* Vol. 5, No. 7, 1965, pp. 193–226.
[16] Garmong, G., "Elastic-Plastic Analysis of Deformation Induced by Thermal Stress in Eutectic Composites: I. Theory," *Metallurgical Transactions A,* Vol. 5, 1974, pp. 2183–2190.
[17] Halford, G. R. and Nachtigall, A. J., "The Strainrange Partitioning Behavior of an Advanced Gas Turbine Disk Alloy, AF2-1DA," *Journal of Aircraft,* Vol. 17, No. 8, 1980, pp. 598–604.
[18] Morrow, J., "Fatigue Properties in Metals," Section 3.2, *Fatigue Design Handbook,* J. A. Graham, Ed., Society of Automotive Engineers, Warrendale, PA, 1968, pp. 21–29.
[19] Manson, S. S. and Halford, G. R., "Treatment of Multiaxial Creep-Fatigue by Strainrange Partitioning," *Symposium on Creep-Fatigue Interaction, MPC-3,* R. M. Curran, Ed., American Society of Mechanical Engineers and the Metals Properties Council, New York, 1976, pp. 299–322.
[20] Manson, S. S. and Halford, G. R., "Discussion of paper by J. J. Blass and S. Y. Zamrik in Symposium on Creep-Fatigue Interaction, ASME, 1976, pp. 129–159," *Journal of Engineering Materials and Technology,* Vol. 99, 1977, pp. 283–286.
[21] Hashin, Z. and Rotem, A., "A Fatigue Failure Criterion for Fiber Reinforced Materials," *Journal of Composite Materials,* Vol. 7, pp. 448–464.
[22] Bigelow, C. A., "The Effect of Uneven Fiber Spacing on Thermal Residual Stresses in a Unidirectional SCS-6Ti-15-3 Laminate," NASA TM 104225, National Aeronautics and Space Administration, Washington, DC, March 1992.
[23] Jansson, S. and Leckie, F. A., "Reduction of Thermal Stresses in Continuous Fiber Reinforced Metal Matrix Composites with Interface Layers," NASA CR-185302, National Aeronautics and Space Administration, Washington, DC, 1990.
[24] Arnold, S. M., Arya, V. K., and Melis, M. E., "Elastic/Plastic Analyses of Advanced Composites Investigating the Use of the Compliant Layer Concept in Reducing Residual Stress Resulting from Processing," NASA TM-103204, National Aeronautics and Space Administration, Washington, DC, 1990.
[25] Chen, E. J. H. and Young, J. C., "The Microdebonding Testing System—A Method of Quantifying Adhesion in Real Composites," *Composites Science and Technology,* Vol. 42, Nos. 1–3, 1991, pp. 189–206.
[26] Ghosn, L. J., Kantzos, P., and Telesman, J., "Modeling of Crack Bridging in a Unidirectional Metal Matrix Composite," NASA TM 104355, National Aeronautics and Space Administration, Washington, DC, May 1991.
[27] Lerch, B. A., Gabb, T. P., and MacKay, R. A., "Heat Treatment Study of the SiC/Ti-15-3 Composite System," NASA TP-2970, National Aeronautics and Space Administration, Washington, DC, Jan. 1990.
[28] Nissley, D. M., Meyer, T. G., and Walker, K. P., "Life Prediction and Constitutive Models for Engine Hot Section Anisotropic Materials Program," Final Report, NASA Contract NAS3-23939, NASA CR-189223, National Aeronautics and Space Administration, Washington, DC, Sept. 1992.
[29] Neu, R. W. and Sehitoglu, H., "Thermomechanical Fatigue, Oxidation, and Creep: Part I. Damage Mechanisms," *Metallurgical Transactions A,* Vol. 20A, 1989, pp. 1755–1767.

[30] Neu, R. W. and Sehitoglu, H., "Thermomechanical Fatigue, Oxidation, and Creep: Part II. Life Prediction," *Metallurgical Transactions A,* Vol. 20A, 1989, pp. 1769–1783.
[31] Miller, M. P., McDowell, D. L., Oehmke, R. L. T., and Antolovich, S. D., "A Life Prediction Model for Thermomechanical Fatigue Based on Microcrack Propagation," *Thermomechanical Fatigue Behavior of Materials, STP 1186,* American Society for Testing and Materials, Philadelphia, 1992, in this publication, pp. 35–49.
[32] Johnson, W. S., Lubowinski, S. J., and Highsmith, A. L., "Mechanical Characterization of Unnotched SCS_6/Ti-15-3 Metal Matrix Composites at Room Temperature," *Thermal and Mechanical Behavior of Metal Matrix and Ceramic Matrix Composites, ASTM STP 1080,* J. M. Kennedy, H. H. Moeller, and W. S. Johnson, Eds., American Society for Testing and Materials, Philadelphia, 1990, pp. 193–218.
[33] Johnson, W. S., "Modeling Stiffness Loss in Boron/Aluminum Laminates Below the Fatigue Limit," *Long-Term Behavior of Composites, ASTM STP 813,* T. K. O'Brien, Ed., American Society for Testing and Materials, Philadelphia, 1983, pp. 160–176.
[34] Verrilli, M. J. and Gabb, T. P., "High Temperature Tension-Compression Fatigue Behavior of a Tungsten Copper Composite," NASA TM 104370, National Aeronautics and Space Administration, Washington, DC, 1990.
[35] Gayda, J., Gabb, T. P., and Freed, A. D., "The Isothermal Fatigue Behavior of a Unidirectional SiC/Ti Composite and the Ti Alloy Matrix," *Fundamental Relationships Between Microstructure & Mechanical Properties of Metal-Matrix Composites,* P. K. Liaw and M. N. Gungor, Eds., Minerals, Metals & Materials Society, 1990, pp. 497–514.
[36] Brindley, P. K., MacKay, R. A., and Bartolotta, P. A., "Thermal and Mechanical Fatigue of a SiC/Ti-24Al-11Nb Composite," *Titanium Aluminide Composites,* P. R. Smith, S. J. Balsone, and T. Nicholas, Eds., WL-TR-91-4020, Wright-Patterson Air Force Base, OH, 1991, pp. 484–496.
[37] Russ, S. M., Nicholas, W. S., Bates, M., and Mall, S., "Thermomechanical Fatigue of SCS-6/Ti-24Al-11Nb Metal Matrix Composite," *Failure Mechanisms in High Temperature Composite Materials,* AD-Vol. 22/AMD-Vol. 122, American Society for Mechanical Engineers, New York, 1991, pp. 37–43.
[38] Conway, J. B., Stentz, R. H., and Berling, J. T., "High Temperature, Low-Cycle Fatigue of Copper-Base Alloys in Argon; Part I—Preliminary Results for 12 Alloys at 1000 °F (538 °C)," NASA CR-121259, National Aeronautics and Space Administration, Washington, DC, 1973.
[39] DeLuca, D. P., Cowles, B. A., Haake, F. K., and Holland, K. P., "Fatigue and Fracture of Titanium Aluminides," WRDC-TR-89-4136, Wright-Patterson Air Force Base, OH, 1990.

Michael G. Castelli[1] *and John R. Ellis*[2]

Improved Techniques for Thermomechanical Testing in Support of Deformation Modeling

REFERENCE: Castelli, M. G. and Ellis, J. R., **"Improved Techniques for Thermomechanical Testing in Support of Deformation Modeling,"** *Thermomechanical Fatigue Behavior of Materials, ASTM STP 1186,* H. Sehitoglu, Ed., American Society for Testing and Materials, Philadelphia, 1993, pp. 195–211.

ABSTRACT: The feasibility of generating precise thermomechanical deformation data to support constitutive model development was investigated. Here, the requirement is for experimental data that is free from anomalies caused by less than ideal equipment and procedures. A series of exploratory tests conducted on Hastelloy X showed that generally accepted techniques for strain controlled tests were lacking in at least three areas. Specifically, problems were encountered with specimen stability, thermal strain compensation and temperature/mechanical strain phasing. The present study was undertaken to identify the source of these difficulties and to develop improved thermomechanical testing techniques to correct them. These goals were achieved by developing improved procedures for measuring and controlling thermal gradients and by designing a specimen specifically for thermomechanical testing. In addition, innovative control strategies were developed to correctly proportion and phase the thermal and mechanical components of strain. Subsequently, the improved techniques were used to generate definitive deformation data for Hastelloy X over the temperature range, 200 to 1000°C.

KEY WORDS: thermomechanical testing, testing techniques, specimen instability, constitutive modeling

Introduction

Overview

The majority of structural components used in high temperature applications experience some form of thermomechanical loading during service. In order to predict performance under such loadings, it is essential that precise thermomechanical tests be conducted on candidate materials. In particular, if thermomechanical testing is performed in support of constitutive model development, the data must be free of anomalies introduced by less than ideal test equipment and control techniques. To this end, recent advances in digital control systems, instrumentation, and mechanical testing equipment have eliminated many of the difficulties encountered in earlier studies. However, maintaining closely controlled conditions during thermomechanical loading remains a relatively complex task, requiring detailed examination.

A state-of-the-art testing facility has been established at NASA Lewis in support of high temperature deformation and fatigue testing of advanced materials. A number of exploratory tests were conducted to investigate the feasibility of generating precise thermomechanical deformation (TMD) data to support constitutive model development. The testing involved proportionally phased values of temperature and mechanical strain under uniaxial strain control.

[1] Structures research engineer, Sverdrup Technology Inc., LeRC Group, Brook Park, OH 44142.
[2] Chief of Fatigue and Fracture Branch, NASA Lewis Research Center, Cleveland, OH 44135.

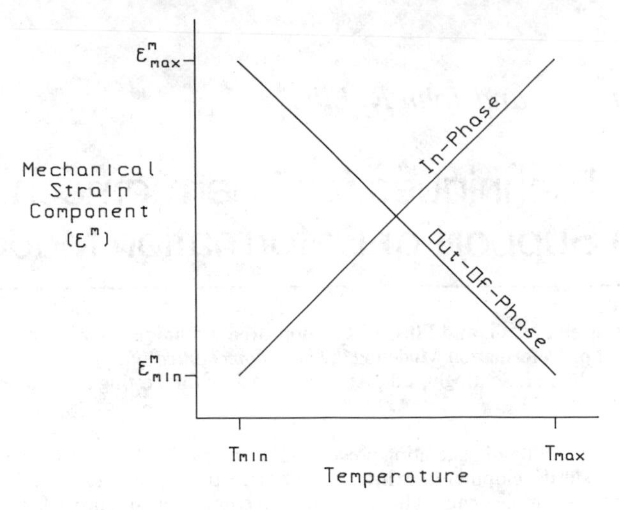

Note: Mechanical Strain (ε^m) = Total Strain (e) - Thermal Strain (ε^{th})

FIG. 1—*Phasing relationship in strain controlled thermomechanical loading.*

The phasing relationships investigated are given in Fig. 1. A triangular waveform was used to control both temperature and mechanical strain as data generated at a constant rate is preferred in constitutive model development. Thermomechanical testing under uniaxial strain control gives rise to complex control conditions because of the first order affect of thermal strain on the control variable [*1–4*]. Because of this added complexity, the techniques detailed in this paper pertain directly to thermomechanical, uniaxial strain controlled testing with triangular command waveforms. However, the techniques are equally applicable to the less complicated mode of load control.

Initial exploratory tests under strain-controlled thermomechanical conditions identified three general areas which were found to be lacking, namely, specimen stability, thermal strain compensation, and temperature-mechanical strain phasing. These three areas were investigated in detail and experimental techniques and procedures were developed to improve the accuracy of each.

Problem Areas/Issues

The first problem area identified pertains to specimen stability during thermomechanical testing. For Hastelloy X®[3] out-of-phase loadings conducted at temperatures greater than 600°C were found to introduce a cyclic specimen instability referred to as barreling. This behavior has been noted in several materials under a variety of loading conditions [*4–11*]. Barreling is a cycle-by-cycle ratchetting phenomenon promoting material flow and localized straining; on a cylindrical specimen, this behavior is evidenced as circumferential bulging. Figure 2 shows a test specimen before and after barreling. In severe cases the barreling is visually detectable very early in cyclic life ($N < 0.2N_f$). The test specimen eventually becomes so severely distorted that initial engineering values of stress and strain lose all significance, complicating interpretation of the data. For Hastelloy X, the barreling did not occur under either isothermal or in-phase loading conditions.

[3] Hastelloy X is a trademark of Haynes International, Inc., Kokomo, IN.

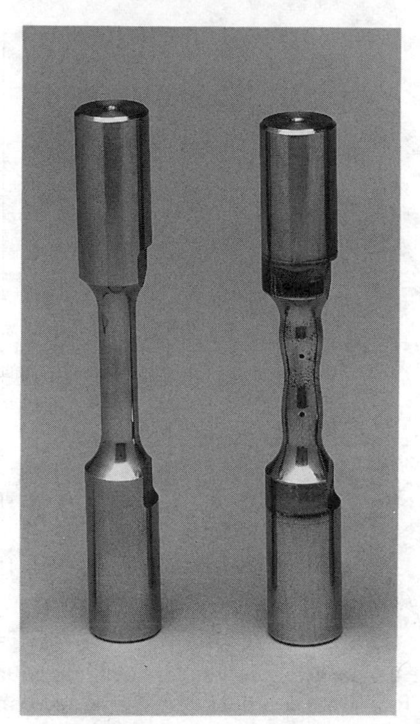

FIG. 2—*Specimen before and after cyclic barreling.*

Although this behavior has been noted on several different materials during high temperature deformation, investigators have only speculated as to its cause [*4–11*] and how it relates to the specific test conditions. The barreling has been attributed in part to several sources including inelastic strain magnitude, strain rate, specimen geometry, and mean stress. However, there has been little or no success at eliminating this problem under conditions where it has been found to occur.

As pictured in Fig. 2, the barreling in the present work often appeared in the form of two bulges occurring above and below the midpoint of the specimen's gage section. After closely monitoring the temperature gradients during the TMD cycle (that is, the dynamic temperature gradients), it was determined that the bulge locations were associated with hot spots ($\approx$ 10°C) at the maximum temperature of the cycle. The correspondence between barreling and the location of a hot dynamic gradient was verified by conducting a test with a dynamic hot spot at the mid-gage position. As suspected, the location of the barreling was mid-gage, coincidental with the location of the hot dynamic gradient. This clearly identified dynamic thermal gradients as a primary source of the material barreling. Thus, to minimize the specimen barreling, effort was placed on refining the techniques used to measure and control dynamic thermal gradients.

The second area identified as being lacking was that regarding thermal strain compensation during a strain-controlled thermomechanical test. If a fixed mechanical strain range is required, an accurate assessment of and compensation for the thermal strain component must be maintained at all times during the TMD cycle.

Two simplified approaches include (1) equal-increment, time-dependent partitioning, and (2) pre-recorded, time-dependent partitioning of thermal strains. The first approach assumes

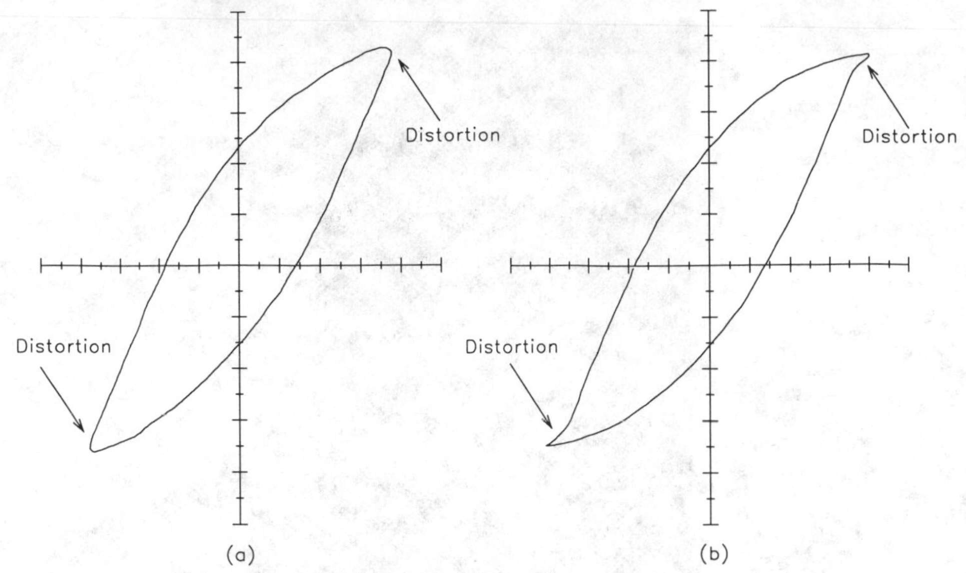

FIG. 3—*Typical distortions found in hysteresis loops where phasing problems exist;* (a) *stress vs. total strain;* (b) *stress vs. mechanical strain.*

that the thermal strain range can be accurately subdivided into equal, time-dependent increments throughout the cycle. Real-time compensation is then accomplished by adding (or subtracting) this constant increment to the desired mechanical strain. This approach can lead to significant errors if the specimen's thermal strain response is not consistently linear (in time), a condition which is typical during temperature reversals. The second approach assumes that the thermal strains can be accurately represented by a single thermal cycle recorded prior to the test. This pre-recorded thermal strain versus time history is subsequently accessed (based on cycle time) to obtain the "real-time" thermal strain component. This approach will promote control errors if small temperature variations are experienced within a cycle or from one cycle to the next. The pre-recorded thermal strains remain fixed in time, while the actual thermal strains are likely to fluctuate.

The third area where problems most often arise during thermomechanical testing is that of command variable phasing. This is a problem because the time lag experienced between the command and actual specimen response is different when comparing temperature to mechanical strain (note that this is also a problem in load-controlled tests). With state-of-the-art equipment, the lag value for the mechanical component of loading is essentially zero. However, the lag value for temperature can be quite large, particularly when a command reversal is issued. Simplified approaches, including simultaneous commands or a mathematical coupling between the mechanical strain command and the temperature response, will usually result in distorted hysteresis loops, such as those shown in Fig. 3. These distortions will be evident, both in the stress − total strain space (Fig. 3a), and after post-processing in the stress − mechanical strain space (Fig. 3b). Such distortions are unacceptable in data intended to support constitutive model development.

Experimental Details

Equipment

A closed-loop, servo-hydraulic test system was used for this study, details of which are shown in Fig. 4. Specimens are installed in water-cooled hydraulic grips which incorporate collets for

FIG. 4—*Experimental setup used for thermomechanical testing.*

gripping a variety of specimen end geometries. During the TMD cycle, the grips serve as the primary mechanism for specimen cooling. The maximum temperature rate is generally governed by the cooling portion of the cycle, particularly when relatively "low" temperatures are involved in the TMD cycle. Forced air can be used as a means of specimen cooling, however, even if the air is diffused, localized cooling is likely to occur where the air streams impinge the surface of the specimen. As the material barreling was known to be promoted by dynamic temperature gradients, this method of cooling was avoided in the present study.

A high temperature, water-cooled extensometer is used to measure axial strains. An axial gage length of 12.7 mm is selected over a length of 25.4 mm because it is easier to control thermal gradients over the shorter test section. Direct induction heating is used for specimen heating. This method is convenient for rapid heating and temperature cycling, and allows relatively free access to the specimen's surface to accommodate the extensometer's probes. Power is provided by a radio frequency (RF) induction heater integrated with a closed loop temperature controller. Type K thermocouple wires are spot-welded onto the specimen's surface to monitor the temperature at several locations.

Typically, induction work coils are constructed with a continuous single segment design, not allowing independent adjustments of the individual windings. Hence, coil adjustments, directly interpreted as temperature changes along the specimen, cannot be pursued in a systematic manner. In contrast, the work coils and support fixture [*12*] used here, shown in Fig. 5, consist of three independent coil segments. Each segment is individually adjustable in vertical and radial directions, allowing a systematic approach to temperature modification at localized positions along the specimen.

A remote minicomputer was used for test control. Control software was written to implement the improved thermomechanical testing techniques developed in this study and manage the data acquisition. The use of RF induction heating often introduces a significant amount of electrical noise which is highly visible in the digital data. An averaging routine was imple-

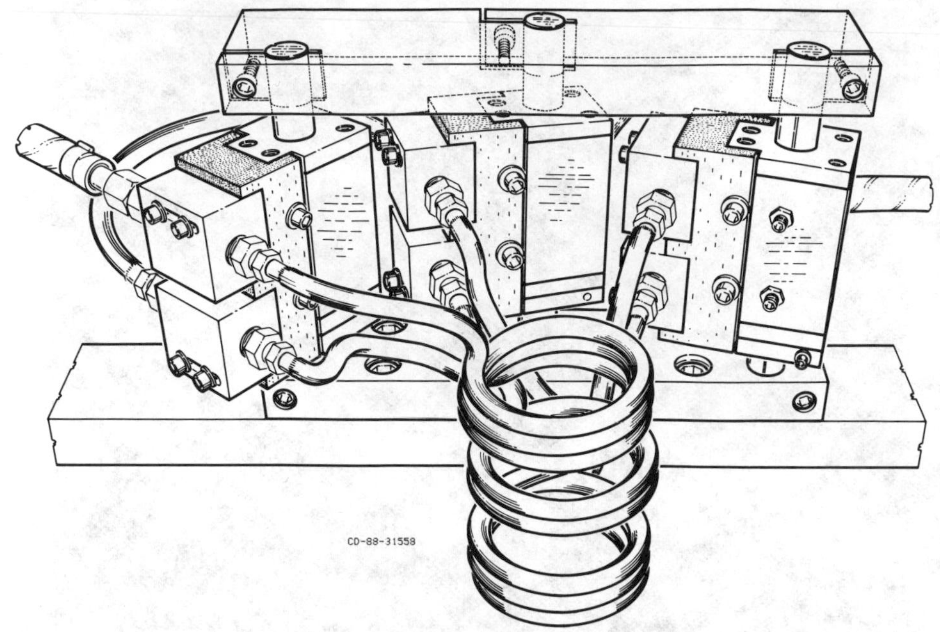

FIG. 5—*Adjustable work coil fixture for direct induction heating.*

mented in the data acquisition routine to collect several data and average them before writing to the disk. This scheme was successful in the presence of unbiased electrical noise.

Specimen Design

Initial TMD tests were conducted on a specimen geometry originally designed for isothermal, low cycle fatigue (LCF) testing. This design is shown in Figs. 2 and 6a. Although specimen geometry was not found to directly affect the material instability, an improved design was established which assisted in reducing the dynamic temperature gradients, and thus, indirectly aided in reducing material barreling. This specimen geometry is shown in Fig. 6b. The improved design is a smooth shank, thin walled tube with an extended parallel working section. In comparison to the solid cross-section specimen design (Fig. 6a), the tubular construction allows for higher heating and cooling rates while maintaining a high degree of waveform control. Also, the tubular geometry inherently lowers the radial temperature gradients present during thermal cycling. Another advantage of this design over the original solid specimen is the extended central parallel section. This feature provides sufficient length to fully accommodate the induction heater work coils, thereby simplifying the relationship between coil adjustment and associated temperature change.

Improved Techniques

Dynamic Thermal Gradients/Specimen Stability

As noted previously, the specimen instability was, in part, promoted by dynamic thermal gradients in the TMD cycle. Gradients of seemingly insignificant quantities were instrumental in promoting barreling under out-of-phase loadings above 600°C. Thus, effort was placed on

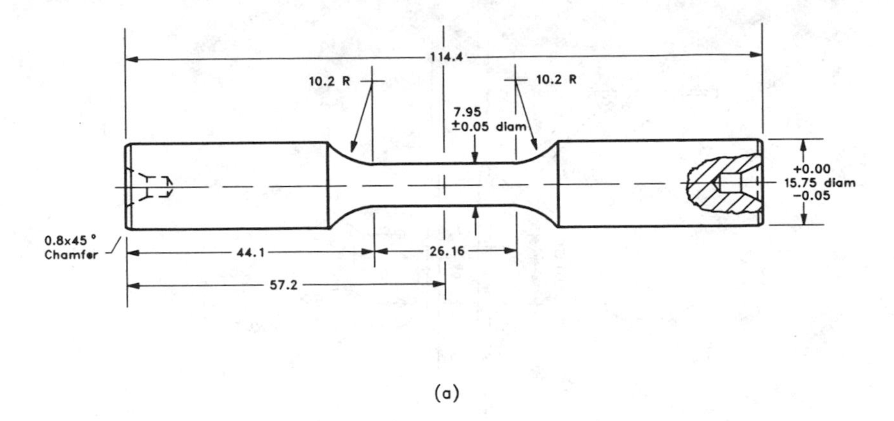

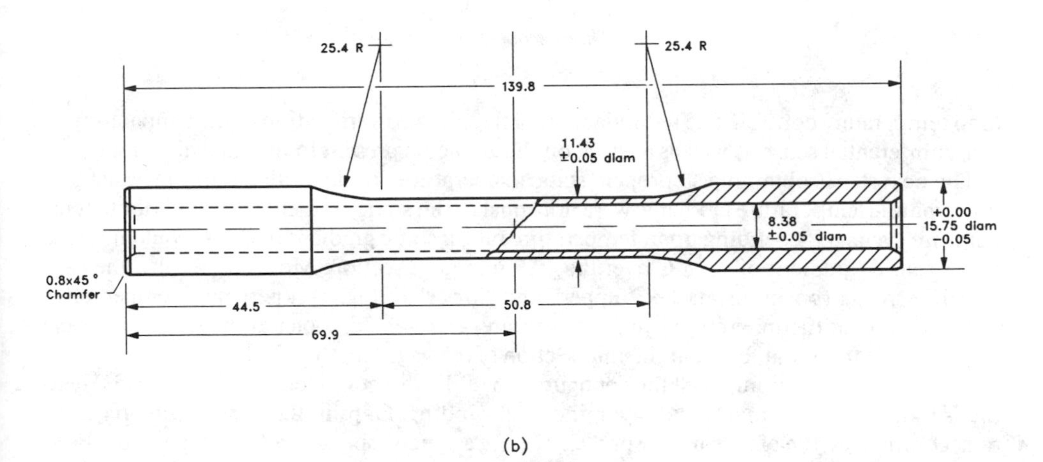

FIG. 6—*Specimen designs;* (a) *solid;* (b) *improved tubular construction for thermomechanical testing.*

the accurate measurement and control of dynamic thermal gradients. Note that the procedures and techniques in the following discussion will be influenced to some degree by specific test material and equipment.

Clearly, close control of thermal gradients requires precise temperature measurement over the specimen's gage section. Thermocouples spot-welded on the specimen's surface were used for temperature monitoring and control. Measurements obtained from thermocouples reflect the temperature at highly localized points (the exact point of physical contact). Experience has shown that direct induction heating can create highly localized "hot" spots. Thus, it is important to place the thermocouples sufficiently close, so as to insure these "hot" spots are detected. After examining several possible configurations, a satisfactory arrangement was established and is shown in Fig. 7. An axial separation of $\approx$6 mm provided adequate resolution for temperature monitoring. This eleven-thermocouple system includes two sets of five on opposite sides of the specimen (180 circumferential separation), and one located on the back-side of the specimen (opposite the extensometer probes) to function as the feedback sensor for closed-

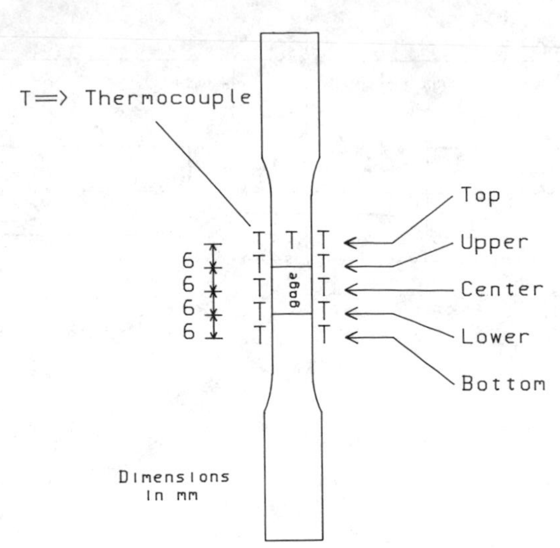

FIG. 7—*Thermocouple configuration.*

loop temperature control. The redundant set of five allowed verification of the temperature in a circumferential sense as well as a check on the gradients present in the axial direction.

The next step is obtaining a "proper" static temperature profile. Initial testing revealed that the dynamic temperature gradients were not equal to the static gradients present prior to temperature cycling. Depending upon temperature rate, a static gradient of $\pm 3°C$ could increase to a dynamic gradient of $\pm 15°C$ over the 12.7 mm gage section. More specifically, the gage length extremes (positions marked "upper" and "lower" in Fig. 7) experienced temperatures $\approx 15°C$ hotter at the maximum cycle temperature, and $\approx 15°C$ cooler at the minimum cycle temperature, than that found at the mid-section of the gage length.

By way of explanation, recall the configuration of the direct induction coils in Fig. 5. Heating is induced by two major and one minor coil winding. Gaps in the windings are required to accommodate the extensometer probes. During an increasing temperature ramp, the locations within the major coils will tend to heat faster than the location encircled by the minor (center) coil. This condition could possibly be avoided through the use of multi-zone control. However, the control complications introduced by such a system under TMD conditions may out-weigh the advantages, as each of the zones will require individual mechanical strain/temperature phasing considerations. Specimen cooling is facilitated by the water-cooled grips. This arrangement will inevitably impose a cold-hot-cold (top-center-bottom) thermal gradient along the full length of the specimen. During a decreasing temperature ramp, a variable cooling rate along the length of the specimen will be realized, as the ends will cool more rapidly than the mid-section. Thus, both heating rates and cooling rates are higher at the gage length extremes, compared to the specimen mid-section.

In addition to these symmetric anomalies, testing in air allows thermal convection currents to play a role in dynamic temperature gradients. A point located within the lower length of the specimen typically requires more energy than its counterpart within the upper length when a uniform temperature is desired. This effect can be countered under static (isothermal) conditions by proper coil adjustments. However, dynamic (thermomechanical) conditions lead to increased heating rates within the upper length during the heating portion of the cycle, and increased cooling rates within the lower length during the cooling portion of the cycle.

Given these constraints, it is impossible to cycle temperature without inducing gradients in

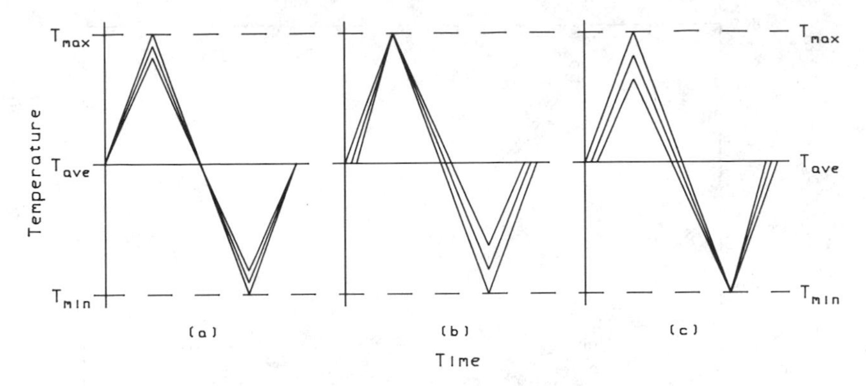

Note: The curves represent temperature histories measured at upper, central, and lower locations within the gage section. Thus, at any given time, curve separations are indicative of thermal gradients over the specimen gage section.

FIG. 8—*Alignment temperatures for dynamic thermal gradients;* (a) *average;* (b) *maximum;* (c) *minimum.*

the gage section. At best, it is possible to eliminate gradients at a specific temperature in the cycle. Three obvious points were considered: (1) average-cycle temperature, (2) maximum cycle temperature and (3) minimum cycle temperature, as schematically represented in Fig. 8. Given that the material barreling increased with increasing temperature, and was associated with the "hot" spots in the gradient, a dynamic temperature profile minimizing the gradients at the maximum cyclic temperature was chosen (for example, Fig. 8b). With this specific dynamic condition required, a corresponding static condition with initial imposed gradients could be established. This technique allowed the hot-end dynamic gradients to be reduced to $\approx \pm 2$°C over the gage section. Note that by choosing this condition, the cold-end gradients increased. Also note that multi-zone temperature control could potentially minimize this problem.

Another related issue concerns the uniformity of the temperature rate. Several axial locations were found to experience a fluctuating temperature rate after reversal. The closed loop temperature control system uses a thermocouple as the feedback sensor, hereafter called the controlling thermocouple (CTC). Strictly speaking, only the CTC temperature rate is controlled; all other points along the specimen are dependent. This is always the case with single-zone control. Typically, the CTC is placed at a central position within the gage section. Shown in Fig. 9 is the dynamic thermal profile of a specimen cycled from 600 to 800°C with the CTC in the "center" position. Here, waveform distortions are revealed at both the "upper" and "lower" gage locations after temperature reversals. This effect is understood by recalling that locations away from the center of the gage section have a higher heating and cooling rate, and hence, these positions respond more quickly to heater adjustments. The solution to this problem is to place the CTC at a "quick" response position. Shown in Fig. 10 is a comparable thermal cycle with the CTC at the "top" position (as defined in Fig. 7). Obvious improvements are visible when controlling from this position. The CTC is now more responsive (relatively) to heater power changes, thereby eliminating conditions of under and over-powering at other locations during reversals. The variation in response with respect to position is now experienced as a slight lingering of the "center" position during reversals, as linearity is not now forced at this position. Utilizing this "top" CTC position reduced dynamic thermal gradients and maintained linear waveforms within the gage section.

The rate of thermal cycling was found to have the greatest influence on all other control parameters; as temperature rate increased, all other aspects became more difficult to control.

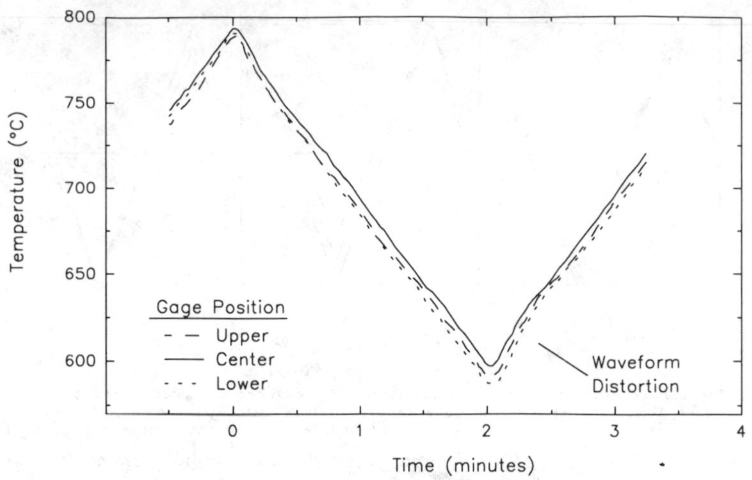

FIG. 9—*Thermal profile with center thermocouple controlling.*

Rates from 25–200°C/min were investigated while monitoring dynamic gradients, temperature rate constancy, and thermal strain compensation. At the higher rates, dynamic temperature gradients became unacceptably high (for example, ±30°C over the gage section) and accurate thermal strain compensation was not possible. To maintain the desired degree of control, a maximum temperature rate of 50°C/min was used.

Thermal Strain Compensation

A strain-controlled environment is preferred in experiments supporting constitutive equation development. However, under TMD, difficulties arise in the test control parameters because of the thermal strain's contribution to the total strain. If a fixed mechanical strain range is required, accurate thermal strain compensation is necessary. After evaluating several techniques, an improved methodology was developed and implemented.

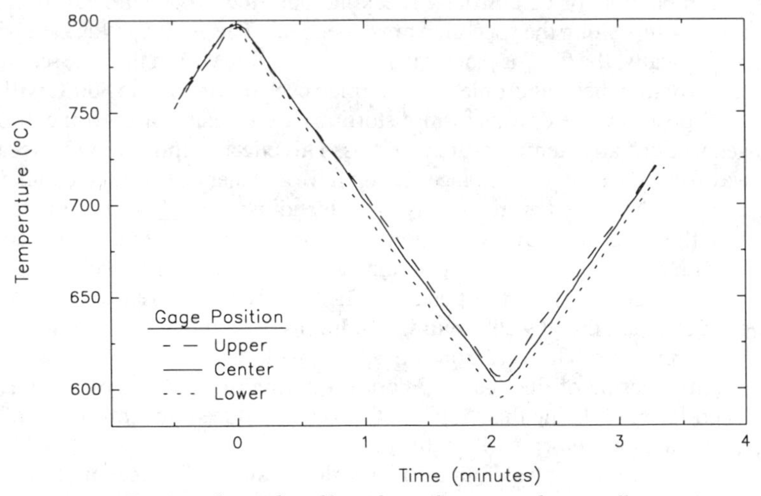

FIG. 10—*Thermal profile with top thermocouple controlling.*

The preferred approach was to relate thermal strain to real-time measurement of specimen temperature. This method relies upon the assumption that a function $f(T)$ is capable of representing the thermal strains using an effective coefficient of thermal expansion (ECTE) which correlates the gage section strains to a single temperature measurement. If the temperature range is relatively small ($\approx$200°C), a linear function will perform adequately. However, this is not usually the case with large temperature ranges. Here the non-linear relationship between temperature and strain is approximated for optimum control using a piece-wise linear function. Higher order polynomials did not perform well at representing thermal strains.

The first step is to determine the physical location of the temperature feedback. After evaluating several axial locations, as well as cases where a temperature feedback was averaged from several locations, the "top" location (Fig. 7) was identified as optimal, particularly during temperature reversal. This location's comparatively quick response characteristics made it most suitable for use in a linear thermal strain equation

$$\varepsilon^{th} = \alpha^* T + C$$

Here, α^* is the ECTE, C is a constant, and T is the real-time specimen temperature at a "top" positioned thermocouple. The specimen is thermally cycled with load held zero to obtain the value for α^*. To insure the accuracy of α^*, the specimen is again thermally cycled, but this time in strain control, where the thermal strain equation is commanding the "compensating" strains.

The possible results from this cycle are shown in Fig. 11, assuming a single linear relation-

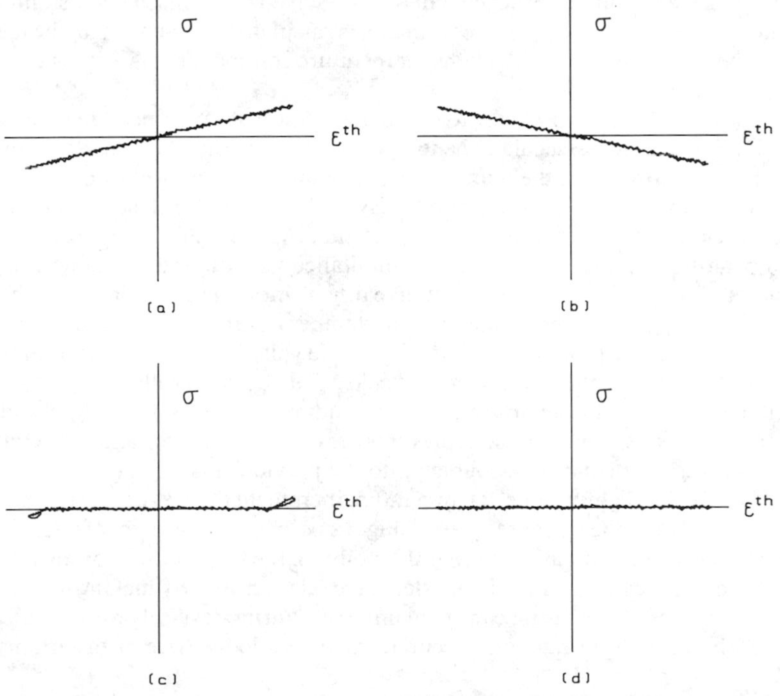

FIG. 11—*Verification of the effective coefficient of thermal expansion (α^*) in the thermal strain compensation equation, $\varepsilon^{th} = \alpha^* T + C$ (α_0 = optimal value for α^*);* (a) $\alpha^* > \alpha_0$; (b) $\alpha^* < \alpha_0$; (c) *temperature rate is too fast; and* (d) $\alpha^* = \alpha_0$.

ship is valid. Figures 11a and 11b represent cases where α^* is greater than and less than the optimal value α_0, respectively. Figure 11c depicts a condition where α^* is accurate during the linear temperature ramp, however, became an invalid relationship during reversals. This condition exists if the temperature rate is too fast and well controlled reversals are not maintained. If the equation is working properly, the specimen is cycled in strain control without inducing a notable stress in the specimen (Fig. 11d). Conditions were considered to be well controlled if the induced stress is contained within 1% of the maximum stress experienced during the test. During the test, the total strain command is calculated by summing two independent functions, thermal strain and mechanical strain. This flexible technique is accurate and capable of representing cycle-to-cycle variations in actual specimen temperature.

Temperature/Mechanical Strain Phasing

Phasing problems arise in thermomechanical loadings because of command/response (C/R) time differences between mechanical strain and temperature. The C/R time is defined as the real time elapsed between issuing a command and obtaining the response on the test specimen. State-of-the-art mechanical loading frames are capable of rapid real-time response, allowing the C/R time of the mechanical strain component to be insignificant. In contrast, the C/R time for temperature is comparatively large, particularly during command reversals.

The temperature/mechanical strain phasing technique was developed with emphasis on maintaining a constant mechanical strain rate. The mechanical strain is calculated by a single, time-based, temperature independent function, ensuring a constant rate throughout the complete cycle. In contrast, two temperature command functions are utilized. The first, referred to as a master/slave command, is dependent on, and slaved to the mechanical strain function. The second, referred to as a decoupled command, is calculated independent of the mechanical strain function. The sequence in which the temperature commands are issued is shown in Fig. 12.

During portions of the cycle which are not in the vicinity of a command reversal, a master/slave relationship is used to calculate the temperature command. At a pre-determined point prior to a command reversal, the temperature command is decoupled from the mechanical strain function and forced to reverse prematurely (Fig. 12, point a). This point is determined by thermal cycle trials (under zero load) prior to the TMD test and is selected such that the actual temperature response reversal occurs simultaneously with the mechanical strain command reversal (Fig. 12, point b). Representative time values between points a and b are 3 to 8 s depending upon specific system configurations, temperature and temperature rate. Note that the value of this interval at the "cold" end of the cycle will likely be different from that established at the "hot" end of the cycle, with the "cold" end value typically greater. As a result of the premature reversal, the temperature command waveform is pushed "ahead" of the mechanical strain waveform. This offset prevents the mechanical strain and temperature commands from being immediately re-coupled into the previous master/slave relationship after reversal. Therefore, a slightly reduced temperature increment ($\approx$95% of the slaved command increment) is used after the response reversal (Fig. 12 point b). At the point of intersection with the master/slave command waveform (Fig. 12, point c), the temperature command is re-coupled, and once again calculated by the master/slave relationship. By making use of this technique, reasonably precise phasing can be maintained during reversals without affecting the constant mechanical strain rate, and accurate hysteresis loops (free of distortions) can be obtained.

Strictly speaking, during the de-coupled portion of the cycle after point b (Fig. 12), the temperature and mechanical strain response are out of proper phasing. However, this slight phase shift occurs during the initial portion of the elastic unload and does not introduce abnormal

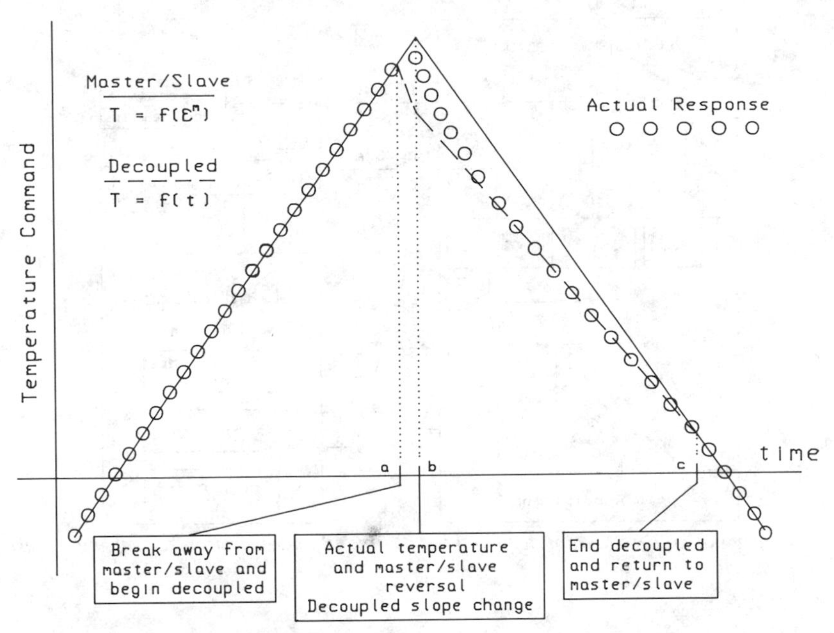

FIG. 12—*Schematic of temperature command sequence.*

distortions in the deformation response. Also, given the real-time calculations for thermal strain compensation (discussed above), no mechanical strain errors are introduced. This methodology is felt to be the best compromise for maximizing accuracy and control.

Discussion

The improved thermomechanical testing techniques and procedures described in this paper were implemented into a real-time test control code and integrated into a state-of-the-art high temperature mechanical testing system. A schematic flow diagram depicting the command and execution sequencing is shown in Fig. 13. As shown, the two control signals, total strain e_2 and temperature T_2, are digitally generated by the microprocessor and converted to analog signals. The thermal component of the total strain is a function of the real-time specimen temperature T_1 and thus requires an incoming signal for control purposes. All remaining control functions are generated as explicit or implicit functions of real-time. Additional analog signals (that is, total strain and load) are sampled for data acquisition.

Given a suitable test system and the control software described above, the sequence for set up and test initiation is as follows. First, temperature ranges and rates are established. Without forced air cooling of the specimen, the cooling portion of the cycle and the desired minimum temperature will dictate the maximum temperature rate. However, from a control standpoint, it is often advantageous to use a temperature rate less than the maximum capable rate, as typically, temperature gradients, phasing control, and thermal strain compensation all improve with decreasing temperature rate. After having established a temperature range and rate, the dynamic temperature gradients over the gage section should be adjusted to desired values. This will require thermal cycling the specimen (under zero load) prior to the TMD test. With all tuning and adjustments to the temperature cycle complete, the next step is to determine appropriate time values for the premature temperature command reversals. Again, this is accom-

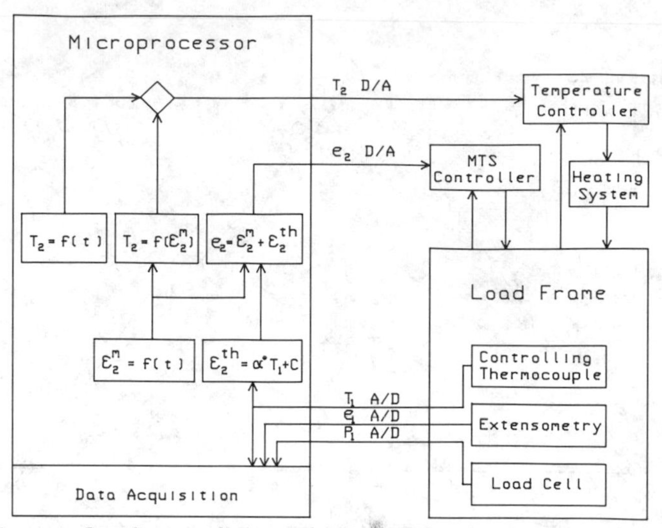

FIG. 13—*Schematic flow diagram depicting the command and execution sequencing for strain controlled thermomechanical testing.*

plished by comparing the temperature commands to the actual temperature response during thermal cycling under zero load. The final step is to calculate an appropriate ECTE under load control and verify the accuracy of the thermal strain compensation function under strain control. This final step will indicate whether or not the various test parameters have been appropriately selected. If this verification does not produce satisfactory results, it will be necessary to adjust one or more of the test conditions (for example, temperature rate) and determine revised test parameters.

Having established the improved techniques, a TMD condition associated with severe material barreling was re-visited. A 600 to 800°C, out-of-phase test with $\varepsilon^m = \pm 0.003$ m/m was conducted; this loading condition consistently promoted severe material barreling after approximately 300 to 500 cycles. The results of this experiment are shown in Fig. 14, alongside an earlier test subjected to identical loading conditions. The specimen tested before implementing the refined techniques was discontinued at 500 cycles. The barreling was visible after only 300 cycles. The specimen tested using the improved techniques experienced over 1100 cycles at which time the test was terminated because of macroscopic cracking. The barreling within the gage section was reduced to an absolute minimum. A small increase in diameter was experienced above the gage section where strict gradient control was not enforced. Since the cross section of the gage remained constant throughout the duration of the test, the deformation data obtained could be used with a high degree of confidence.

Although the material barreling was clearly associated with temperature gradients present during thermomechanical loading, it is likely that other factors contribute to this phenomenon. As this is a ratchetting/flow type process, the magnitude of inelastic strain will likely have an influence, as suggested by Coffin in Ref *6*. A series of 600 to 800°C out-of-phase tests were conducted with similar initial static gradients (± 3°C) over the gage section. Each test was performed with a different mechanical strain range including 0 (thermal cycle only), ± 0.002, ± 0.003 and ± 0.004 m/m. The thermomechanical cycles were periodically interrupted under controlled conditions for careful diameter measurements. Results are shown in Fig. 15 where the diameter increase at the location of most pronounced barreling was reported. The tests

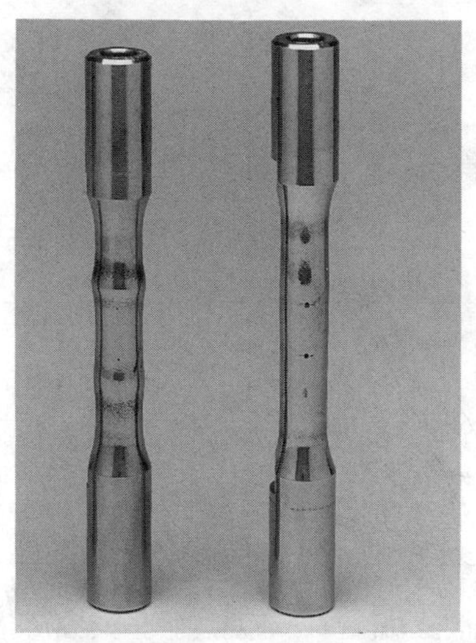

FIG. 14—*Specimen barreling;* (a) *after 500 cycles, before improved techniques;* (b) *after 1100 cycles with improved control techniques.*

reveal that the material barreling is highly influenced by degree of inelastic strain. As previously observed, the barreling was located at dynamic hot spots. It is noteworthy that the material did not barrel under thermal cycle only conditions, but rather a small mechanical component was necessary to promote the behavior.

The improved techniques resulted in the ability to perform well controlled, accurate, TMD experiments. These techniques were used to conduct a series of strain-controlled TMD tests on the nickel-base superalloy, Hastelloy X, at temperatures from 200 to 1000°C. A temperature range of 200°C with a mechanical strain range of ± 0.003 m/m was used. In-phase hys-

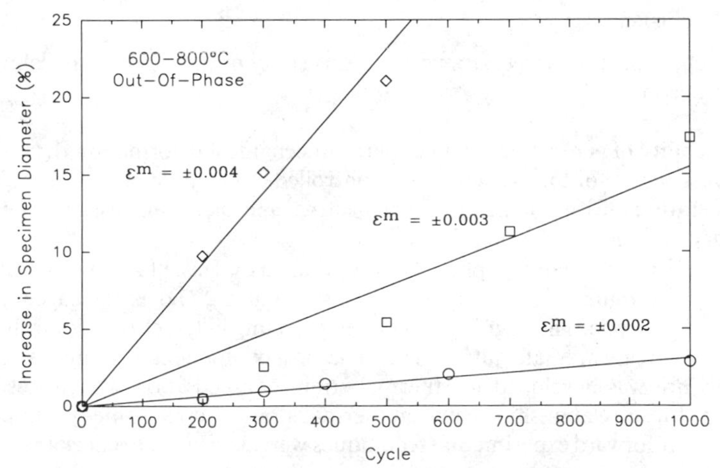

FIG. 15—*Effect of mechanical strain range on specimen barreling for 600 to 800°C out-of-phase loading conditions.*

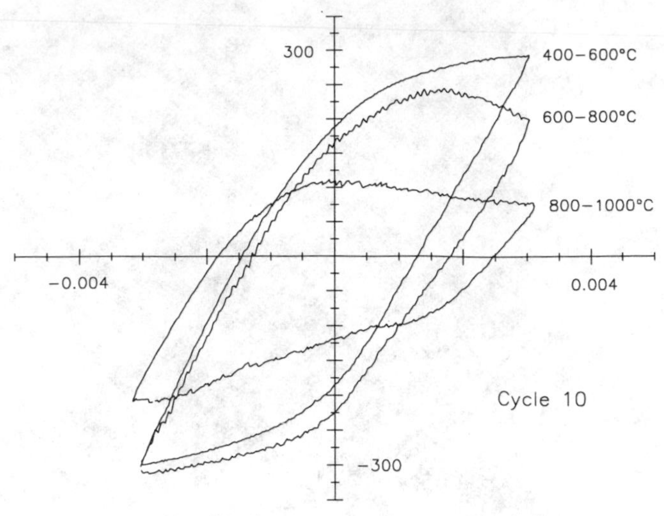

FIG. 16—*In-phase hysteresis loops for Hastelloy X exhibiting minimal distortions.*

teresis loops taken from three of the tests are shown in Fig. 16. Note the absence of abnormal distortions at reversal locations and the distinctive thermomechanical slope changes which occur, most notable in the compressive 800–1000°C deformation. These subtle variations are easily lost if well controlled conditions are not maintained.

Performing this series of definitive tests enabled the thermomechanical hardening mechanisms and behavior of Hastelloy X to be studied in detail, leading to the identification of behavior unique to thermomechanical loading conditions [*13,14*]. Most recently, these data were used to provide guidance for the development of thermoviscoplastic constitutive theories.

Summary/Conclusions

The following conclusions were drawn from this study of thermomechanical testing techniques and procedures.

1. The feasibility of generating reliable thermomechanical deformation data was demonstrated using state-of-the-art, computer controlled test equipment. Some improvements in temperature measurement, specimen heating, and specimen design were found to be necessary in achieving this goal.
2. Specimen barreling in out-of-phase tests conducted on Hastelloy X was effectively eliminated by controlling axial thermal gradients within ± 2°C over the gage section at the hot end of the cycle. The degree of specimen barreling in less closely controlled tests was shown to increase dramatically as the mechanical component of strain was increased.
3. A procedure was developed for thermal strain compensation which uses linear and piecewise linear relationships between thermal strain and real-time specimen temperature. Straightforward experimental techniques were developed to check the effectiveness of the thermal strain compensation prior to testing.
4. A technique was developed to ensure proper phasing between temperature and the mechanical component of strain. Temperature is slaved to mechanical strain over the

major part of the cycle and decoupled during reversals. Reasonably precise phasing is achieved during reversals by tailoring the temperature command waveform to the thermal characteristics of the test system.

5. The techniques and procedures developed in this study were used successfully to generate definitive thermomechanical deformation data for Hastelloy X.

References

[1] Hopkins, S. W., "Low-Cycle Thermal Mechanical Fatigue Testing," *Thermal Fatigue of Materials and Components, STP 612,* D. A. Spera and D. F. Mowbray, Eds., American Society for Testing and Materials, Philadelphia, 1976, pp. 157–169.

[2] Jones, W. B., Schmale, D. T., and Bourcier, R. J., "A Test System for Computer-Controlled Thermomechanical Fatigue Testing," SAND-88-2183C, Sandia National Laboratory, Albuquerque, NM, 1988.

[3] Ellis, J. R. and Robinson, R. N., "Some Advances in Experimentation Supporting Development of Viscoplastic Constitutive Models," NASA CR-174855, LeRC, Cleveland, OH, April 1985.

[4] Bakis, C. E., Castelli, M. G., and Ellis, J. R. "Thermomechanical Testing in Torsion: Test Control and Deformation Behavior," *Multiaxial Fatigue Behavior STP 1191,* D. L. McDowell and J. R. Ellis, Eds., American Society for Testing and Materials, Philadelphia, 1993.

[5] Carden, A. E. and Slade, T. B., "High-temperature Low-cycle Fatigue Experiments on Hastelloy X, *Fatigue at High Temperature, ASTM STP 459,* American Society for Testing and Materials, Philadelphia, 1969, pp. 111–129.

[6] Coffin, L. F., Jr., "The Stability of Metals Under Cyclic Plastic Strain," Transactions of the ASME, *Journal of Basic Engineering,* Vol. 82D, Sept. 1960, pp. 671–682.

[7] Li, P., Marchand, N. J., and Ilschner, B., "Inhomogeneous Cyclic Deformation Behavior of Polycrystalline Aluminum," In: Low-Cycle Fatigue and Elasto-Plastic Behaviour of Materials, Elsevier Applied Science, London and New York, 1988, pp. 55–64.

[8] Conway, J. B., Stentz, R. H., and Berling, J. T., "High Temperature, Low-Cycle Fatigue of Copper-Base Allows in Argon: Part II—Zirconium-Copper at 482, 538 and 593 C," NASA LeRC CR-121260, Aug. 1973.

[9] Sheffler, K. D. and Doble, G. S., "Influence of Creep Damage on the Low-Cycle Thermal-Mechanical Fatigue Behavior of Two Tantalum Base Alloys," NASA LeRC CR-121001, May 1972.

[10] Avery, L. R., Carayanis, G. S., and Michky, G. L., "Thermal-Fatigue Tests of Restrained Combustor-Cooling Tubes," *Experimental Mechanics,* Vol. 7, No. 6, June 1967, pp. 256–264.

[11] Verrilli, M. J., Kim, Y. S. and Gabb, T. P., "High Temperature Fatigue Behavior of Tungsten Copper Composites," NASA TM-102404, LeRC, Cleveland, OH, 1989.

[12] Ellis, J. R. and Bartolotta, P. A., "Adjustable Induction Heating Coil," NASA Tech Brief, Vol. 14, No. 11, Nov. 1990, p. 50.

[13] Miner, R. V. and Castelli, M. G., "Hardening Mechanisms in a Dynamic Strain Aging Alloy, Hastelloy X, During Isothermal and Thermomechanical Cyclic Deformation," *Metall. Trans.,* Vol. 23A, Feb. 1992, pp. 551–562.

[14] Castelli, M. G., Miner, R. V., and Robinson, D. N., "Thermomechanical Deformation Behavior of a Dynamic Strain Aging Alloy, Hastelloy X," *ThermoMechanical Fatigue Behavior of Materials, STP 1186,* H. Sehitoglu Ed., American Society for Testing and Materials, Philadelphia, 1993, pp. 106–125, this publication.

Henry L. Bernstein,[1] *Timothy S. Grant,*[1] *R. Craig McClung,*[1] *and James M. Allen*[2]

Prediction of Thermal-Mechanical Fatigue Life for Gas Turbine Blades in Electric Power Generation

REFERENCE: Bernstein, H. L., Grant, T. S., McClung, R. C., and Allen, J. M., **"Prediction of Thermal-Mechanical Fatigue Life for Gas Turbine Blades in Electric Power Generation,"** *Thermomechanical Fatigue Behavior of Materials, ASTM STP 1186,* H. Sehitoglu, Ed., American Society for Testing and Materials, Philadelphia, 1993, pp. 212–238.

ABSTRACT: Thermal-mechanical fatigue (TMF) cracking is the primary mode of mechanical failure of first stage blades in gas turbines used for electric power generation. Because of the high replacement costs of these superalloy components, it is important that accurate predictions of TMF life be made so that blades can be reconditioned before cracks initiate and propagate beyond repair limits. This paper describes the development of a TMF life prediction methodology for first stage blades, which is part of a Life Management System developed by the Electric Power Research Institute (EPRI).

The TMF life model is based on laboratory TMF tests of the superalloy IN-738LC in both the coated and uncoated conditions. Results for TMF crack initiation life are presented and modeled. The life model is based upon the strain range, the dwell time, and the strain *A*-ratio. The dwell time term accounts for the time dependence of the fatigue life due to environmental attack and creep. The strain ratio term helps account for the effect of mean stress in the cycle. The ability of the life model to predict the fatigue lives of the current test data and the fatigue lives of IN-738 test samples for various test conditions reported in the literature is presented. The effect of coatings and environment upon the TMF life is also discussed.

The strain-temperature response of the blade was obtained from the manufacturer. The life prediction methodology is presented and applied to the prediction of TMF life of blades from a number of engines. Miner's rule is used to sum cumulative damage from the different engine cycles that occur. Good agreement is obtained between the service experience and the life predictions.

KEYWORDS: thermal-mechanical fatigue, nickel-base superalloys, IN-738LC, gas turbine blades, life prediction, high-temperature coatings

Introduction

Industrial gas turbines, also called combustion turbines, are used by utilities and private industrial companies to generate electricity. In the 1960s and 1970s utilities primarily used them to provide peaking power, that is, electricity at the highest demand periods of the day. In the 1970s and especially in the 1980s, combined cycle application became widespread in which the exhaust gases were used to make steam in a boiler, and that steam was used to gen-

[1] Principal engineer, engineer, and senior research engineer, respectively, Southwest Research Institute, P.O. Drawer 28510, San Antonio, TX 78228-0510.

[2] Formerly, project manager, Electric Power Research Institute, 3412 Hillview Ave., Palo Alto, CA 94303; currently, consulting engineer, 7982 Sunderland Drive, Cupertino, CA 95014.

erate additional electricity with a steam turbine or was used in industrial process plants. In Europe, steam or hot water is used for district heating purposes. Today, many simple and combined cycle gas turbine plants are being built to provide additional power and to replace aging power plants both in the United States and abroad. Combined cycle operation can provide thermal efficiencies greater than 50% and can be purchased in modular units to reduce the financial risk associated with building large, conventional coal-fired steam plants or nuclear plants. Smaller industrial gas turbines are used as mechanical drives for numerous applications and to provide electricity in remote locations.

The durability of a gas turbine is principally limited by those components operating at high temperatures in the turbine section. The first stage nozzle and blades usually are the first parts to need repair, apart from the combustion components. The first stage blades, also called buckets, experience creep, fatigue, and high-temperature attack from oxidation and hot corrosion. Because of the conservative design of the heavy section buckets, creep has not been an operational problem. The buckets are usually removed from service for repair because of fatigue cracking or coating degradation.

Heavy Wall Buckets

The type of bucket, or blade, considered in this paper is called a heavy wall bucket because of its solid cross section, as shown in Fig. 1. This additional mass provides greater resistance to fracture from impact by foreign objects and provides the bucket with more resistance to environmental attack, which can compromise its structural integrity. The bucket has a cooling scheme consisting of holes drilled through the span of the bucket. The new, highest tempera-

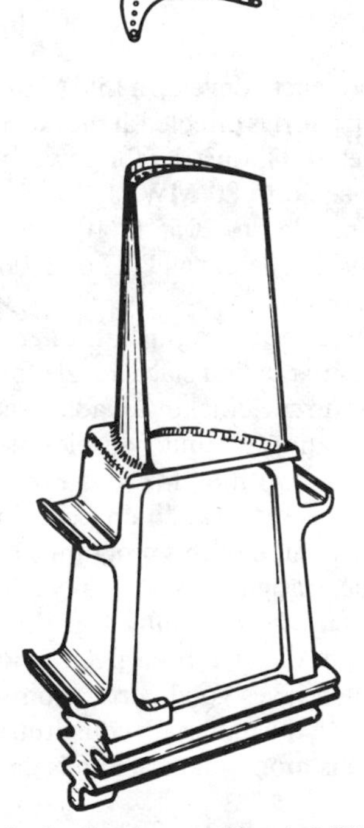

FIG. 1—*"Heavy wall" first stage bucket for General Electric MS7001E industrial gas turbine. Bucket is 35 cm (14 in.) long, weighs 6.8 kg (15 lb), and is cooled by through holes.*

ture, gas turbines employ more advanced cooling designs similar to those used in aircraft engines, consisting of serpentine passages and film cooling.

Because of the high cost of a new bucket set, which is roughly 0.5 million U.S. dollars, the ability to refurbish the buckets is highly desirable. Buckets cannot be refurbished, however, when fatigue cracks are present in the base metal or when too much base metal attack has occurred from environmental attack or foreign object damage. The amount of allowable damage is defined by the manufacturer. Operators of gas turbines wish to obtain 24 000 to 48 000 h of operation before removing the buckets for refurbishment, which is typically three to six years of continuous operation.

If fatigue cracks penetrate into the base metal beyond the blending limits, the bucket cannot be refurbished. The industry practice is to refurbish before fatigue cracks occur, because fatigue cracking to lengths much greater than 0.2 mm (a few mils) is cause to reject the bucket. Consequently, fracture mechanics approaches are not useful because they assume the existence of cracks, which would quickly lead to the removal of the buckets for refurbishment. Furthermore, metallurgical examinations of many buckets at high magnification have not found pre-existing cracks. It should be noted that IN-738 heavy wall buckets can operate with leading edge cracks of 5 mm (0.2 in.) length without fracture provided that vibratory stresses are very low and there is no embrittlement of the metal.

Coating degradation is currently under study, and models for describing this degradation are under development. This is a very complex area and very important because it can determine the bucket life in both the current and the newer, high-temperature engines being built today. Little in the way of life prediction can be done to account for foreign object damage. Most foreign object damage in the turbine section is caused by pieces of metal that originate in the combustion section.

Life Management of Gas Turbines

This study is part of a larger project to develop a Life Management System (LMS) for industrial gas turbines [*1*]. The current effort is directed at the General Electric MS7001 gas turbines, which are used widely by utilities and industries for power generation. These gas turbines are physically very large and produce 60 to 80 MW of electricity. They operate at turbine inlet temperatures (TIT) ranging from 1850 to 2055°F (1010 to 1125°C). Advanced gas turbines, just beginning operation as of 1991, operate at TITs of 2300°F (1260°C) and produce around 150 MW of electricity.

The LMS project concentrates on the components of greatest interest to gas turbine operators and manufacturers, which are the first stage nozzle (or vanes) and the first stage blades. Models have been developed for nozzle cracking, blade creep, and blade thermal-mechanical fatigue. Models for coating degradation are under development. The amount of life consumed and the remaining life are calculated with REMLIF, a Fortran-based computer program that tracks the engine and component history in addition to performing life predictions, and REMLIF EXPERT, an expert system version of this program.

A model for thermal-mechanical fatigue was originally developed for this project by General Electric, and that model has been used successfully. A preliminary discussion of this effort and a presentation of some of the fatigue data has been given by Russell [*2*]. The GE model is based on maximum tensile stress, which gives good correlation with life, but is difficult to apply because of its strong dependency on the material's constitutive behavior, which is not that well known. A strain-based approach is more practical and is the focus of this paper.[3]

[3] Approaches requiring an accurate calculation of the stresses are difficult to use because inelastic behavior caused by stress-relaxation, cyclic hardening, and softening, and so forth, makes the determination of the stresses quite difficult and uncertain.

High-Temperature Fatigue of Nickel-Base Superalloys

The isothermal fatigue of nickel-base superalloys at high temperature has been widely studied. Summaries of the literature have been given by Miner [*3*] and Skelton [*4*], among others. Conclusions of significance to this paper are that compressive hold times decrease life by elevating the mean stress, and that tensile hold times have the opposite effect, both conclusions being valid when inelastic strains are limited. Environmental attack caused by high-temperature oxidation has been found to be detrimental to IN-738 by Nazmy [*5*] and Gabrielli et al. [*6*]. Hot corrosion has also been shown to shorten the life of Udimet 720 by Allen and Withlow [*7*] and Withlow et al. [*8*].

The TMF of nickel-base superalloys has not been as widely reported. Leverant et al. [*9*] have summarized many of the important issues. TMF data for IN-738 have been given by Russell [*2*] and by Kuwabara et al. [*10*], both of which are discussed in this paper. Some TMF data have also been given by Viswanathan [*11*] for Udimet 710, which is also discussed in the text, and tests on Mar-M247 have been reported by Boismier and Sehitoglu [*12*].

Information on the high-temperature fatigue of coated superalloys is also not widely available. Grünling et al. [*13*] have summarized much of what is known about the effect of coatings on mechanical properties. Bernstein [*14*] has discussed how coatings can affect the fatigue life of industrial gas turbine blades. Fairbanks and Hecht [*15*] have shown that the fatigue life is reduced by coatings and that different coatings can affect the fatigue life differently. Heine et al. [*16*] conducted TMF tests on a coated nickel-base single crystal superalloy and found that coatings reduced the fatigue life. They contended that the coating and the substrate must be considered as a single system in order to predict the TMF life of coated turbine blades.

TMF modelling of nickel-base superalloys has mostly attempted to use models developed for isothermal fatigue. Kuwabara et al. [*10*] claimed that TMF tests gave lives longer than isothermal tests at the maximum TMF temperature, and consequently, that TMF lives could be conservatively predicted from isothermal test results. However, Heine et al., at Pratt and Whitney [*16*] stated that the use of isothermal data has not been successful in predicting the fatigue life of blades and that TMF data must be used. Halford and Manson [*17*] have attempted to extend strain range partitioning to TMF, and Heine et al. [*16*] have used a tensile hysteretic energy approach that requires an accurate constitutive model. Neu and Sehitoglu [*18*] have developed a general model for high-temperature fatigue, including thermal-mechanical fatigue, and Sehitoglu and Boismier [*19*] have applied it to Mar-M247. Their model accounts for damage from fatigue, environment, and creep as separate terms, which are linearly summed, and makes use of a constitutive equation to compute the creep damage.

Finally, it should be noted that the large body of high-temperature fatigue data on steels and stainless steels is of limited value for applications to nickel-base superalloys because of the large differences in ductility and strength between these materials. In general, the nickel-base superalloys are more sensitive to mean stress and environment than the stainless steels.

Scope of Paper

The main purpose of the work reported in this paper is to develop a model to predict the TMF life of industrial gas turbine blades. The model developed is a semi-empirical model, similar to most engineering models that are actually used to predict low-cycle fatigue. (The model does not attempt to separately model the different mechanisms of damage, that is, fatigue, creep, and environmental attack.) To this end, a presentation is made of what is involved in actually predicting the life of a component, and how the material data on fatigue properties fit into this process. A secondary objective of this work is to better understand the TMF behavior of IN-738.

The body of the paper is divided into four sections. The first section describes the experimental program, the fatigue results, and the development of a model to predict the TMF of blades. The second section describes the structural analysis of a cooled blade. The third section discusses the life predictions for service exposed blades and includes metallurgical analyses. The last section is a discussion of the results from the paper.

Experimental Program

Materials

The buckets and test specimens are made from IN-738LC, which is an investment cast nickel-base superalloy used worldwide for industrial gas turbine hot section components. It has greater high-temperature corrosion resistance than superalloys used in aircraft engines because industrial gas turbines are expected to operate for long times on low-quality fuels. IN-738LC has good high-temperature creep strength, limited ductility, and good room-temperature strength. It is strengthened primarily by a high fraction of gamma-prime particles, and secondarily by grain boundary carbides and solid solution strengthening. The nominal composition (in wt%) of the material is 16% Cr, 3.4% Ti, 3.4% Al, 8.3% Co, 2.6% W, 1.7% Mo, 0.9% Cb, 0.10% C, 0.01% B, and 1.75% Ta.

In service, the buckets are coated to protect against high-temperature environmental attack. The test specimens were coated on the outer surface with GT-29, which is a proprietary overlay coating applied by a low-pressure plasma-spray process. The coating is a CoCrAlY type, having a composition (in wt%) of 29% Cr, 6% Al, and 0.3% Y, with the balance being Co. The coating consists of a Co + Cr matrix containing particles of CoAl. The inner surface of the hollow specimen was not coated because it is not possible to plasma spray this coating on the inner surface.

Specimen blanks for this testing program were cast under carefully controlled conditions to obtain a grain structure representative of gas turbine buckets. Following casting, the blanks were hot isostatically pressed (HIPd) in an argon atmosphere. Following HIPing, the castings were solution-treated in vacuum at 1121°C (2050°F) for 2 h followed by air cooling and subsequent aging at 843°C (1550°F) for 24 h with air cooling.

Final specimen geometry was a hollow cylinder with a gage section outer diameter of 10.2 mm (0.400 in.), a wall thickness of 1.52 mm (0.060 in.), and a gage length of 25.4 mm (1 in.). The specimens were finish machined using low-stress grinding on the outside and longitudinal honing on the inside, resulting in a very smooth finish. Most of the specimens were coated with GT-29. The coating thickness ranged from 0.117 to 0.155 mm (4.5 to 6 mils), which is somewhat less than the typical thickness of a GT-29 bucket coating, which is 0.2 to 0.25 mm (8 to 10 mils). Following the coating application, the specimens were given a surface treatment involving peening and heat treatment in order to meet the standard bucket coating requirements.

Test Program

The tests were conducted at General Electric laboratories using servo-hydraulic test machines under strain control. Specimens were induction heated and cooled by forced air through the hollow tube. Strains were measured with an axial extensometer over a 15-mm (0.6-in.) gage section.[4] The temperature gradient over the gage section was maintained to within ±2°C. The test program comprised three groups of experiments: isothermal low-cycle fatigue, thermal-mechanical fatigue, and "bucket" thermal-mechanical fatigue tests.

[4] The knife edges of the extensometer were lightly pulled into the specimen by springs, and no dimpling of the surface was used.

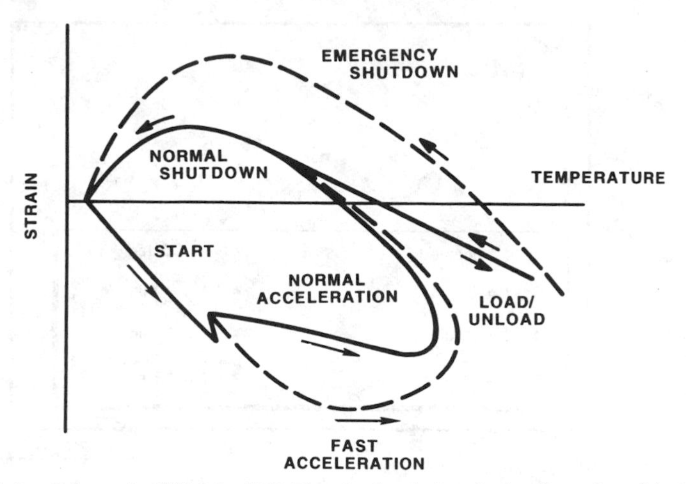

FIG. 2—*Schematic of "bucket" TMF cycle simulating the leading edge of the bucket.*

Isothermal low-cycle fatigue (LCF) tests were conducted at 760, 871, and 982°C (1400, 1600, and 1800°F). The tests were run with compressive holds of 0, 0.33, 2, and 15 min. Total strain ranges[5] varied from 0.2 to 0.8%, and the cycling frequency was 0.33 Hz. Strain amplitude ratio $A = \varepsilon_{amp}/\varepsilon_{mean}$ was held constant at -1.[6] All but three specimens were coated.

Thermal-mechanical fatigue (TMF) tests were on linear out-of-phase (LOP) cycles in which the maximum strain and minimum temperature occurred simultaneously. This cycle is representative of industrial gas turbine buckets in which the highest temperatures occur while the surface is in compression. The minimum temperature was held constant at 427°C (800°F) and the maximum temperature was either 871, 916, or 982°C (1600, 1680, or 1800°F). Compressive hold times of 0, 2, and 15 min were applied with both coated and uncoated specimens. The strain range varied from 0.4% to 0.8% with $A = \infty$ in most cases. Four tests were conducted at $A = -1$ to investigate the significance of mean strain effects.

The "bucket" thermal-mechanical fatigue tests followed a strain-temperature history representative of the leading edge of the first stage bucket in service. This bucket history was developed by GE from their component analysis; a schematic is shown in Fig. 2. Tests were carried out at maximum temperatures of 916 and 960°C (1680 and 1760°F), which are in the range of application. The hold time was held constant as a 2-min compressive dwell (at maximum temperature), and the strain amplitude ratio was maintained at $A = -2.3$. The only free parameters in these tests were the strain range and the presence or absence of coatings.

Test Results

The current study focused on a parametric analysis of the test data using a strain-based approach. To accomplish this task, total strain ranges versus fatigue life were plotted holding all but one parameter constant. This facilitated identification of the most important factors to be included in a life prediction model. The total strain range was used because the cyclic inelas-

[5] All strains reported are mechanical strains, in which the thermal strain (or displacement) has been subtracted from the strain (or displacement) measured by the extensometer. Thus, the "total strain" is the "total mechanical strain," the "inelastic strain" is the "inelastic mechanical strain," and so forth.

[6] $\varepsilon_{amp} = (\varepsilon_{max} - \varepsilon_{min})/2$ and $\varepsilon_{mean} = (\varepsilon_{max} + \varepsilon_{min})/2$ where ε_{max} and ε_{min} are the maximum and minimum values of the strain, respectively.

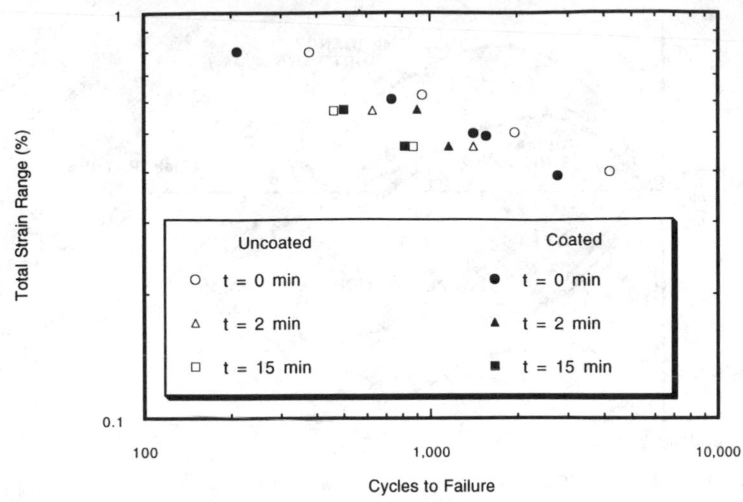

FIG. 3—*LOP test results at* T_{max} = *871°C (1600°F) and* A = ∞, *showing the effect of compressive hold time and coatings.*

tic strains are negligible in the range of application, and inelastic strain was reported as zero for a significant fraction of the test data.[7] The failure life N_f corresponded to 50% load drop.

The five parameters investigated, apart from strain range, were the compressive hold time, the test temperature, the presence or absence of a coating, the strain *A*-ratio, and the cycle type—isothermal, LOP and bucket.

Effect of Compressive Hold Time—The effect of compressive hold time upon fatigue life in the LOP tests is shown in Fig. 3 for both coated and uncoated specimens. With the exception of one test, there was a clear decrease in life as the hold time increased. Similar decreases in life with hold time were observed for the LOP tests at 982°C (1800°F) and the isothermal tests at 871°C (1600°F).

Effect of Temperature—The effect of temperature upon the fatigue life in the LOP tests with a 2-min hold is shown in Fig. 4*a* for coated specimens and Fig. 4*b* for uncoated specimens. The life at 982°C (1800°F) was consistently shorter than at 871°C (1600°F) for both the coated and uncoated specimens. Similar behavior was seen for the uncoated LOP tests with no hold time. The isothermal tests of coated specimens with $A = -1$ and a 2-min hold time showed decreases in life as the temperature was increased from 760 to 871 to 982°C (1400 to 1600 to 1800°F). For the tests in Fig. 4*b*, the lives at 916°C (1680°F) appeared to be similar to those at 982°C (1800°F). There also was no difference in the fatigue life between the 916 and 960°C (1680 and 1760°F) bucket tests for both coated and uncoated specimens.

Effect of Coatings—The effect of coatings upon the fatigue life is more difficult to interpret because the inner surface of the specimen was not coated. Furthermore, the tensile properties of the coated specimens showed somewhat greater ductility and lower strength than the uncoated specimens. The reason for this behavior is not known. As shown in Fig. 3 for LOP tests at an *A*-ratio of infinity, the coating reduced the life for 0 hold time tests at both 871 and 982°C (1600 and 1800°F). For 2- and 15-min holds, the coated and uncoated specimens had similar lives. For the bucket specimens, which had a 2-min hold time, the coating caused a significant reduction in the fatigue life of the specimens, as shown in Fig. 5 for the 916°C (1680°F) data. The effect of coatings on the bucket tests at 960°C (1760°F) was similar. For the

[7] The extent of the inelastic strain depended upon the strain range of the test and was negligible at low strain ranges. In addition, the material initially hardened and then softened during the test.

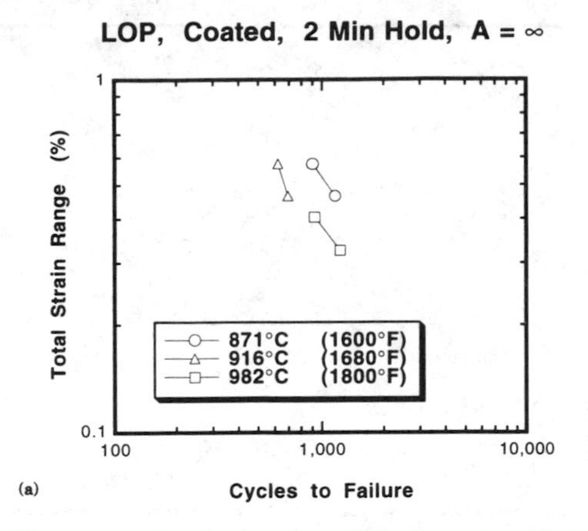

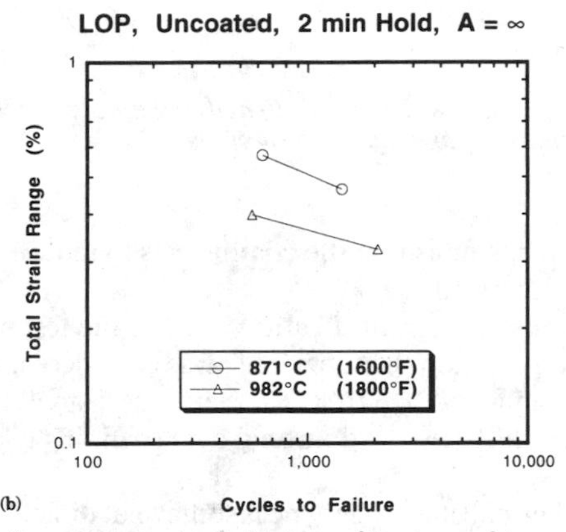

FIG. 4—*LOP test results for a compressive hold time of 2 min, and* A = ∞, *showing the effect of temperature:* (a) *coated specimens and* (b) *uncoated specimens.*

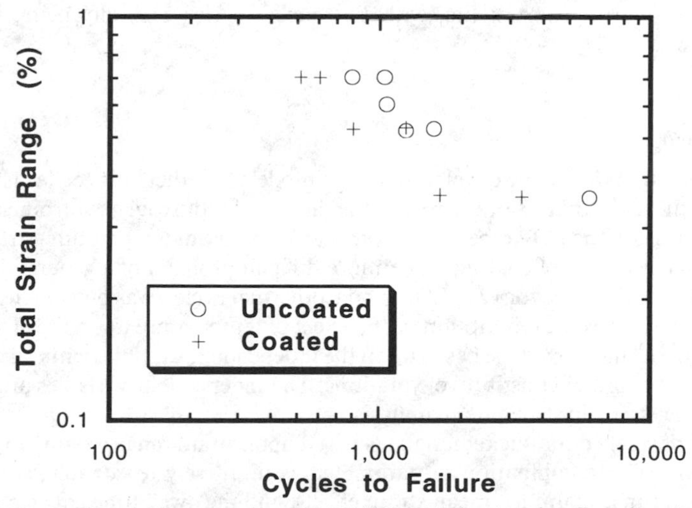

FIG. 5—*"Bucket" test results at 916°C (1680°F) showing the effect of coatings.*

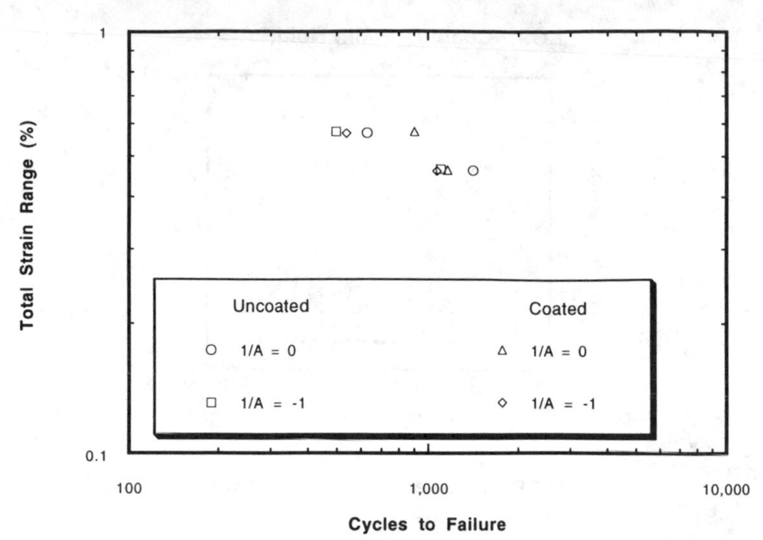

FIG. 6—*LOP test results at* T_{max} = *871°C (1600°F) and a compressive hold time of 2 min showing the effect of strain* A-*ratio for both coated and uncoated specimens.*

TMF tests, the general conclusion is that the coating acts to reduce the life. More will be said about this subject in the discussion.

Effect of A-Ratio—Because the strain *A*-ratio varied in the test program and in the application, the effect of strain *A*-ratio on the fatigue life was considered. As shown in Fig. 6 for the LOP tests with a 2-min hold at 871°C (1600°F), there was a small but consistent decrease in life for $A = -1$ compared to $A = \infty$. Additional data spanning a larger range in *A* were not available.

Effect of Cycle Type—The effect of cycle type is somewhat difficult to distinguish in the data because each cycle type was run at a different combination of temperature and *A*-ratio. The data for coated specimens with a 2-min compressive hold are shown in Fig. 7 for the isothermal, LOP, and bucket cycles. It can be seen that the isothermal tests gave the longest life, and the LOP tests gave the shortest life, with the bucket cycles generally being intermediate between the two.

TMF Life Model

These observations led to the development of a model to predict fatigue life. The model was optimized to those conditions of interest to the fatigue of industrial gas turbine blades: a few well-defined cycles of the LOP type; a relatively narrow range of temperature, strain range and *A*-ratio; and the presence of coatings. The more difficult problem of a general description of thermal-mechanical fatigue for LIP, LOP, and isothermal cycles, was outside the scope of the modelling effort, as was the description of the effect of temperature upon life. The goal of the effort was to model the fatigue life based upon the independent variables without complicating the analysis by the use of constitutive equations. The independent variables are those determined from the elastic finite-element analysis.

The model developed for bucket fatigue is based upon strain range, strain amplitude ratio, and dwell time. This combination of parameters is intuitively reasonable, since the strain amplitude ratio can account for mean strain effects, and the dwell time can account for both

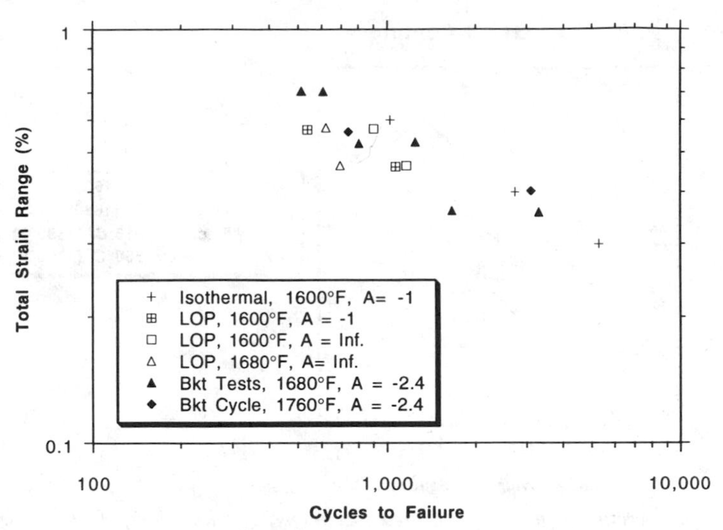

FIG. 7—*Test results on coated specimens with a 2-min compressive hold time showing the effect of different cycle types.*

environmental effects and mean stress effects caused by steady-state stress relaxation, both of which can affect the fatigue life. The general form of the model is

$$N_f = C_0(\Delta\varepsilon)^{C_1}(t_h)^{C_2} \exp\left(\frac{C_3}{A}\right) \tag{1}$$

where N_f is the fatigue life, A is the strain A-ratio, $\Delta\varepsilon$ is the total strain range (expressed in percent strain), and t_h is the dwell time (in minutes). Two equations for fatigue life were developed, one for coated and one for uncoated specimens. The empirical constants C_i in Eq 1 were determined from multiple regression of the LOP data at 871°C (1600°F), and are given in Table 1. As shown in Fig. 8, the model adequately described the behavior of IN-738LC for the LOP test conditions at 871°C (1600°F). This model was then used to predict the results of the bucket cycle tests, which had *not* been utilized to generate the model constants. Figure 8 shows that the model did indeed adequately predict the bucket cycle lives for the coated specimens. Similar results were obtained for the uncoated specimens.

Prediction of Literature Data—The model was further evaluated using IN-738 data available in the literature. Only isothermal test conditions were found, except for one source of TMF data, and all the literature data were taken from uncoated specimens at an A-ratio of infinity. To predict the TMF data, the constants in the model were developed using the uncoated specimen results from the present test program. Consequently, the comparisons with these literature data are genuinely independent predictions. For the isothermal tests in the lit-

TABLE 1—*Constants in Eq 1.*

Type	C_0	C_1	C_2	C_3
Coated	125	−3.40	−0.217	0.247
Uncoated	171	−3.45	−0.381	0.388

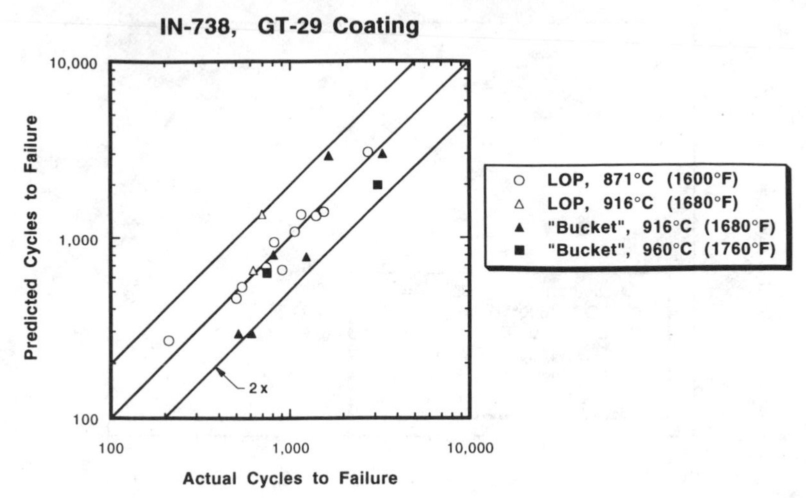

FIG. 8—*Predicted versus actual life for coated specimens. The linear out-of-phase tests at 871°C (1600°F) were correlated, and the other tests were predicted.*

erature, the literature data were fitted to the model, but without the *A*-ratio term.[8] In order to account for strain rate effects in the literature data, the "dwell time" in the life model was interpreted as the sum of both hold time and time in the compression-going portion of a single cycle.

The TMF data were from Kuwabara et al. [*10*] for LOP cycles at T_{min} = 300°C (572°F) and T_{max} = 900°C (1652°F). Total strain ranges varied from 0.3 to 1.0%, and cycling frequencies were held constant at 0.0056 Hz. The predictions of the TMF model using the uncoated data in the current test program are shown in Fig. 9. Although the data were predicted to within a factor of 2 (worst point is 2.65), all the tests had shorter lives than predicted. Isothermal data from this same source, which are presented in the next paragraph, had shorter lives than other isothermal data in the literature for IN-738.

Four different sets of isothermal data in the literature were used, along with the three uncoated isothermal tests in the current test data. The literature data were from Kuwabara et al. [*10*], Gabrielli et al. [*6*], Nazmy [*20*], and Ostergren [*21*]. Kuwabara et al. reported isothermal LCF data at 900°C (1652°F). Gabrielli et al. conducted isothermal LCF tests on cast IN-738LC specimens at 850°C (1562°F) with total strain ranges between 0.5 and 2.5%. These tests were conducted at constant strain rates of 10^{-3} or 10^{-2} s^{-1} or with compressive hold times of 60 s. Nazmy published isothermal LCF results at 850°C (1562°F) for seven different strain wave shapes, including constant strain rates between 10^{-2} and 10^{-5} s^{-1}, fast-slow and slow-fast waveforms, and hold times in tension or compression, or both. Ostergren reported isothermal LCF data at 871°C (1600°F) with and without hold times in tension or compression. Cycling frequency was held constant at 0.33 Hz, and total strain ranges varied from 0.3% to 0.8%.

The isothermal test conditions were correlated from the TMF model, but without the *A*-ratio term. The results of the correlation are shown in Fig. 10, and the majority of the data are correlated to within a factor of 2. The worst prediction was a factor of 2.6.

[8] There were an insufficient number of uncoated isothermal tests in the current data set, that is, 3 tests, to develop the constants.

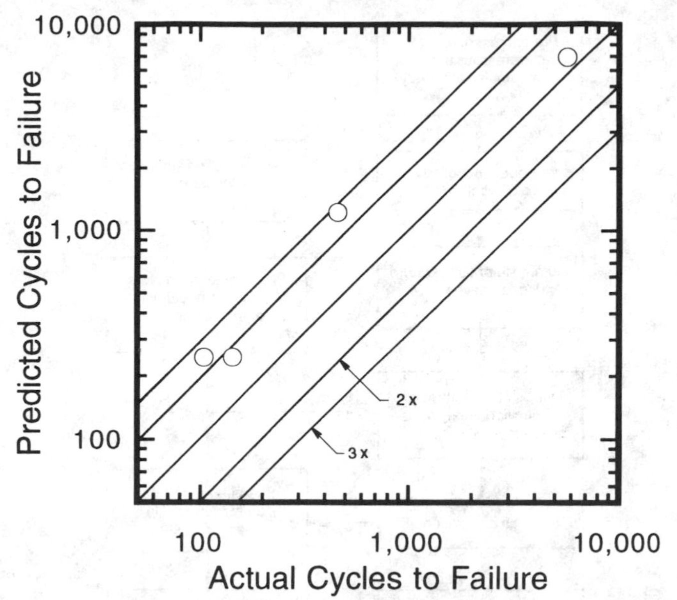

FIG. 9—*Predicted versus actual life for data from Kuwabara et al.* [10] *using the model and constants given in this paper.*

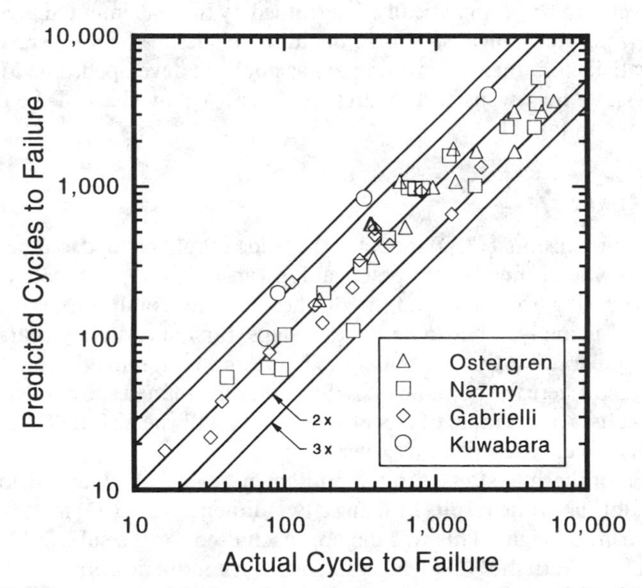

FIG. 10—*Predicted versus actual life from data in the literature using the model in this paper to correlate the data.*

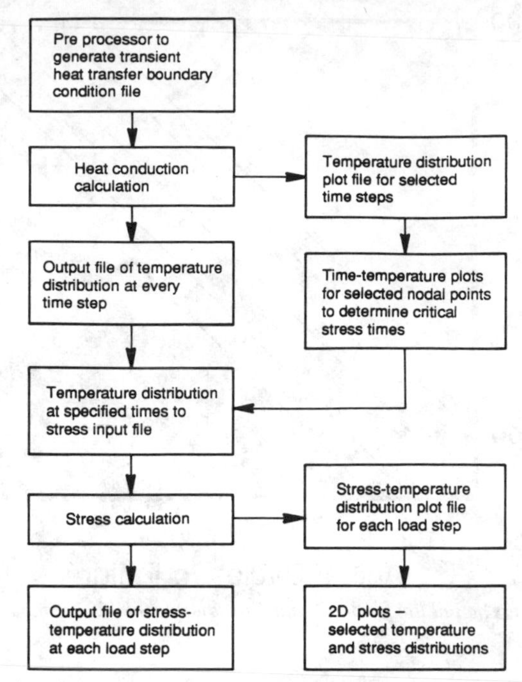

FIG. 11—*Flow chart for a transient finite-element analysis of a cooled gas turbine blade.*

Structural Analysis

Blade TMF cycles must be analytically determined by finite-element analysis (FEA). Measurement of thermal-mechanical strains in operating blades at the temperatures involved is difficult, and a satisfactory means for doing so has not been developed. The FEA entails transient and steady-state heat transfer and stress analyses. A flow chart of the general analysis procedure is shown in Fig. 11.

Finite-Element Model

Ideally, a three-dimensional (3-D) model of the blade should be used, but because 3-D models are so large, they are generally not practical for transient analysis if detailed modelling of features, such as cooling holes and leading edge heat transfer conditions, is required. In these circumstances, a 3-D analysis has to be done in steps. First, a relatively coarse 3-D model is used to obtain a general solution for the entire blade and then refined sub-models of critical regions are analyzed where the boundary conditions for the refined sub-models come from the coarse model results. An example of a coarse 3-D model of the airfoil of a blade of the type considered in this paper is described by Allen [*22*].

For calculation of thermal stresses in the mid-height region of a blade airfoil, it is possible to obtain reasonably accurate results utilizing a two-dimensional (2-D) model of a blade cross section at or near mid-height. This was the approach used for the subject blade. The model used is shown in Fig. 12. In the 2-D idealization, no heat conduction is allowed in the direction normal to the cross section, and stresses are computed assuming that the blade is in a state of generalized plane strain [*22*]. The generalized plane strain formulation leads to a quasi 3-D solution in which strains normal to the cross section are allowed to vary linearly with the

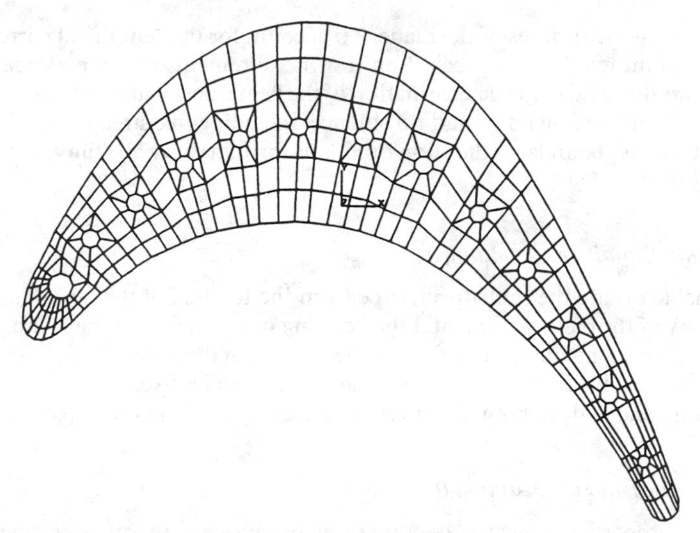

FIG. 12—*Finite-element model of the MS7001E first stage bucket.*

dimensions of the cross section (plane sections remain plane after deformation) and the out-of-plane shearing strains are assumed to be zero. This amounts to superposing a uniform strain (translation) and two bending strain distributions (rotations about the x and y axes) to the classical plane strain (zero normal strain) solution [*23*]. The magnitude of these superposed strains are such that static equilibrium of the external force in the normal direction (centrifugal force acting on the top half of the blade) and external bending moments (from the aerodynamic loads on the top half of the blade) about the x and y axes are satisfied.

A plane strain or generalized plane strain solution yields exact results for cross sections well removed from the ends (on the order of a chord length) of prismatic bodies with constant or linear thermal expansion in its lengthwise direction [*23*]. Obviously, these conditions are never completely satisfied with any turbine blade, but if the aspect ratio (length to chord ratio) approaches two, the taper and twist are gradual, and the spanwise temperature profile is roughly linear, then reasonably accurate results can be expected. Applications of the generalized plane strain model and some additional FEA modelling considerations are described in Refs *22* and *24*.

Heat Transfer Boundary Conditions

A major part of the analysis effort is the calculation of heat transfer film coefficient and gas temperature distributions for starting, steady-state, and shutdown operating conditions. This requires numerous turbine performance, aerodynamic flow field, heat-transfer boundary layer, and cooling airflow network calculations to define the time-dependent boundary condition functions for each element face of the model that is exposed to the blade path or cooling air flow streams. For the cooling air temperature and heat transfer coefficient calculations, iteration is required to get the cooling air temperatures correct, because of significant heatup of the cooling air as it flows through the blade. The aerodynamic flow field calculations are generally made assuming that the flow is circumferentially uniform as it enters the blade flow passage, but this is not the case because the total temperature pattern exiting the combustor is nonuniform. In addition, wakes in the flow are introduced by the trailing edge of the upstream stator. Thus, the flow is unsteady relative to the rotating blade, and its time average temperature will vary along the surfaces of the blade. To account for this, empirical corrections must be made to the film temperatures, based on test measurements and experience.[9] Empirical relationships are also used in the determination of the heat transfer film coefficients to account for the effects of surface roughness and turbulence, including free stream turbulence. Representative heat transfer boundary condition functions for an industrial/utility gas turbine blade are given in Ref *24*.

Stress Boundary Conditions (Loadings)

The external loads are the centrifugal force from the top half of the airfoil, applied at the center of gravity of the cross section, and the bending moments about the x and y axes of the model from the aerodynamic loads acting on the top half of the airfoil. The stresses computed from the model must balance these applied loads. The model is fixed against plane rigid body motion by fixing one node in both directions and a second node in one direction.

Material Properties and Constitutive Behavior

The material properties of thermal conductivity, specific heat, thermal coefficient of expansion, and Young's modulus vary significantly with temperature and are therefore specified as functions of temperature in the finite-element model input. For an inelastic stress analysis, the inelastic properties are even more temperature-dependent and must be accurately specified if satisfactory stress results are to be obtained. However, because the inelastic time-temperature-cyclic constitutive behavior of these alloys has not been well characterized and because the blade TMF is essentially a strain-controlled process, as opposed to a load-controlled process, the usual practice is to perform only an elastic analysis and to use the resulting elastic strains in conjunction with strain-controlled TMF data to predict TMF life.

The assumption that total mechanical strains from an elastic solution are approximately the same as from an elastic-plastic-creep solution for this class of problem has been shown to be reasonable [*26*]. Furthermore, it is the cyclic strain that needs to be accurately determined for TMF life prediction purposes, and the cyclic strain in this case will shake down (stabilize) to an elastic cycle over the bulk of the blade. Thus, the elastic solution should be quite acceptable for the intended purpose. Where holes or notches are involved, Neuber's rule should be used if elastic shakedown does not occur.

It should be noted that creep (overall permanent extension of the blade) is generally limited by design to be negligible in gas turbine blades of this type, particularly in peaking machines where TMF is the life-limiting mode of mechanical failure. For this particular blade, the creep design limit is 0.5% creep strain in 70 000 h for one blade in 10 000 [*27*]. This means that for blades run in cyclic (peak load) service, virtually no creep will ever occur over the useful lifetime of the blades. Thus, neglecting mean creep in the TMF life models is justified.

Blade Strain-Temperature Cycles

Strain-temperature cycles for the leading edge of the subject blade are depicted in Fig. 2. The leading edge is the most responsive part of the blade to transient heating and cooling owing to the high heat transfer coefficients there relative to the rest of the airfoil [*24*]. The response can be minimized by increasing the radius of the leading edge, since the heat transfer coefficients

[9] With the advanced computational fluid dynamics (CFD) codes that have been developed in recent years, it is now possible to model these effects, but because of the enormous computer capacity and run times required, this is not yet practical for general design and analysis purposes [*25*].

there depend on the radius of the leading edge. It should be noted that, for cooled blades, the maximum compressive strain does not necessarily occur during the transient heatup of the blade. It is possible for it to occur at steady state because of the cooling. For this reason, large steady-state compressive strains also occur on the suction and pressure surfaces of the airfoil [*24*], and consequently cyclic strains can be of the same order as for the leading edge. The highest magnitude cyclic strains in the airfoil actually occur tangentially to the mid-chord cooling holes due to the large through-the-thickness thermal gradient and stress concentration effect of the holes, but because the metal temperatures at these holes are so much lower than on the airfoil surface, there is no history of TMF cracking of these holes. For the subject blade, experience has shown that the leading edge is the most vulnerable region to cracking [*28*]. Substantial cracking has also been observed on the suction and pressure surfaces [*28*].

Effect of Engine Startup-Shutdown Cycle—As shown in Fig. 2, compressive peaks occur after ignition, during acceleration, and at steady state. The tensile peak occurs during shutdown. The ignition and acceleration compressive peaks occur at low and intermediate temperatures while the steady-state (hold time) compressive peak occurs at maximum temperature. The shutdown peak occurs at low temperature. Thus, the linear-out-of-phase TMF cycle best simulates the blade TMF cycle among the simplified cycles that have been studied.

It can be seen from Fig. 2 that the magnitude of the strain cycle depends on how the machine is started and shut down. The normal mode of operation is the normal startup-normal acceleration-normal loading to base load-fired shutdown path. The fast start option increases the strain range by about 20%. This option is for emergency demands for electricity and is therefore seldom used. The full load trip (or emergency shutdown) increases the strain range by about 35% and is a major reason for premature TMF cracking in service. Full load trips occur because of vibration or temperature limits being exceeded or by sudden disconnection from the electrical grid.

Life Prediction

Field Data

The field data are from General Electric MS7001E (Frame 7E) gas turbines operating base load and intermediate duty cycles. (Base load is running 24 h a day for weeks or months at a time. Intermediate duty is running 6 to 12 h a day.) The data represent 16 inspections from 9 engines at 6 sites. The engines are geographically dispersed, located in the east, west, southwest, and north. The engines run at turbine inlet temperatures of 1085 to 1104°C (1985 to 2020°F) and mostly burn natural gas, although there is some use of fuel oils.

The data on bucket cracking come from inspection reports and metallurgical investigations performed by SwRI and GE. Inspection reports are based upon visual observations and fluorescent dye penetrant inspections. When fatigue cracking is found by nondestructive evaluation (NDE), fatigue cracks are usually present on most, if not all, of the buckets.

The engine history obtained from the field consisted of the number of starts, fast starts, trips, and fired hours. The number of hours of operation between each start was not available, nor was the load level at which a trip occurred. (Trips from full load are more damaging than trips from part load.) Furthermore, the number of hours at any one load level was not available. The bucket history consisted of when the part was installed in the engine, when it was removed, the type of coating, and whether the part was new or repaired.

All of the buckets considered in this study were original equipment and had not been repaired. They were all coated with a platinum-aluminide coating. The platinum-aluminide coating is primarily nickel-aluminide (NiAl) with platinum present as a solid solution element. When manufactured, platinum-aluminide particles ($PtAl_2$) are present in the outer layer of the coating, but during service, the $PtAl_2$ particles dissolve into the NiAl matrix. Platinum increases the oxidation resistance of the NiAl by 2 to 4 times. The reason for this is not known [*14*].

Prediction of Field Data

The TMF model presented previously was used to predict the fatigue life of the buckets in service. The leading edge of the bucket is where fatigue cracks first initiate and is the area that the predictions are for. The strains and temperatures at this location for each cycle type were obtained from the manufacturer, who used a process similar to that described in the section on Structural Analysis.

The data used to develop the TMF model were for specimens coated with GT-29, since similar data were not available on specimens coated with platinum-aluminide. Another series of TMF tests (in a bending configuration) were conducted on specimens removed from leading edges of new buckets that had either GT-29 or platinum-aluminide coatings. These test lives were similar for the two coatings: 388 cycles for the GT-29 at $\Delta\varepsilon = 0.7\%$, and 470 cycles for the platinum-aluminide at $\Delta\varepsilon = 0.6\%$.

The dwell time used was an average value determined by dividing the total number of hours by the total number of starts. Precise values of operating times between starts were not available. The sensitivity of the predictions to different methods of determining the dwell time was investigated and was found to be of second order.

It was necessary to assume the sequence of starting and shutdown events because records of the exact sequence were not available. The assumptions were deliberately conservative. A normal start was followed by either a normal shutdown or a trip from full load. A fast start was followed by a normal shutdown. Since the number of fast starts was small, this assumption did not introduce significant error. Only for a single site was it possible to determine the number of trips from full speed-no load. (Full speed-no load is the idle condition.)

The engines were operated at full power, except for two engines that ran at part load. Since the fatigue data were independent of temperature in the range of metal temperatures expected at full power, temperature was not considered in the predictions. Part load operation, in which the metal temperatures are lower, was not considered.

A linear damage rule was used to account for the different cycle types

$$D = \Sigma \frac{N_i}{N_{fi}}$$

where the damage ratio D was assumed to be one at 100% life consumed. Here N_i is the number of cycles of type i, and N_{fi} is the number of cycles to failure of cycle type i.

The predictions of the field data are shown in Fig. 13 as the percentage of life consumed versus the number of starts. The predictions account for all cycle types and dwell times. Below 100% life, cracks are not expected to be present.[10] Above 100% life, cracks are expected in the leading edge. As the percentage of life consumed becomes greater than 100%, the extent of cracking is expected to be greater, but the model does not make a quantitative prediction of the crack length. Around 100% life, cracks are expected to be present. Shown in Fig. 13 are the field data and the nature of the cracking observed. The next section on metallurgical analysis shows examples of each of these types of cracking. It should be noted that a metallurgical analysis is required to determine if cracks are just beginning or not.

[10] Cracks are defined for the purposes of this work as engineering size cracks of the order of 0.2 to 0.4 mm (5 to 10 mils) in depth. Certainly, smaller cracks exist in both the test specimens and the buckets before this point, but it is not the intent of this work to describe the formation and growth of these smaller cracks.

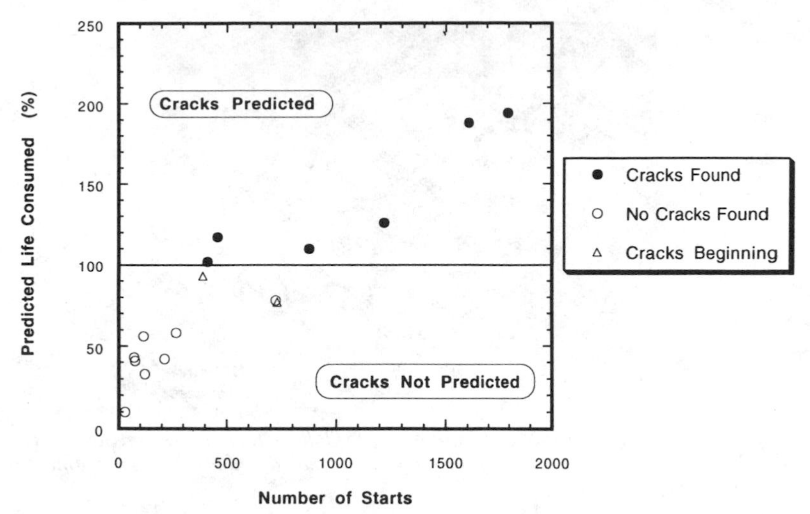

FIG. 13—*Predicted life consumed for field data and presence of cracks. At 100% life consumed, cracks are predicted to initiate.*

As can be seen in Fig. 13, the agreement of the field data with the predictions was remarkably good. All points below the 100% line were not predicted to have any cracking, and no cracking was found. All points above the 100% line were predicted to have cracking, and cracks were found. The points in which cracks were just beginning to form were predicted at 77 and 93% life consumed.

Metallurgical Analysis

A number of the buckets for which predictions are shown have been metallurgically analyzed to verify the predictions and to better understand the cracking process. Figure 14 shows

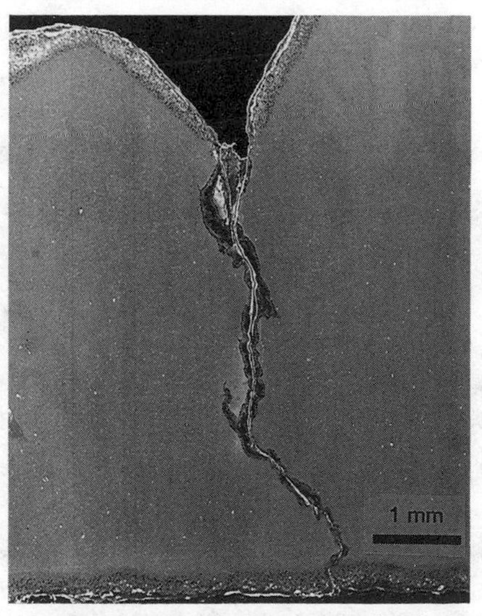

FIG. 14—*Crack in leading edge that has grown to the first cooling hole, a distance of 5 mm (0.2 in.). It was predicted that 194% of the life was consumed. (unetched).*

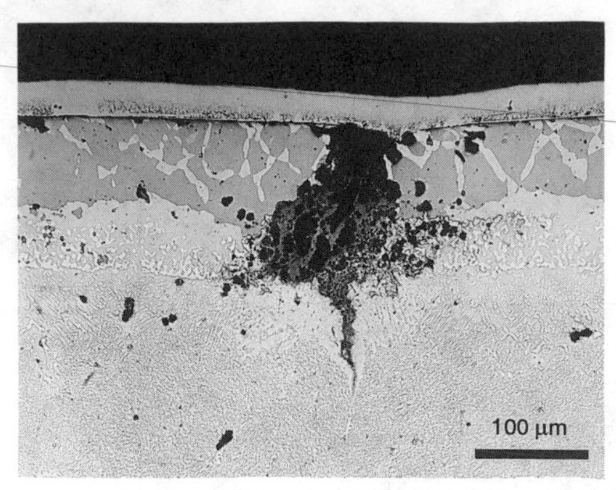

FIG. 15—*Crack in leading edge that is 0.13 mm (5 mils) long. Other cracks were found as deep as 0.5 mm (20 mils). It was predicted that 110% of the life was consumed. Outer layer is a platinum aluminide coating. (lactic acid etch).*

a crack in the leading edge of a bucket that has grown to the first cooling hole, a distance of 5 mm (0.2 in.), which is very severe. It was predicted that 194% of the life of this bucket was consumed, which is in agreement with the condition of the bucket. Figure 15 shows a crack beginning in the leading edge of another bucket from a sister engine. Other cracks on the leading edge were found as deep as 0.5 mm (20 mils), which is in reasonable agreement with the predicted life consumed of 110%. Additional details about the metallurgical analysis of these two buckets have been presented by Bernstein and Allen [*28*].

Micrographs of buckets without any cracks do not show any crack-like features. They have a continuous, uncracked coating, as shown in Fig. 16 for a bucket that has only 41% of its life consumed.

Cracks, which are just beginning in the leading edge, are shown in Fig. 17 for a bucket from a northern location that was predicted to have 93% of its life consumed. The coating is cracked, environmental attack is occurring at the interdiffusion zone between the coating and the base metal, and environmental attack is taking place in the grain boundary of the base metal under the crack. This oxide spike in the grain boundary is the precursor to the formation of a crack.

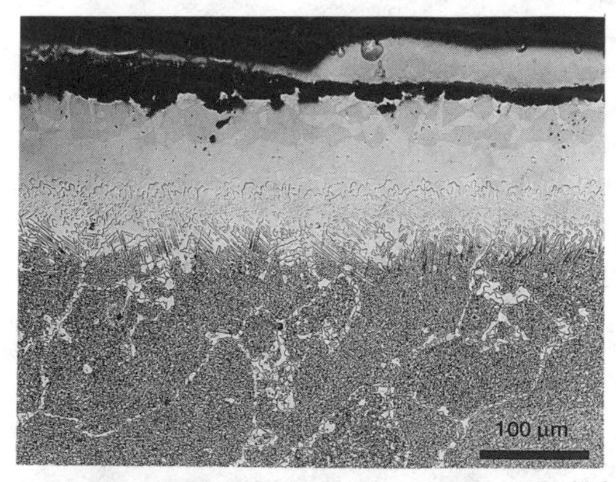

FIG. 16—*Micrograph of the leading edge of a bucket that had a predicted life consumption of 41%. No cracking was found. Outer layer is a platinum aluminide coating. (lactic acid etch).*

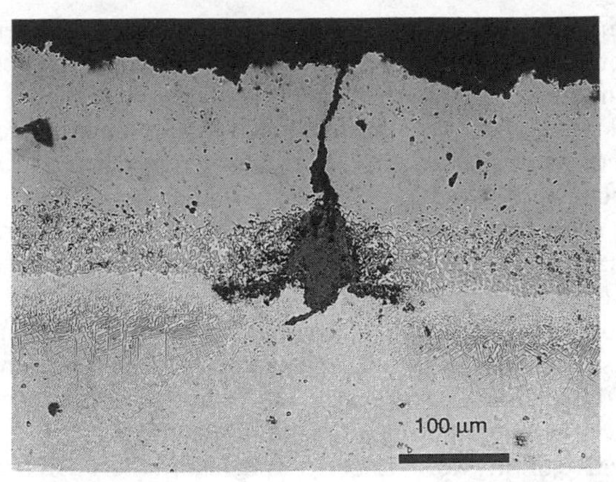

FIG. 17—*Cracks beginning by grain boundary oxidation in the leading edge of a bucket which had a predicted life consumption of 93%. Outer layer is a platinum aluminide coating. (lactic acid etch).*

As can be seen in the micrographs, environmental attack is present in all cases of fatigue cracking. This phenomenon is shown vividly in Fig. 18, where three cracks in various stages of formation are shown. On the right is the start of the crack formation. The coating has cracked and environmental attack is proceeding along the interdiffusion zone between the coating and base metal, as well as into the base metal. In the middle, the environmental attack has deepened into the base metal. On the left, a crack about 0.5 mm (20 mils) deep has formed within this environmental attack. Thus, the formation of fatigue cracks is a strongly environmentally assisted process. First, the coating cracks. Second, the base metal is attacked by the environment. Third, a crack forms in the attacked region of the base metal. And fourth, the crack grows preceded by environmental attack of the crack tip.

Discussion

The analysis of thermal-mechanical fatigue in gas turbine blades is a complex problem involving aerothermal, fluid mechanics, heat transfer, and stress analyses, in addition to materials analysis. Each of these analyses contributes its own unique set of difficulties and uncer-

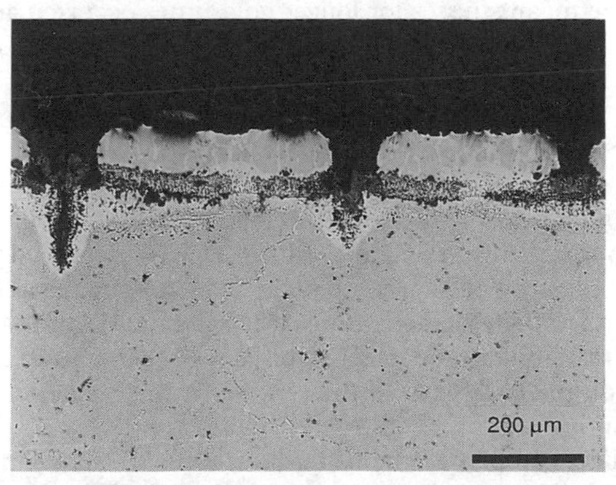

FIG. 18—*Stages in the formation of a crack showing how initial environmental attack leads to cracking. Micrograph is from the convex side (suction side) of a field run bucket. (lactic acid etch).*

tainties. From the standpoint of material analysis, TMF of coated superalloys is one of the more complex forms of material behavior involving the interaction of stress, strain, time, temperature, and environment.

Despite these complexities, very good predictions have been obtained for the fatigue lives of blades in service. No cracks were found in blades that were predicted to have less than 100% life consumed, and cracks were found in all blades that were predicted to have consumed more than 100% life. In two cases in which 77 to 93% life consumption was predicted, cracks were beginning by the formation of grain boundary oxide spikes. This degree of accuracy is fairly remarkable in light of the usual material scatter in fatigue of 2 to 3 times.

Thus, the principal objective of this work, to predict the TMF life of blades in service in GE MS7001E gas turbines, has been achieved. The next objective, to extend this model to the MS7001B and MS7001EA engines, is in progress. Further work is expected to address other materials, and possibly other engines, that are used by electric utilities.

The secondary objective of this work, to better understand the TMF behavior of coated IN-738, has only been partially achieved. The effects of time, temperature, coatings, *A*-ratio, and environment upon the TMF life were sought. Part of the difficulty of understanding these effects, apart from their complexity, was the practical limitations of performing all of the necessary tests to elucidate these effects. Brief discussions of each of these effects follow below.

Effect of Hold Time

The effect of compressive hold time upon the fatigue life of IN-738 was to decrease the life when considered on a total strain range basis. The life became shorter as the hold time was increased from 2 to 15 min, which was the longest time studied. The test matrix was not sufficient to determine if the hold time effect would saturate at longer times. TMF data for uncoated Udimet 710 show that the life continues to decrease for hold times of 14 to 162 to 300 min [*29*].

Compressive hold time can affect the TMF life for at least three reasons. First, the inelastic strain increases because of stress relaxation, which reduces the life as given by the Coffin-Manson relation. Second, there is a shift to tensile mean stresses for compressive hold times on nickel-base superalloys, which reduces the fatigue life [*30*]. And third, the environment has more time to interact with the fatigue process during longer hold periods. Which of these effects is most important for IN-738 cannot be easily identified. The inelastic strain increased with longer hold times, but this change occurred concurrently with the environmental attack. A shift to higher tensile mean stresses for longer hold times occurred at 1800°F (982°C) but not at 1600°F (871°C). The most that can be said at this point is that the independent variable of compressive hold time is important to predicting the TMF life of IN-738.

Effect of Temperature

The general effect of temperature observed from the data is that the fatigue life decreased as the temperature was increased from 760 to 871 to 982°C (1400 to 1600 to 1800°F), when the data were analyzed on the basis of total strain range. This behavior is not surprising, since the inelastic strains increase at higher temperatures when the total strain range is kept constant. What is surprising is the similar fatigue behavior at 916, 960, and 982°C (1680, 1760, and 1800°F). Although additional testing is necessary to verify this behavior, the data do not show a temperature effect in this range.

The temperature effect can perhaps be better understood from a simple Universal Slopes calculation of the fatigue life based upon the measured tensile behavior of the material. For a strain range of 0.5%, the difference in fatigue life between 871 and 927°C (1600 and 1700°F), as well as between 927 and 982°C (1700 and 1800°F), is a factor of 1.6.[11] However, the difference in life between 871 and 982°C (1600 and 1800°F) is a factor of 2.5. Scatter in fatigue life may be masking the effect of temperature upon life for a 56°C (100°F) change in temperature, but not for a 111°C (200°F) change. Environmental attack must also be considered because it is greater at higher temperatures.

Effect of Coatings

One of the objectives of this study was to gain a better understanding of the effect of coatings upon the TMF life of IN-738, because this alloy is coated in service. Unfortunately, the specimen design did not allow an unequivocal answer to the effect of coatings, because only the outer surface of the specimen was coated, and the inner surface was uncoated. Furthermore, the coated specimens underwent a secondary heat treatment after coating, in order to refine the coating microstructure and to improve the adherence of the coating to the base metal. This processing may have been responsible for the slightly reduced strength and increased ductility of the coated specimens compared to the uncoated specimens.

The TMF data clearly showed that the coatings reduced the life of the LOP tests without any hold time and reduced the life of the bucket tests with a 2-min hold time. However, the LOP tests with 2- and 15-min hold times showed similar lives for both coated and uncoated specimens. It is hard to reconcile these apparently contradictory results. One could argue that the uncoated inner surface was attacked by the environment in the LOP test with a hold time, but this same attack would have occurred in the bucket tests. The only difference is that in the bucket tests, the maximum temperature occurred at somewhat less than the maximum strain.

When the model is extrapolated to longer hold times, beyond the range of available data, it predicts that the coated specimens have longer TMF lives than the uncoated specimens. This is not surprising since the LOP tests on coated specimens gave shorter lives for no hold time and similar lives for 2- and 15-min hold times. The only way of determining if these predications are realistic is additional testing at significantly longer hold times.

The effect of coatings reported in the literature is to reduce the TMF life, as discussed in the Introduction. This is because the coatings are much less ductile than the base metal and begin to crack before the base metal. Diffusion coatings, which are nickel or cobalt aluminides, NiAl or CoAl, go through a ductile-to-brittle transition (DBTT) at about 593°C (1100°F), in which the strain to crack in the brittle region is around 0.35%, as shown in Fig. 19. Overlay coatings, such as the type considered in this paper, can be tailored to be relatively ductile or brittle.

Because of the DBTT, the effect of coatings on fatigue can be radically changed by altering the strain-temperature cycle. For isothermal cycles at high temperatures, the coating is very ductile and may have a beneficial effect upon the life by reducing environmental attack. For TMF cycles that impose a high strain below the DBTT, coatings can have a very detrimental effect, but may not be that harmful if the strain or temperature is kept below the strain to crack or the DBTT. The design of blades in engines must consider the DBTT behavior of the coatings and limit the tensile strains for temperatures below the DBTT.

In order to obtain a better understanding of the role of coatings on the TMF of IN-738, additional testing is needed on specimens that are coated on all exposed surfaces. In addition, tests are needed on coatings having different ductilities.

[11] These somewhat small effects of temperature are due to the compensating effects of increased ductility and decreased strength as the temperature increases.

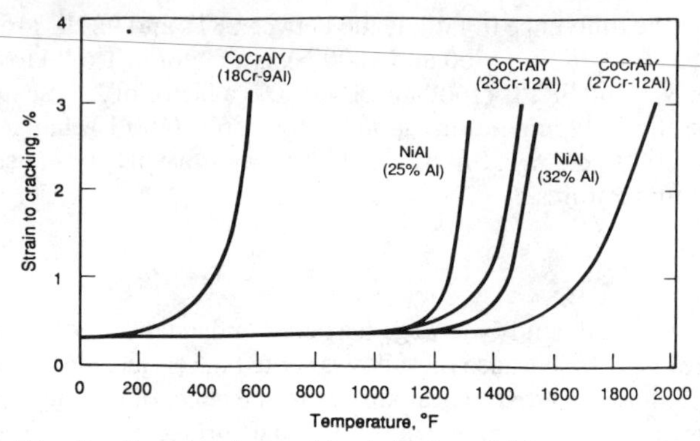

FIG. 19—*Ductile-to-brittle transition behavior of high-temperature coatings.*

Effect of Strain A*-Ratio*

The effect of the strain *A*-ratio on the LOP tests with a 2-min hold time was to decrease the life slightly as *A* changed from infinity to -1. A broader range of *A* is needed to properly evaluate the effect of the *A*-ratio upon the life. Such a range of *A* was run on René 95 in the forged condition at 650°C (1200°F) at a $\Delta\varepsilon = 1.2\%$ with no hold time, as shown in Fig. 20 [*31*]. For this case, the life was increased by a negative *A* and decreased by a positive *A*, which is opposite to the effect observed in the current data. For the René 95 data, the change in life can be explained by the change in mean stress, which was negative for negative *A* and positive for positive *A*. Thus, the effect of the *A*-ratio upon the life should be understood in terms of the changes in mean stress and inelastic strain caused by *A*-ratio, time, and temperature.

Effect of Cycle Type

The effect of the different types of cycles was difficult to judge because each cycle was run with a different combination of temperature and *A*-ratio. The overall trend of the data was

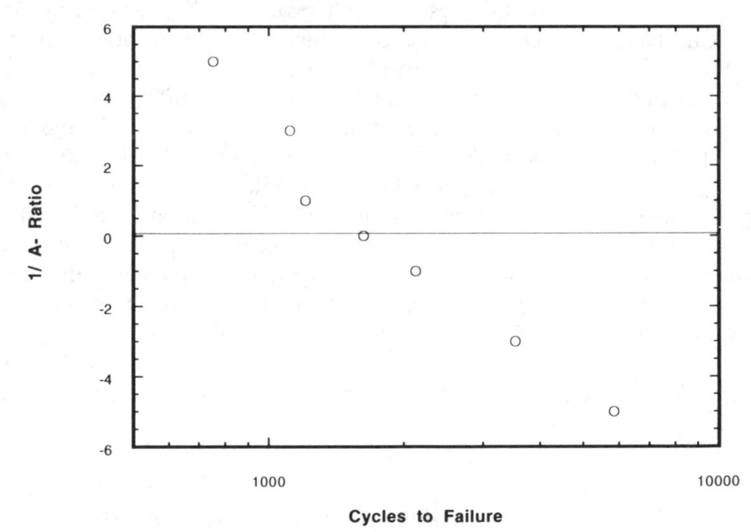

FIG. 20—*Isothermal fatigue tests on René 95 at 650°C (1200°F) at a strain range of 1.2% showing the effect of strain* A*-ratio.*

that the isothermal tests gave the longest lives, and the LOP tests gave the shortest lives. This difference is clearest for the $A = -1$ tests, which varied only in cycle type. The bucket cycles gave lives intermediate between the LOP and isothermal cycles. Additional testing with the same temperatures and A-ratios is required to better define the differences in cycle type. Understanding the effect of cycle types requires looking into the inelastic strain and mean stress developed during the cycle.

Kuwabara et al. [*10*] conducted TMF tests on uncoated IN-738 for LOP, LIP (linear in-phase), and isothermal cycles at a cyclic frequency of 0.0056 Hz (3 min/cycle). They found that in terms of total strain range, the LIP tests had the longest lives, and the isothermal tests had the shortest lives. The LOP tests had lives somewhat longer than the isothermal tests. These results are opposite to the current findings of the LOP tests having the shortest lives. The difference in test conditions is that the current tests employed a 2-min hold at the minimum strain and maximum temperature, whereas the data of Kuwabara were for continuous cycling conditions.

Effect of Environment

Environmental effects on the high-temperature fatigue of nickel-base superalloys, including IN-738, are well documented for the increase in life for vacuum testing, as discussed in the Introduction. One of the reasons given for an effect of hold time upon fatigue life is the increased environmental attack by high-temperature oxidation that occurs for longer hold times. For industrial gas turbine blades, hold times from a few hours to thousands of hours occur. This provides ample opportunity for environmental attack, and it is difficult, if not impossible, to simulate this attack in the laboratory.

TMF Model

The TMF model developed for this application was optimized to the conditions of interest for industrial gas turbine blades. It is not a general purpose model, and it is not claimed to be applicable for conditions of tensile hold or LIP cycles. The aim of the modeling effort was to predict the field data using the variables that can be determined from the engine operation and an elastic stress analysis. These variables are the total strain range, the strain A-ratio, and the hold time, which are also the independent variables in the fatigue tests. Using these variables, an excellent correlation of the 871°C (1600°F) LOP data was obtained, and the bucket tests were predicted to within a factor of 2.

The hold time term is thought to account for the time-dependent processes of environmental attack, microstructural aging and stress relaxation, which can affect the mean stresses and inelastic strains. No attempt was made to account for each of these time effects separately. Previous studies have shown that inelastic strain, time, and stress terms are needed to adequately correlate and predict isothermal data [*32*].

The strain A-ratio term was used to account for the different A-ratios in the data. This term, along with the hold time, would account for changes in mean stresses and inelastic strain. Although the effect of strain A-ratio was small in the model, it provided a meaningful improvement to the model.

The bucket tests conducted at the highest strain range of 0.7% were uniformly underpredicted by a factor of 2, although the lives of LOP tests at strain ranges of 0.6 to 0.8% were well correlated. This difference may be due to the use of a straight-line approximation to the strain-life curve. At sufficiently high strain ranges, the straight-line approximation would underpredict the data.

The one term missing from the TMF model is temperature. Although the TMF data for temperatures in the range of 871 to 916°C (1600 to 1680°F) are well correlated and predicted by the model, data at 760 and 982°C (1400 and 1800°F) show a significant difference in life compared to the 871°C (1600°F) data.

Finally, to gain a better understanding of the fatigue process, the constitutive behavior of the material must be considered. Effects of time, temperature, strain range, and strain *A*-ratio must be considered in relation to how they affect the inelastic strains and the stresses, since these "developed" variables have a direct effect upon the fatigue life. One may not want to predict the fatigue life of a blade from these "developed" variables because of the difficulties and uncertainties of calculating them, as well as the inherent scatter in the material's constitutive response. As shown in this paper, fatigue life can be computed accurately by considering only the independent variables. However, a better understanding of the TMF behavior of a material requires understanding its constitutive response.

Summary

Excellent predictions of the thermal-mechanical fatigue life of first stage turbine blades from the GE MS7001E industrial gas turbine were obtained using a TMF model consisting of a power-law equation with strain range, time, and strain *A*-ratio terms. The constants in the model were obtained from linear out-of-phase thermal-mechanical fatigue tests on IN-738LC specimens coated with GT-29. The model was able to correlate the test data to within a factor of 2 and was able to predict tests run under simulated service conditions to within a factor of 2.1.

The effect of various variables upon the fatigue life was studied on the basis of total strain range. Compressive hold times reduced the fatigue life, with the life becoming lower for longer holds. Higher temperatures reduced the life, although differences because of temperature changes of 56°C (100°F) were not always apparent. Coatings acted to reduce the fatigue life, but the data were confounded by the uncoated inner surface of the hollow specimen. Negative strain *A*-ratio acted to reduce the life, but the effect probably depends upon both temperature and compressive hold time. For 2-min compressive hold times, the isothermal tests gave the longest lives and the out-of-phase tests the shortest lives. The bucket simulated service condition tests yielded lives intermediate between these two test types.

Acknowledgments

The authors thank the Electric Power Research Institute for supporting this work under contract RP3064-1 and the General Electric Company for supplying the test data and bucket cycle data under EPRI contract RP2421-1, 2. Thanks are also extended to Ms. Loretta Mesa for preparing the manuscript and Mr. Harold Saldana for the metallurgical sample preparation.

References

[*1*] Bernstein, H. L., "Life Management System for Frame 7E Gas Turbine," *Life Assessment and Repair Technology for Combustion Turbine Hot Section Components,* R. Viswanathan and J. M. Allen, Eds., ASM International, Materials Park, OH, 1990, pp. 111–118.

[*2*] Russell, E. S., "Practical Life Prediction Methods for Thermal-Mechanical Fatigue of Gas Turbine Buckets," *Proceedings: Conference on Life Prediction for High-Temperature Gas Turbine Materials,* V. Weiss and W. T. Bakker, Eds., EPRI AP-4477, Electric Power Research Institute, Palo, Alta, CA, 1986, p. 3-1–3-39.

[*3*] Miner, R. V., "Fatigue," *Superalloys II,* C. T. Sims, N. S. Stoloff, and W. C. Hagel, Eds., Wiley, New York, 1987, pp. 263–287.

[4] Skelton, R. P., *High Temperature Fatigue: Properties and Prediction,* Elsevier Applied Science, New York, 1987.
[5] Nazmy, M. Y., "The Effect of Environment on the High Temperature Low Cycle Fatigue Behaviour of Cast Nickel-base IN-738 Alloy," *Materials Science and Engineering,* Vol. 55, 1982, pp. 231–237.
[6] Gabrielli, F., Marchionni, M., and Onofrio, G., "Time Dependent Effects on High Temperature Low Cycle Fatigue and Fatigue Crack Propagation of Nickel Base Superalloys," *Advances in Fracture Research,* Proceedings of the 7th International Conference on Fracture (ICF7), Vol. 2, Houston, TX, March 1989, pp. 1149–1163.
[7] Allen, J. M. and Whitlow, G. A., "Observations on the Interaction of High Mean Stress and Type II Hot Corrosion on the Fatigue Behavior of a Nickel Base Superalloy," *ASME Journal of Engineering for Gas Turbines and Power,* Vol. 107, 1985, pp. 220–224.
[8] Whitlow, G. A., Johnson, R. L., Pridemore, W. H., and Allen, J. M., "Intermediate Temperature, Low-Cycle Fatigue Behavior of Coated and Uncoated Nickel Base Superalloys in Air and Corrosive Sulfate Environments," *ASME Journal of Engineering Materials and Technology,* Vol. 106, 1984, pp. 43–49.
[9] Leverant, G. R., Strangman, T. E., and Langer, B. S., "Parameters Controlling the Thermal Fatigue Properties of Conventionally-Cast and Directionally-Solidified Turbine Alloys," *Superalloys: Metallurgy and Manufacture, 3rd International Symposium,* Seven Springs, PA, 1976.
[10] Kuwabara, K., Nitta, A., and Kitamura, T., "Thermal-Mechanical Fatigue Life Prediction in High-Temperature Component Materials for Power Plant," *Proceedings of the ASME International Conference on Advances in Life Prediction Methods,* Albany, NY. April 1983, pp. 131–141.
[11] Viswanathan, R., *Damage Mechanisms and Life Assessment of High-Temperature Components,* ASM International, Metals Park, OH, 1989.
[12] Boismier, D. A. and Sehitoglu, H., "Thermo-Mechanical Fatigue of Mar-M247: Part 1—Experiments," *ASME Journal of Engineering Materials and Technology,* Vol. 112, 1990, pp. 68–79.
[13] Grünling, H. W., Schneider, K., and Singheiser, L., "Mechanical Properties of Coated Systems," *Materials Science and Engineering,* Vol. 88, 1987, pp. 177–189.
[14] Bernstein, H. L., "Gas Turbine Blade and Vane Coatings Technology," *Guidebook and Software for Specifying High-Temperature Coatings for Combustion Turbines,* EPRI GS-7334-L, Electric Power Research Institute, Palo Alto, CA, 1991.
[15] Fairbanks, J. W. and Hecht, R. J., "The Durability and Performance of Coatings in Gas Turbine and Diesel Engines," *Materials Science and Engineering,* Vol. 88, 1987, pp. 321–330.
[16] Heine, J. E., Warren, J. R., and Cowles, B. A., *Thermal Mechanical Fatigue of Coated Blade Materials,* WRDC-TR-89-4027, Wright Research and Development Center, Wright-Patterson AFB, OH, 1989.
[17] Halford, G. R. and Manson, S. S., "Life Prediction of Thermal-Mechanical Fatigue Using Strain-range Partitioning," *Thermal Fatigue of Materials and Components, STP 612,* D. A. Spera and D. F. Mowbray, Eds., American Society for Testing and Materials, Philadelphia, 1976, pp. 239–254.
[18] Neu, R. W. and Sehitoglu, H., "Thermomechanical Fatigue, Oxidation, and Creep," *Metallurgical Transactions A,* Vol. 20A, Sept. 1989, pp. 1755–1783.
[19] Sehitoglu, H. and Boismier, D. A., "Thermo-Mechanical Fatigue of Mar-M247: Part 2—Life Prediction," *ASME Journal of Engineering Materials and Technology,* Vol. 112, 1990, pp. 80–89.
[20] Nazmy, M. Y., "High Temperature Low Cycle Fatigue of IN 738 and Application of Strain Range Partitioning," *Metallurgical Transactions A,* Vol. 14A, 1983, pp. 449–461.
[21] Ostergren, W. J., "A Damage Function and Associated Failure Equations for Predicting Hold Time and Frequency Effects in Elevated Temperature, Low Cycle Fatigue," *Journal of Testing and Evaluation,* Vol. 4, 1976, pp. 327–339.
[22] Allen, J. M., "Effect of Temperature Dependent Mechanical Properties on Thermal Stress in a Cooled Gas Turbine Blade," *ASME Journal of Engineering for Power,* Vol. 104, 1982, pp. 349–353.
[23] Timoshenko, S. and Goodier, J. N., *Theory of Elasticity,* McGraw-Hill, New York, 1951, pp. 11–27, 33, 52, 169–170, 399–403.
[24] Allen, J. M. and Munson, K. E., "Advanced Ceramic Coating Development for Industrial/Utility Gas Turbine Applications, Section 5.0, Design Analysis," NASA CR-165619, 1982, pp. 85–118.
[25] Sharma, O. P., Pickett, G. F., and Ni, R. H., "Assessment of Unsteady Flows in Turbines," *Journal of Turbomachinery,* Vol. 114, Jan. 1992, p. 79.
[26] Manson, S. S., *Thermal Stress and Low Cycle Fatigue,* McGraw-Hill, New York, 1966.
[27] Foster, A. D., "First-Stage Bucket Coatings-Description, Life and Refurbishment," General Electric Company, GE Turbine Reference Library, GER-3632, 1990.
[28] Bernstein, H. L. and Allen, J. M., "Analysis of Cracked Gas Turbine Blades," ASME Paper 91-GT-16, IGTI Conference, Orlando, FL, June 1991; also Bernstein, H. L. and Allen, J. M., "Analysis of

Cracked Gas Turbine Blades," *Journal of Engineering for Gas Turbines and Power,* Vol. 114, April 1992, pp. 293–301.

[*29*] Viswanathan, op. cit., p. 128.

[*30*] Bernstein, H. L., "An Evaluation of Four Creep-Fatigue Models for a Nickel-Base Superalloy," *Low-Cycle Fatigue and Life Prediction, STP 770,* C. Amzallag, B. N. Leis, and P. Rabbe, Eds., American Society for Testing and Materials, Philadelphia, 1982, pp. 105–134.

[*31*] Bernstein, H. L., unpublished data, Southwest Research Institute, San Antonio, TX.

[*32*] Bernstein, H. L., "A Stress-Strain-Time Model (SST) for High-Temperature, Low-Cycle Fatigue," *Methods for Predicting Material Life in Fatigue,* W. J. Ostergren and J. R. Whitehead, Eds., ASME, New York, pp. 89–100.

Toshio Sakon,[1] Masaharu Fujihara,[1] and Tetsuo Sada[2]

Residual Life Assessment of Pump Casing Considering Thermal Fatigue Crack Propagation

REFERENCE: Sakon, T., Masaharu, F., and Sada, T., **"Residual Life Assessment of Pump Casing Considering Thermal Fatigue Crack Propagation,"** *Thermomechanical Fatigue Behavior of Materials, STP 1186,* H. Sehitoglu, Ed., American Society for Testing and Materials, Philadelphia, 1993, pp. 239–252.

ABSTRACT: Thermal fatigue crack propagation is the main concern on the life assessment of the boiler feed-water circulation pump casings under cyclic operation. This paper describes the experimental and analytical studies on the thermal fatigue crack growth behavior, and the remaining life prediction of the casings.

First, it was confirmed by low-cycle fatigue test that the propagation behavior of surface cracks accompanied with the coalescence could be estimated by the da/dN-ΔJ relationship for long cracks. Secondly, the simplified J-integral estimation method for the thermal stress problem was discussed based on some bench mark calculations. Finally, the crack growth behavior in the casings was estimated and the results were compared with the field inspection data. Good agreement was found between them and, hence, these type of pump casings had adequate remaining lives to the final failure.

KEYWORDS: thermal fatigue, crack growth, stress intensity factor, cyclic J-integral, remaining life, boiler water pump casings

Introduction

In recent years, the small or medium size fossile power plants have been operated with frequent start/stops. Such cyclic operation gives rise to a thermal fatigue damage due to the high transient thermal stress. Consequently, the evaluation of thermal fatigue damage has become one of the great concerns in the remaining life assessment for some components.

One such component is the casing of boiler feed-water circulation pumps whose inner surface is subjected to high thermal shock at start-up. Since the casings are carbon steel castings, many thermal fatigue cracks are initiated from casting porosities at the inner surface of casings. Although grinding of the cracks might be an idea for extending the life of the casings, it is not an easy task; besides, it may not be effective since new cracks will be initiated in early cycles from the porosities, which were inside but appeared on the new surface. Hence, it will be a better choice to leave the cracks alone, provided that they will not propagate to the critical size leading to the final catastrophe of the casing.

The purpose of this study is the evaluation of crack growth behavior in the casing. Since thermal fatigue in general is the inelastic problem, nonlinear fracture mechanics is required

[1] Assistant chief research engineer, Materials and Strength Laboratory, Takasago R&D center, Mitsubishi Heavy Industries, Ltd., Takasago, Hyogo, Japan.

[2] Manager, Boiler design section, Kobe Shipyard & Machinery Works, Mitsubishi Heavy Industries, Ltd., Kobe, Hyogo, Japan.

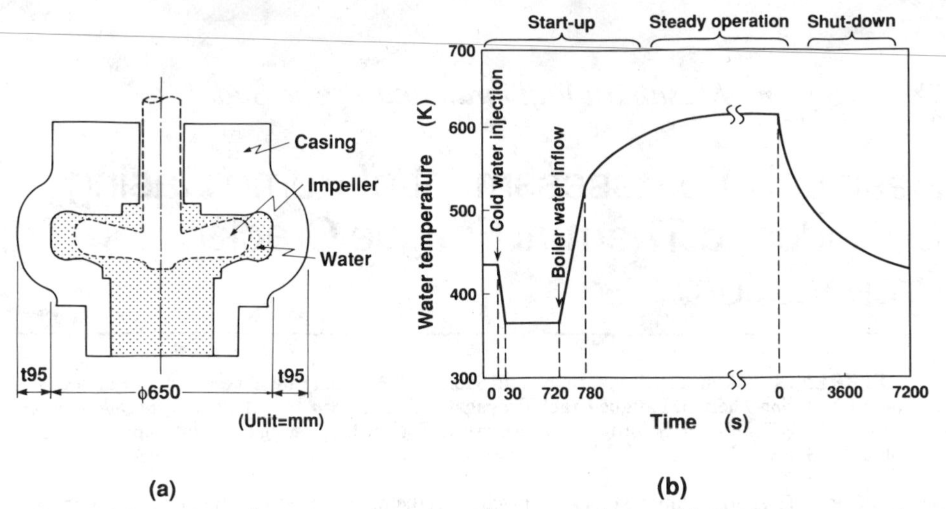

FIG. 1—*Schematic configuration and water temperature variation of boiler feed-water circulation pump.*

for crack growth evaluation. At present the *J*-integral proposed by Rice [*1*] is widely used as the nonlinear fracture mechanics parameter. For the application to fatigue crack growth, Dowling [*2*] proposed cyclic *J*-integral ΔJ which is also used frequently. For the application of *J*-integral, Kumar et al. [*3,4*] presented solutions of the *J*-integral for various kind of cracked bodies subject to mechanical loads. But it is difficult to apply these solutions to cracked components subject to the deformation-controlled stress such as thermal stress. Therefore, the *J*-integral is estimated from the strain-basis intensity factor in this study, and its effectiveness for the evaluation of short crack growth behavior is confirmed by low-cycle fatigue test of casing material.

Moreover, it is supposed that the strain-based intensity factor in the thermal stress field will decrease with crack growth due to the reduction of constraint. Hellen et al. [*5*] proposed the equation for stress intensity factor under linear temperature gradient. Since the temperature gradient through the casing wall is nonlinear, the authors use Hellen et al.'s equation with the linearized stress. The method of stress linearization is confirmed by a bench mark calculation.

A series of the procedures mentioned above is applied to the evaluation of crack growth in a casing to estimate its residual life.

Thermal Stress in Casings

As shown in Fig. 1, cold water is injected into the casing at the beginning of start-up. The hot boiler feed-water then comes in and the water temperature in the casing goes up rapidly at this time. Spontaneously, high compressive thermal stress is generated in the inner surface region. The steady operation temperature is not high (about 620 K). At shutdown, the water temperature drops slowly to about 420 K.

The metal temperature changes and the induced thermal stress in this operation cycle were calculated by using a finite element code, namely, MARC. The finite element model in the heat transfer analysis consisted of the casing body and insulation around the outer wall. The change of water temperature in the casing was specified according to Fig. 1*b*. The calculated metal temperature changes at representative locations on inner and outer surfaces are shown in Fig. 2*a* and Fig. 2*b*. Thermal stress analyses were conducted after 850 s from the beginning of start-up and after 1800 from the beginning of shutdown, when the temperature differences

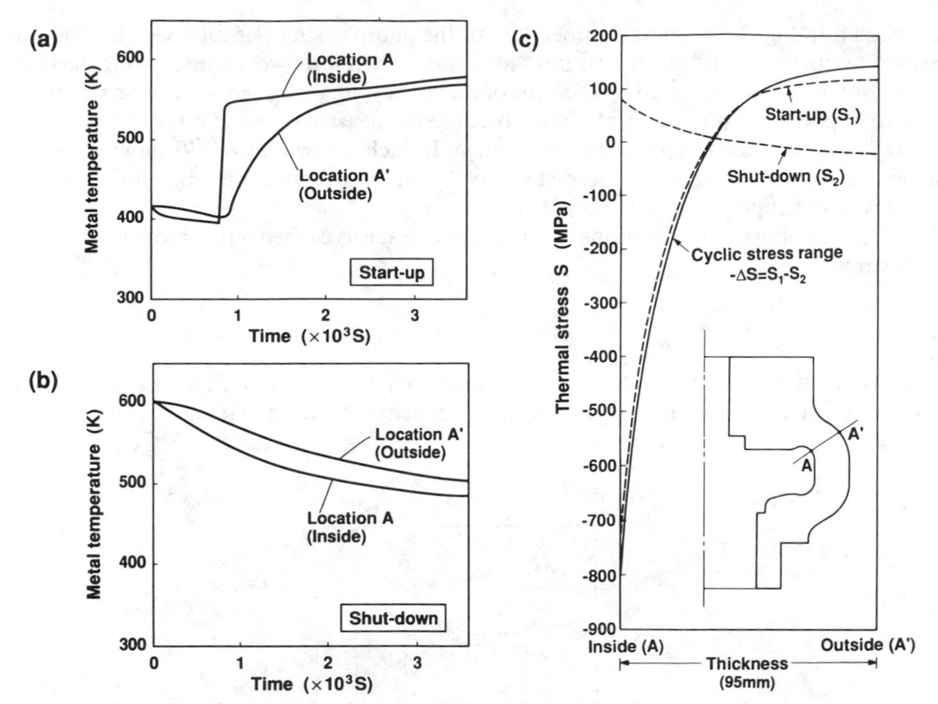

FIG. 2—*Metal temperature history and thermal stress distribution at critical location.*

between inner and outer surfaces become maximum in the respective stages. The broken lines in Fig. 2*c* show the results. The stress range determined from these extremes is also shown by solid line in the figure. The inner surface, thus, is subjected to the highest thermal stress range of 800 MPa which is equivalent to the strain range of about 0.4%. Consequently, the thermal fatigue crack will be nucleated within a limited number of cycles.

Low-Cycle Fatigue Crack Growth Tests

Objectives

According to the field inspections, many short cracks were found on the inner surface of casings within early operation cycles and most of them were initiated from casting porosities. This suggests that the understanding of short crack growth behavior will be very important for the life assessment of casings rather than the evaluation of crack initiation life. Hence, low-cycle fatigue crack growth tests were conducted using the specimens sampled from a retired pump casing.

Test Temperature

It is known that the relationship between low-cycle fatigue crack growth rate da/dN and cyclic J-integral ΔJ does not strongly depend on temperature or material. For instance, Ohtani [*6*] showed da/dN-ΔJ relations for the various kind of steels and test temperatures fell into the scatter of factor about two. Ogata et al. [*7*] also showed the thermal fatigue crack growth rate plotted against ΔJ for Type 304 stainless steel coincided with isothermal fatigue data except the "in-phase" temperature-strain wave with very high maximum temperature (973 K). Additionally, Taira et al. [*8*] showed that the life of "out of phase" thermal fatigue was longer than that of isothermal fatigue at the temperature range of 373 ~ 673 K for a carbon steel.

As shown in Fig. 2, the metal temperature of the pump casing was relatively low and the phase of temperature and strain was not "in-phase." Therefore, we assumed no remarkable acceleration in da/dN occurred by this type of thermal cycling. The temperature-strain phase in the casing was not pure "out of phase" because the peak tensile and compressive strains occurred at the middle of the temperature range. In such a case, Russel [9] showed that the fatigue life was similar to the isothermal fatigue life at the minimum temperature in a cycle for a nickel base superalloy.

Based on the above considerations, a crack growth test was carried out at room temperature in this study.

Test Procedure

Two types of specimens, compact type and plate-type as shown in Fig. 3*a* and *b*, respectively, were employed in this study. Compact specimens were used to obtain the relationship

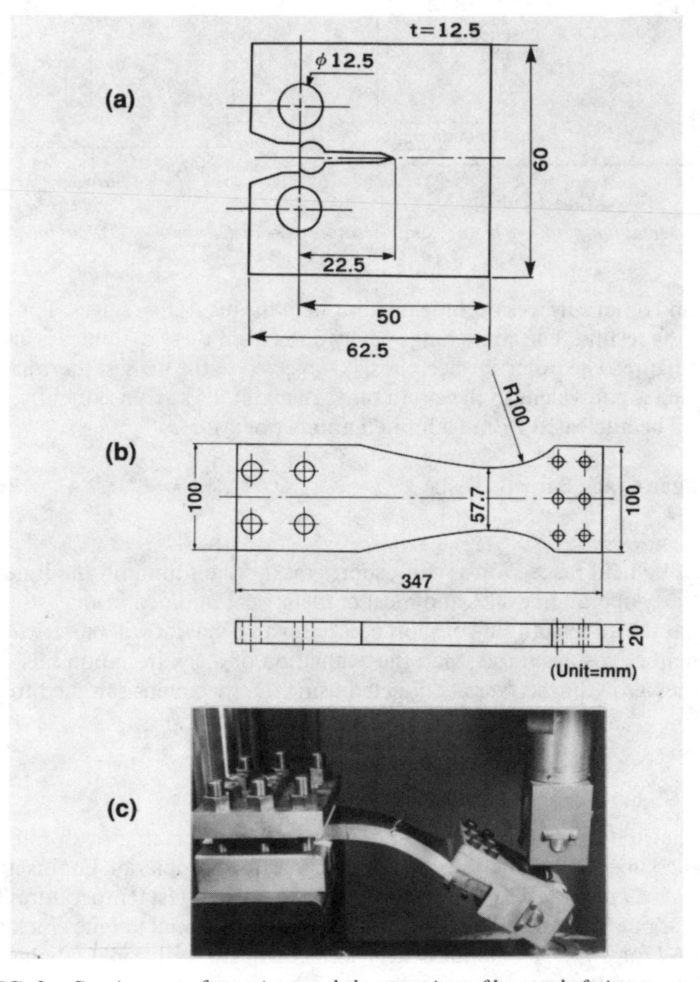

FIG. 3—*Specimen configurations and close-up view of low-cycle fatigue test section.*

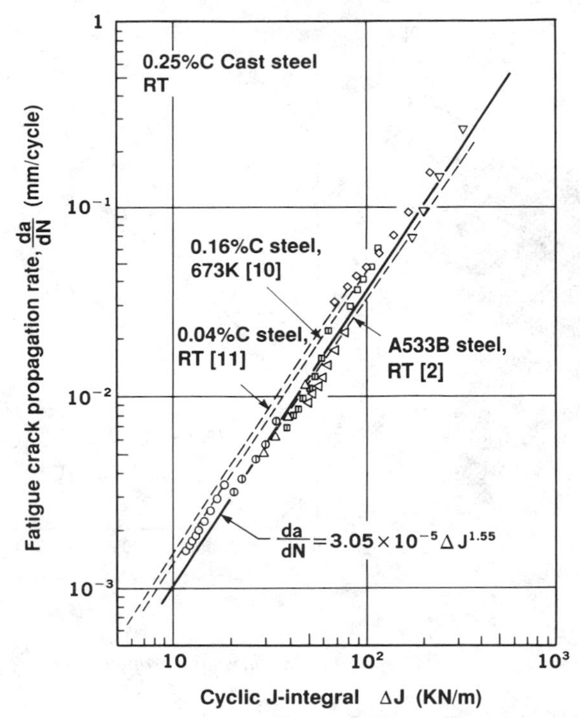

FIG. 4—*Low-cycle fatigue crack propagation rate of casing material at room temperature.*

between ΔJ and da/dN for a long crack. This test was carried out by controlling the load-line displacement which was measured by an extensometer clamped in the mouth of the slit. Constant displacement ranges from 0.24 to 3.1 mm were applied to eight specimens. Frequency was 0.1 Hz. ΔJ was calculated from the load-displacement hysteresis loop according to Dowling's method [*2*].

Plate specimens were used for observing the short crack growth behavior. Bending load was applied to the end of the specimen through a universal joint. The test section view is shown in Fig. 3*c*. The section profile of the specimen was adjusted to make the strain distribution uniform in the gage section. The test was conducted by deflection control and the strain-range was measured by the strain gage on the specimen surface. The strain ranges applied were 0.5, 1.0, 2.1, 4.2%, with a frequency of 0.1 Hz. The crack growth behavior was observed by taking the replica of the specimen surface intermittently during the tests.

Test Results

The relationship between da/dN and ΔJ obtained by compact specimens is shown in Fig. 4. The test results obtained by other investigators [*2,10,11*] on carbon steels and A533B were also shown in this figure. The data of casing material were fairly close to those reported data.

An example of the failed plate specimen is shown in Fig. 5*a*. As shown in the photo, a lot of surface cracks were nucleated on the specimen, and the growth and coalescence of them resulted in the final fatigue failure. Figure 5*b* shows the change of the surface crack lengths with the number of cycles. It was found that short cracks nucleated in the early stage of the fatigue life and that most of the life period corresponded to the crack growth process.

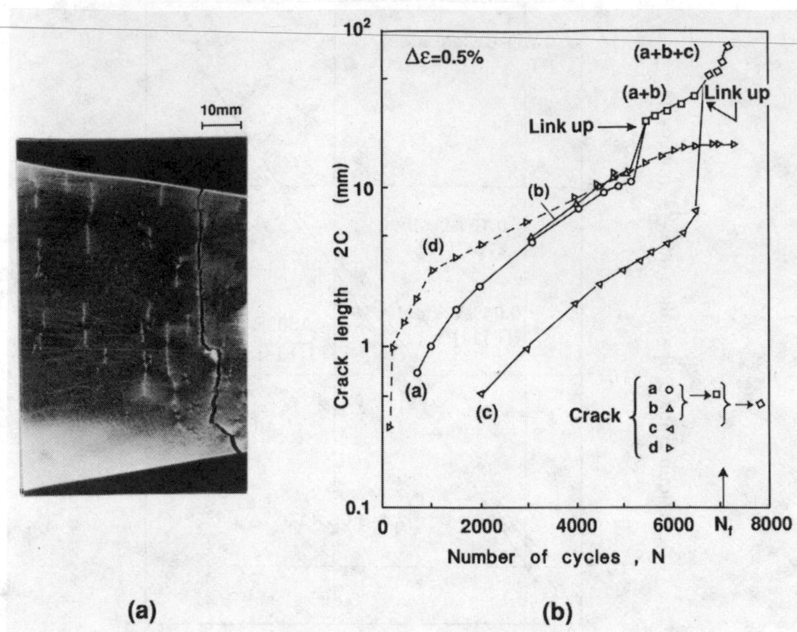

FIG. 5—*Photo of the surface of plate specimen fractured and examples of crack growth data.*

Estimation of Cyclic J-Integral for Surface Crack

Sakon and Kaneko [*12*] have shown analytically that the J-integral of a crack under deformation-controlled loading could be estimated approximately from the strain-based intensity factor using the formula

$$\left.\begin{aligned} J &= EK_e^2 \\ K_e &= \varepsilon_n\sqrt{\pi a f} \end{aligned}\right\} \quad (1)$$

where E and ε_n are the Young's modulus and the nominal strain, respectively, and f is the shape factor for the elastic stress intensity factor. As shown in Fig. 6 [*12*], the solid lines with the slopes of two which were obtained by elastic FEM analyses almost agreed with the open symbols which were calculated by the inelastic FEM analyses. As a result the *J*-integral under the deformation controlled loading can be estimated approximately by the elastic analysis. In other words, Eq 1 will be effective for such a case. Other investigators [*13,14*] showed that the strain-based intensity factor could be applied to the evaluation of low-cycle fatigue crack growth rate. An example of these results was shown in Fig. 7 [*13*]. Moreover, Rau et al. [*15*] applied it successfully to thermal fatigue crack growth.

Based on these investigations, the cyclic *J*-integral of the surface crack in a plate specimen was estimated by

$$\Delta J = E\Delta\varepsilon^2\pi a f^2 \quad (2)$$

where $\Delta\varepsilon$ is the strain range measured by strain gage, and f is the shape factor obtained from the Newman-Raju equation [*16*]. Values of aspect ratio used for the Newman-Raju equation were determined based on those measured by observing the fracture surfaces of specimens.

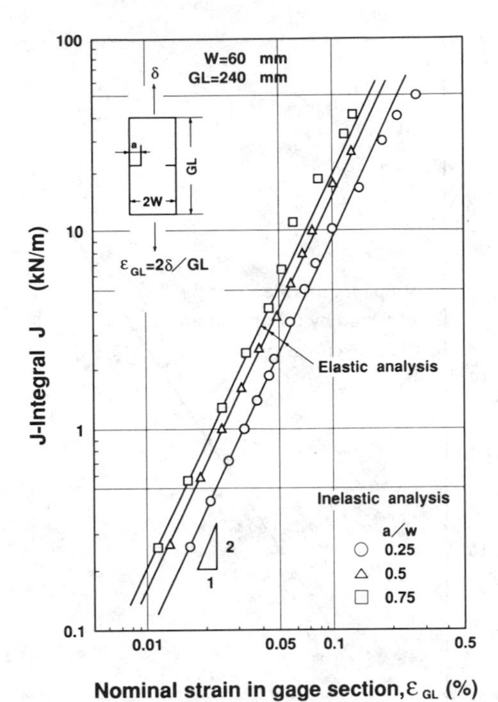

FIG. 6—*Examples of J-integral analyses under deformation-controlled loading.*

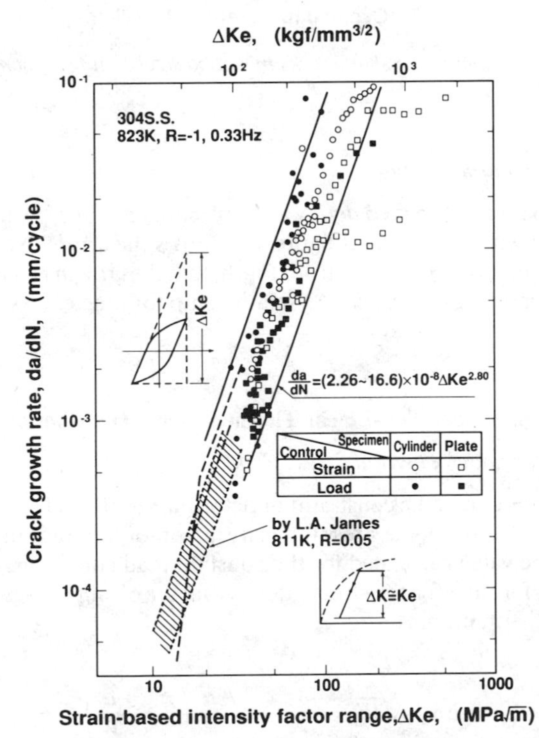

FIG. 7—*Relation between strain-based intensity factor and fatigue crack growth rate.*

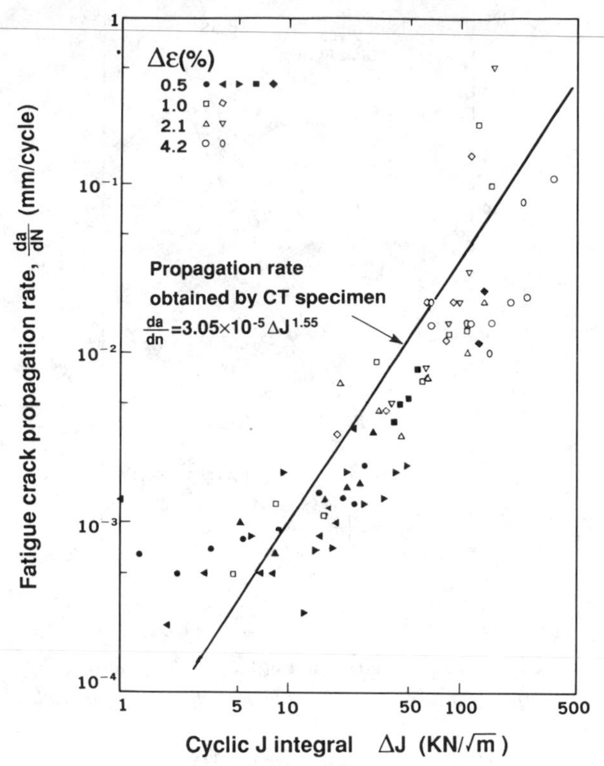

FIG. 8—$\frac{dc}{dn}$-ΔJ *relationship for short cracks in carbon steel casting at room temperature.*

Short Crack Growth Characteristics

The relationship between observed dc/dN and estimated ΔJ for typical surface cracks in the plate specimen is shown in Fig. 8. This figure also shows the da/dN-ΔJ relation obtained by the compact specimen. It was confirmed from Fig. 8, that the growth behavior of surface cracks could be estimated based on the da/dN-ΔJ relationship for long cracks, where the ΔJ is estimated from Eq 2.

Estimation of J-Integral under Non-Linear Thermal Stress Distribution

Stress Intensity Factor under Thermal Stress Field

Thermal stress arises due to the constraint of deformation. Therefore, due to the reduction of constraint with crack growth, the stress intensity factor for the crack in a thermal stress field will be lower than the value calculated for the constant load condition. Hellen et al. [5] presented the following formula for the stress intensity factor of single edge-cracked plate under linear temperature distribution.

$$K_I = E\alpha T_0\sqrt{W}\,\frac{4.48}{\sqrt{\pi}}\,\sqrt{\frac{a}{W}}\left(\frac{\pi}{4} - \frac{a}{W}\right)f\left(\frac{a}{W}\right) \tag{3}$$

Considering the thermal stress in a non-cracked plate as

$$\sigma = E\alpha T_0 \tag{4}$$

Equation 3 yields

$$K_I = 1.12\sigma\sqrt{\pi a}\left(1 - \frac{4}{\pi}\cdot\frac{a}{W}\right)f\left(\frac{a}{W}\right) \tag{5}$$

In this equation the term $(1 - 4/\pi \cdot a/W)$ characterizes the thermally stressed crack.

Linearization of Stress Distribution

Thermal stress in the pump casing is not distributed linearly as shown in Fig. 2. For calculating the stress intensity factor conveniently in such a non-linear stress distribution, ASME Sec. XI recommends linearizing the stress distribution as shown in Fig. 9*a*. Using the linearized stress components, K_I is obtained by

$$K_I = (M_m\sigma_m + M_b\sigma_b)\sqrt{\pi a} \tag{6}$$

where σ_m and σ_b are membrane and bending stress components, respectively, and M_m, M_b are the shape correction factors.

However, in the case of a thermal stress field, it is supposed that the high peak stress near the surface may be relieved by the crack growth. Consequently, the stress intensity factor may not be so high as estimated by the above method. Taking such a possibility into account, the authors propose the alternative method of stress linearization as shown in Fig. 9*b*, where the tangent line of the stress distribution at the point corresponding to the crack tip is used for the stress intensity calculation.

A Bench Mark Calculation

A bench mark calculation was conducted in order to clarify the applicability of the stress intensity estimation method using the linearized stress and the thermal stress correction term mentioned above, which is represented by

$$K_I = (M_m\sigma_m + M_b\sigma_b)\sqrt{\pi a}\left(1 - \frac{4}{\pi}\cdot\frac{a}{W}\right) \tag{7}$$

where, σ_m and σ_b are determined by linearizing the stress distribution according to Fig. 9*a* or Fig. 9*b*.

The bench mark model is an axially cracked cylinder under radial temperature distribution as shown in Fig. 10*a*, whose numerical solution of the stress intensity factors has been obtained by FEM [*17*]. The solution of FEM and the estimates from Eqs 6 and 7 are shown in Fig. 10*b*. Using the linearizing method of Fig. 9*a*, the stress intensity factor was overestimated in the region of $a/t > 0.2$, even though the thermal stress correction term was used. On the other hand, the combination of the linearization method shown in Fig. 9*b* and the thermal stress correction term yielded a fairly good agreement with the FEM result up to $a/t \simeq 0.4$. Based on these results, the latter method was adopted in this study.

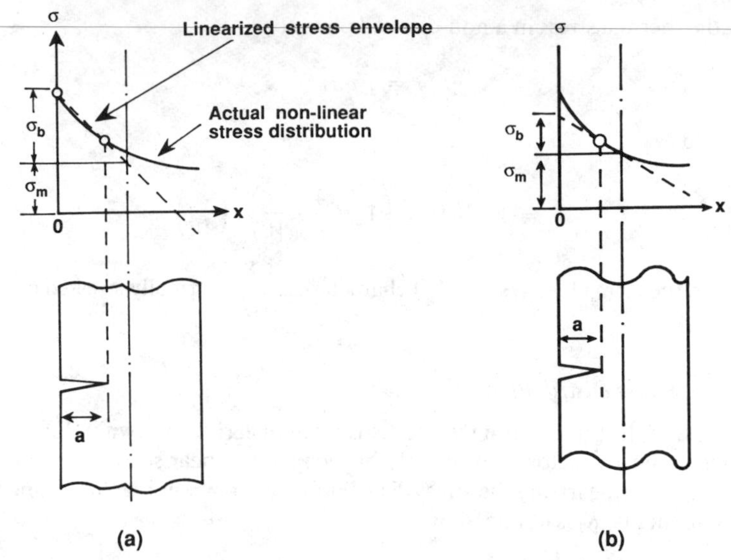

FIG. 9—*Stress linearization procedures:* (a) *recommended in ASME Sec. XI; and* (b) *proposed in the present investigation.*

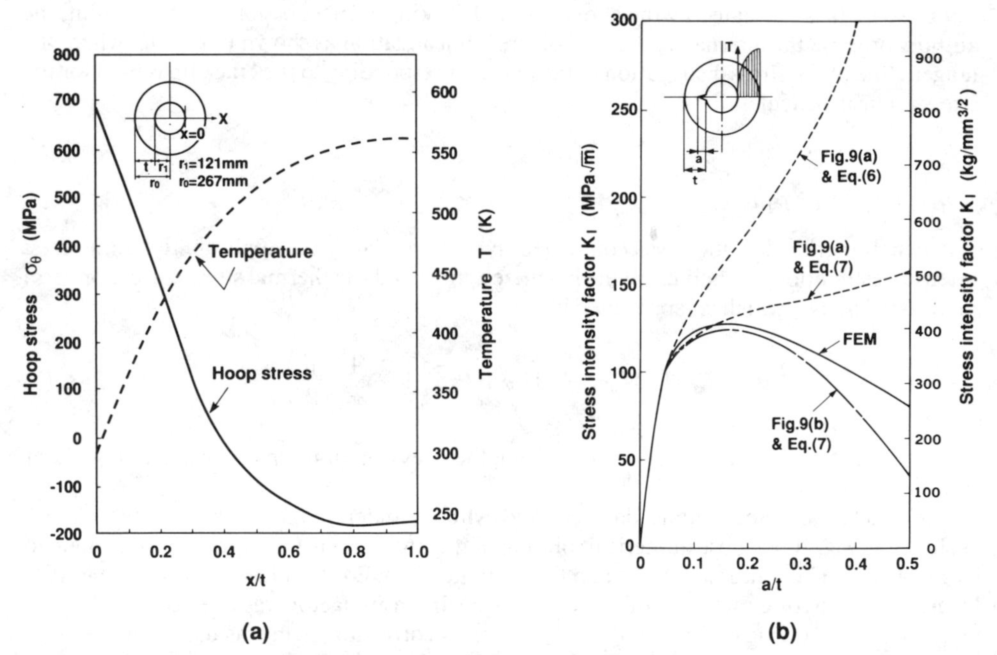

FIG. 10—*Boundary conditions* (a) *and estimated stress intensity factors* (b) *for axially cracked cylinder with non-linear temperature distribution.*

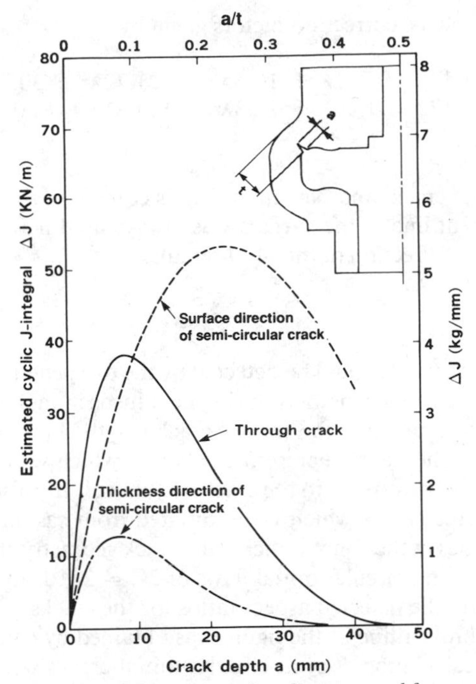

FIG. 11—*Examples of the estimated cyclic J-integral for pump casing.*

Conversion of K_I *to J-Integral*

As mentioned earlier, the *J*-integral under deformation-controlled loading could be simply estimated from the strain-based intensity factor by using Eq 1. In the case of thermal stress, the nominal strain in Eq 1 can be given by

$$\varepsilon_n = S/E \tag{8}$$

where S is the elastically calculated nominal stress. Using the cyclic stress range, the cyclic *J*-integral is given by

$$\Delta J = \frac{1}{E}(M_m \Delta S_m + M_b \Delta S_b)^2 \pi a \left(1 - \frac{4}{\pi} \cdot \frac{a}{W}\right)^2 \tag{9}$$

Remaining Life Estimation of Casings

Calculation of Cyclic J-Integral

The cyclic J-integral for a crack in the casing can be calculated according to the distribution of the thermal stress range in Fig. 2 and Eq 9 in the previous paragraph. Examples of the calculated results are shown in Fig. 11, where the solid line shows the value for a grooved crack through the inner surface, and the broken lines correspond to the values for the deepest point and surface side of a semi-circular crack.

In this calculation, the shape correction factors given by

$$\left.\begin{aligned} M_m &= 1.12 - 0.231\lambda + 10.55\lambda^2 - 21.72\lambda^3 + 30.39\lambda^4 \\ M_b &= 1.122 - 1.40\lambda + 7.33\lambda^2 - 13.08\lambda^3 + 14.0\lambda^4 \\ \lambda &= a/t \end{aligned}\right\} \tag{10}$$

were used for the through crack, and Newman-Raju's equation [*16*] was used for the semi-circular crack. The tangent line at the surface was always used here as the linearized stress distribution for the surface direction of the semi-circular crack.

Estimation of Crack Geometry

In the case of long cracks which could be detected by the dye penetrant test, their depths are measured by the electric resistance method during field inspection. Figure 12 shows the relationship between the aspect ratio and the surface crack length obtained through such inspections. The open circles in the figure represent the cracks which were initiated from casting porosites and the solid ones correspond to the cracks without clearly distinguishable porosities. According to the figure, the cracks which were initiated from casting porosities have larger aspect ratios. The solid lines in the figure indicate the crack geometry changes which were estimated by postulating the semi-circular initial flaws of 2C = 2, 10, and 40 mm. The trend of the estimation agreed with the detected aspect ratios for the cracks initiated from porosities. On the other hand, the broken line in the figure was obtained by considering the link-up of multiple cracks. In this calculation, 7 semi-circular initial cracks were postulated, and their

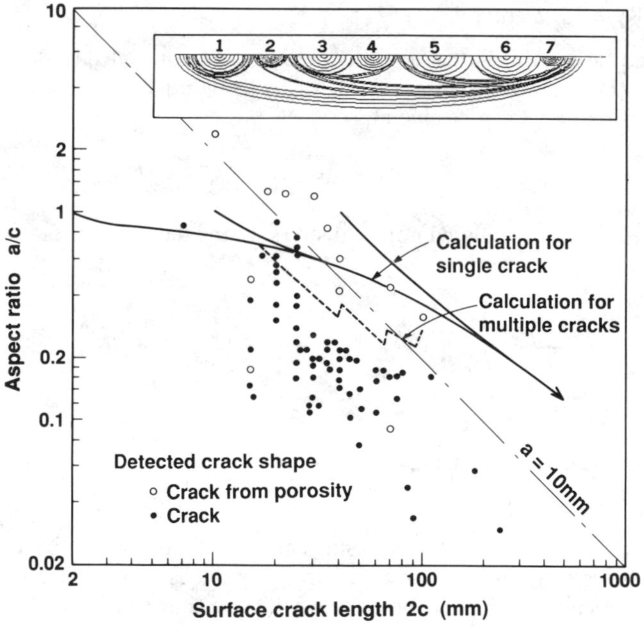

FIG. 12—*Comparison between the estimated crack geometry change and those detected by field inspection.*

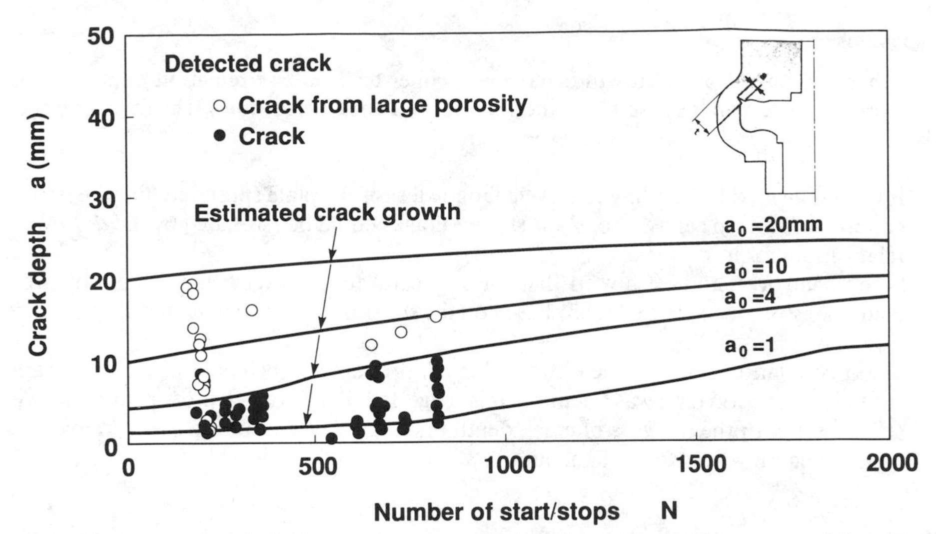

FIG. 13—*Detected crack depth in casing versus number of start/stops of the pumps and the estimated crack growth behavior.*

surface lengths and pitches were given by the random number. The link-up of them was assumed to occur when two cracks contact each other. As shown in the figure, the result of this calculation coincides better with the field inspection data.

These simulation analyses and the inspection data suggest that crack growth was dominated in the surface direction due to linking-up, and deep cracks ($a \geq 10$ mm) might be initiated from large porosities.

Estimation of Crack Growth in Thickness Direction

The important issue on the remaining life evaluation of casings may not be the crack growth in the surface direction but in the thickness direction. Since it was suggested above that the crack geometry would become shallow with the crack growth, it will be conservative and rational to postulate a through crack in the case of estimating the increases in the crack depth direction.

Figure 13 shows the estimated crack growth curves assuming single-cracks with initial depths of 1 to 20 mm. In the estimation, the cyclic J-integral shown by the solid line in Fig. 11 was used. This result suggests that the extension of a shallow crack may be comparatively severe, but a deep crack can hardly extend. The data points in the figure indicate the crack depths detected by the field inspections in the operation cycles for several pump casings. Deep cracks shown by the open circles which were initiated from porosities have no correlation with the number of operation cycles. However, in the shallow crack cases shown by the solid symbols, the crack depths tend to increase with the operation cycles. These features agreed well with the estimated results. Hence, we conclude that this estimation procedure can be applied to the remaining life evaluation of the casings.

The critical flaw size for the casing was estimated as 50 mm from the fracture toughness of the casing material and the applied stress due to the internal pressure. Figure 13 also suggests that a great number of operating cycles will be needed for the cracks extending up to the critical size, so that it will be harmless to leave the cracks as they are.

Conclusions

The thermal fatigue crack growth behavior in boiler feed-water circulation pump casings were estimated based on the results of experimental and analytical studies. The following conclusions were derived through this investigation.

1. It was confirmed by bending low-cycle fatigue test of the plate specimen that the propagation and link-up behavior of short surface cracks could be estimated by the da/dN-ΔJ relationship for long cracks.
2. Some analytical studies showed that the J-integral for the crack under thermal stress could be estimated approximately based on the stress distribution obtained by the elastic stress analysis.
3. Good correlation between the estimated crack propagation behavior of the casing and the field inspection data was obtained. It is suggested that the crack propagation rate will slow down with the increase of crack depth thus, these type of pump casings have adequate remaining lives to the final failure.

References

[1] Rice, J. R., *Journal of Applied Mechanics,* Series D, *Transactions,* American Society of Mechanical Engineers, Vol. 35, June 1968, pp. 379–386.

[2] Dowling, N. E., *Cracks and Fracture,* ASTM STP 601, American Society for Testing and Materials, 1976, pp. 19–32.

[3] Kumar, V., German, M. D., and Shih, C. F., NP-1931, Electric Power Research Institute, July 1981.

[4] Kumar, V., German, M. D., Wipkening, W. W., Andrews, W. R., deLorenzi, H. G., and Mowbray, D. F., NP-3607, Electric Power Research Institute, August 1984.

[5] Hellen, T. K., Cesari, F., and Maitan, A., *International Journal of Pressure Vessels and Piping,* Vol. 10, 1982, pp. 181–204.

[6] Ohtani, R., *Transactions of Japan Society of Mechanical Engineers,* Vol. 52, Series A, June 1986, pp. 1461–1467.

[7] Ogata, T., Nitta, A., and Kuwabara, K., Report No. T86061, Central Research Institute of Electric Power Industry, May 1987.

[8] Taira, S., Fujino, M., and Higaki, H., *Journal of The Society of Materials Science, Japan,* Vol. 25, No. 271, April 1976, pp. 375–381.

[9] Russell, E. S., Proceedings of Conference on Life Prediction for High-Temperature Gas Turbine Materials, AP-4477, Electric Power Research Institute, April 1986, p. 3-1.

[10] Ohtani, R., Proceedings of International Conference on Engineering Aspects of Creep, The Institution of Mechanical Engineers, Vol. 2, September 1980, pp. 17–22.

[11] Taira, S., Tanaka, K., and Ogawa, S., *Journal of The Society of Materials Science, Japan,* Vol. 26, No. 280, 1977, pp. 93–98.

[12] Sakon, T. and Kaneko, H., Proceedings of the Fifth International Conference on Creep of Materials, ASM International, May 1992, pp. 227–233.

[13] Asada, Y., Yuuki, R., and Sunamoto, D., Proceedings of International Conference on Engineering Aspects of Creep, The Institution of Mechanical Engineers, Vol. 2, September 1980, pp. 23–28.

[14] EL Haddad, M. H., Smith, K. N., and Topper, T. H., in *Fracture Mechanics,* ASTM STP 677, American Society for Testing and Materials, 1979, pp. 274–289.

[15] Rau, C. A., Jr., Gemma, A. E., and Leverant, G. R., in *Fatigue at Elevated Temperatures,* ASTM STP 520, American Society for Testing and Materials, 1973, pp. 166–178.

[16] Newman, J. C. and Raju, I. S., NASA Technical Paper 1578, 1979.

[17] Urabe, Y., Mitsubishi Juko Giho, Mitsubishi Heavy Industries, Ltd., Japan, Vol. 19, 1982, pp. 243–248.

11-17-93

SUBJECT